Topics in Structural Graph Theory

The rapidly expanding area of structural graph theory uses ideas of connectivity to explore various aspects of graph theory, and vice versa. It has links with other areas of mathematics, such as design theory, and is increasingly used in such areas as computer networks where connectivity algorithms are an important feature.

Although other books cover parts of this material, none has a similarly wide scope. Ortrud R. Oellermann (Winnipeg), internationally recognized for her substantial contributions to structural graph theory, acted as academic consultant for this volume, helping to shape its coverage of key topics. The result is a collection of 13 expository chapters, each written by acknowledged experts. These contributions have been carefully edited to enhance readability and to standardize the chapter structure, terminology and notation throughout. An introductory chapter details the background material in graph theory and network flows, and each chapter concludes with an extensive list of references.

LOWELL W. BEINEKE is Schrey Professor of Mathematics at Indiana University–Purdue University Fort Wayne, where he has been since receiving his Ph.D. from the University of Michigan under the guidance of Frank Harary. His graph theory interests are broad, and include topological graph theory, line graphs, tournaments, decompositions and vulnerability. With Robin Wilson he edited *Selected Topics in Graph Theory* (three volumes), *Applications of Graph Theory*, *Graph Connections*, *Topics in Algebraic Graph Theory* and *Topics in Topological Graph Theory*. Until recently he was editor of the *College Mathematics Journal*.

ROBIN J. WILSON is Emeritus Professor of Pure Mathematics at the Open University, UK. He was recently Gresham Professor of Geometry, London, and a Fellow in Mathematics at Keble College, Oxford University, and now teaches at Pembroke College, Oxford. He graduated in mathematics from Oxford University, and received his Ph.D. in number theory from the University of Pennsylvania. He has written and edited many books on graph theory and the history of mathematics, including *Introduction to Graph Theory, Four Colours Suffice* and *Lewis Carroll in Numberland*, and his research interests include graph colourings and the history of combinatorics. He is currently President of the British Society for the History of Mathematics.

ENCYCLOPEDIA OF MATHEMATICS AND ITS APPLICATIONS

All the titles listed below can be obtained from good booksellers or from Cambridge University Press. For a complete series listing visit www.cambridge.org/mathematics.

97 M. C. Pedicchio and W. Tholen (eds.) *Categorical Foundations*
98 M. E. H. Ismail *Classical and Quantum Orthogonal Polynomials in One Variable*
99 T. Mora *Solving Polynomial Equation Systems II*
100 E. Olivieri and M. Eulália Vares *Large Deviations and Metastability*
101 A. Kushner, V. Lychagin and V. Rubtsov *Contact Geometry and Nonlinear Differential Equations*
102 L. W. Beineke and R. J. Wilson (eds.) with P. J. Cameron *Topics in Algebraic Graph Theory*
103 O. J. Staffans *Well-Posed Linear Systems*
104 J. M. Lewis, S. Lakshmivarahan and S. K. Dhall *Dynamic Data Assimilation*
105 M. Lothaire *Applied Combinatorics on Words*
106 A. Markoe *Analytic Tomography*
107 P. A. Martin *Multiple Scattering*
108 R. A. Brualdi *Combinatorial Matrix Classes*
109 J. M. Borwein and J. D. Vanderwerff *Convex Functions*
110 M.-J. Lai and L. L. Schumaker *Spline Functions on Triangulations*
111 R. T. Curtis *Symmetric Generation of Groups*
112 H. Salzmann *et al. The Classical Fields*
113 S. Peszat and J. Zabczyk *Stochastic Partial Differential Equations with Lévy Noise*
114 J. Beck *Combinatorial Games*
115 L. Barreira and Y. Pesin *Nonuniform Hyperbolicity*
116 D. Z. Arov and H. Dym *J-Contractive Matrix Valued Functions and Related Topics*
117 R. Glowinski, J.-L. Lions and J. He *Exact and Approximate Controllability for Distributed Parameter Systems*
118 A. A. Borovkov and K. A. Borovkov *Asymptotic Analysis of Random Walks*
119 M. Deza and M. Dutour Sikirić *Geometry of Chemical Graphs*
120 T. Nishiura *Absolute Measurable Spaces*
121 M. Prest *Purity, Spectra and Localisation*
122 S. Khrushchev *Orthogonal Polynomials and Continued Fractions*
123 H. Nagamochi and T. Ibaraki *Algorithmic Aspects of Graph Connectivity*
124 F. W. King *Hilbert Transforms I*
125 F. W. King *Hilbert Transforms II*
126 O. Calin and D.-C. Chang *Sub-Riemannian Geometry*
127 M. Grabisch *et al. Aggregation Functions*
128 L. W. Beineke and R. J. Wilson (eds.) with J. L. Gross and T. W. Tucker *Topics in Topological Graph Theory*
129 J. Berstel, D. Perrin and C. Reutenauer *Codes and Automata*
130 T. G. Faticoni *Modules over Endomorphism Rings*
131 H. Morimoto *Stochastic Control and Mathematical Modeling*
132 G. Schmidt *Relational Mathematics*
133 P. Kornerup and D. W. Matula *Finite Precision Number Systems and Arithmetic*
134 Y. Crama and P. L. Hammer (eds.) *Boolean Models and Methods in Mathematics, Computer Science, and Engineering*
135 V. Berthé and M. Rigo (eds.) *Combinatorics, Automata and Number Theory*
136 A. Kristály, V. D. Rădulescu and C. Varga *Variational Principles in Mathematical Physics, Geometry, and Economics*
137 J. Berstel and C. Reutenauer *Noncommutative Rational Series with Applications*
138 B. Courcelle and J. Engelfriet *Graph Structure and Monadic Second-Order Logic*
139 M. Fiedler *Matrices and Graphs in Geometry*
140 N. Vakil Real *Analysis through Modern Infinitesimals*
141 R. B. Paris *Hadamard Expansions and Hyperasymptotic Evaluation*
142 Y. Crama and P. L. Hammer *Boolean Functions*
143 A. Arapostathis, V. S. Borkar and M. K. Ghosh *Ergodic Control of Diffusion Processes*
144 N. Caspard, B. Leclerc and B. Monjardet *Finite Ordered Sets*
145 D. Z. Arov and H. Dym Bitangential *Direct and Inverse Problems for Systems of Integral and Differential Equations*
146 G. Dassios *Ellipsoidal Harmonics*
147 L. W. Beineke and R. J. Wilson (eds.) with O. R. Oellermann *Topics in Structural Graph Theory*
148 L. Berlyand, A. G. Kolpakov and A. Novikov *Introduction to the Network Approximation Method for Materials Modeling*

Karl Menger (1902–1985),
the founder of structural graph theory.

Topics in Structural Graph Theory

Edited by

LOWELL W. BEINEKE
Indiana University–Purdue University
Fort Wayne

ROBIN J. WILSON
The Open University
and Pembroke College, Oxford University

Academic Consultant

ORTRUD R. OELLERMANN
University of Winnipeg

CAMBRIDGE UNIVERSITY PRESS
Cambridge, New York, Melbourne, Madrid, Cape Town,
Singapore, São Paulo, Delhi, Mexico City

Cambridge University Press
The Edinburgh Building, Cambridge CB2 8RU, UK

Published in the United States of America by Cambridge University Press, New York

www.cambridge.org
Information on this title: www.cambridge.org/9780521802314

First published 2013

Printed and bound in the United Kingdom by the MPG Books Group

A catalogue record for this publication is available from the British Library

Library of Congress Cataloging in Publication Data

Topics in structural graph theory / edited by Lowell W. Beineke, Robin J. Wilson.
p. cm. – (Encyclopedia of mathematics and its applications ; 147)
ISBN 978-0-521-80231-4 (Hardback)
1. Graph theory–Data processing. I. Beineke, Lowell W. II. Wilson, Robin J.

QA166.T645 2013
511′.5–dc23

2012022109

ISBN 978-0-521-80231-4 Hardback

Contents

Foreword

Ortrud R. Oellermann

The overriding theme of this volume is connectedness in graphs. In its simplest form a graph is connected if every two vertices are connected by some path. Karl Menger's celebrated theorem changed the way we think about connectedness in graphs. The best-known version of Menger's theorem states that the maximum number of internally disjoint paths between a given pair of non-adjacent vertices in a graph equals the minimum number of vertices that separate the pair.

The connectivity of a graph is the minimum number of vertices whose deletion disconnects the graph. For a given integer k, a graph is k-connected if its connectivity is at least k. Menger's theorem can be used to establish Whitney's characterization of k-connected graphs: 'a graph is k-connected if and only if any two vertices are connected by at least k internally disjoint paths'. Another well-known result that follows from Menger's theorem is Dirac's cycle theorem: 'in a k-connected graph every set of k vertices lie on a common cycle'. Connectivity in graphs has given rise to a substantial body of work on minimally and critically k-connected graphs which is largely due to W. Mader.

An alternative formulation of Menger's theorem states: 'for given sets V and W of vertices in a graph G and a given integer k, there are k disjoint V–W paths in G if and only if every V–W separating set contains at least k vertices'; this is true in particular if V and W are disjoint sets of k vertices. So if V and W cannot be separated by fewer than k vertices, then there exist k disjoint paths, where each path has one end in V and the other end in W. If, for all such choices of V and W, one is able to specify the ends for each path in the collection of disjoint paths, then the graph is said to be *k-linked.*

W. T. Tutte's wheel theorem states: 'every 3-connected graph can be constructed from a wheel graph by repeatedly either splitting a vertex or by adding an edge between a pair of non-adjacent vertices'. Equivalently, 'every 3-connected graph, other than a wheel, can be reduced to a smaller 3-connected graph by either deleting or contracting an edge'. Thus every 3-connected graph has a wheel as a minor. As a result of Wagner's famous conjecture, the theory of graph minors was developed by

Neil Robertson and Paul Seymour who settled this conjecture in a series of papers. Since then, graph minors have played a fundamental role in many areas of graph theory, connectivity being no exception.

The theory of random graphs began in the 1960s, in a series of papers by P. Erdős and A. Rényi. One of their best-known and influential results in this area deals with the phase transition of the component structure in a typical random graph as the number of edges grows from less than half of the number of vertices to more than half.

Menger's theorem, Dirac's cycle theorem, Tutte's wheel theorem, graph minors, and Erdős and Renyi's work on phase transitions underpin many of the chapters in this volume. Menger's theorem is the basis for Chapter 1. Graphs whose connectivity equals the maximum degree are the subject of Chapter 2. Minimally and critically k-connected graphs and reduction methods for 3-connected graphs, first introduced by Tutte, serve as a platform for the material of Chapter 3, and contractible edges in k-connected graphs are further explored in Chapter 4. Dirac's cycle theorem serves as motivation for Chapter 5. The stronger version of the alternative formulation of Menger's theorem (k-linked graphs) and the work on graph minors play a fundamental role in Chapters 6 and 7. Measures of connectedness other than the connectivity are explored in Chapters 1, 7, 8 and 9. The work of Erdős and Renyi on random graphs inspired the results presented in Chapter 10. Network reliability depends on a probabilistic approach to connectedness in graphs and is the subject of Chapter 11. In Chapter 12 the evolution of deterministic algorithms for finding the connectivity of a graph are surveyed. The final chapter describes how different structures of graphs play a fundamental role in finding the best block designs.

Preface

The field of graph theory has undergone tremendous growth during the past century. As recently as the 1950s, the graph theory community had few members and most were in Europe and North America; today there are hundreds of graph theorists and they span the globe. By the mid-1970s, the subject had reached the point where we perceived a need for a collection of surveys of various areas of graph theory: the result was our three-volume series *Selected Topics in Graph Theory*, comprising articles written by distinguished experts and then edited into a common style. Since then, the transformation of the subject has continued, with individual branches (such as graph connectivity) expanding to the point of having important subdivisions themselves. This inspired us to conceive of a new series of books, each a collection of articles within a particular topics of graph theory written by experts within that area. The first two of these books were the companion volumes to the present one, on algebraic graph theory and on topological graph theory. This is thus the third volume in the series.

A special feature of these books is the engagement of academic consultants (here, Ortrud R. Oellermann) to advise us on topics to be included and authors to be invited. We believe that this has been successful, with the result being that the chapters of each book cover the full range of topics within the given area. In the present case, the area is connectivity, also called structural graph theory, with chapters written by authors from around the world. Another important feature is that, to the extent possible, we have imposed uniform terminology and notation throughout, in the belief that this will aid readers in going from one chapter to another. For a similar reason, we have not tried to remove a small amount of material common to some of the chapters.

We hope that these features will facilitate usage of the book in advanced courses and seminars. We sincerely thank the authors for cooperating in these efforts, even though it sometimes required their abandoning some of their favourite conventions – for example, computer scientists commonly use the term *node*, whereas graph theorists use *vertex*; not surprisingly, the graph theorists prevailed on this one. We also asked our contributors to endure the ordeal of having their early versions subjected to detailed critical reading. We believe that as a result the final product is thereby significantly better than it would otherwise have been (as a collection of individual chapters with differing styles and terminology). We want to express our heartfelt appreciation to all of our contributors for their cooperation in these endeavours.

We extend special thanks to Ortrud Oellermann for her service as Academic Consultant – her advice has been invaluable. We are also grateful to Cambridge University Press for publishing these volumes; in particular, we thank Roger Astley for his advice, support, patience and cooperation. Finally we extend our appreciation to several universities for the ways in which they have assisted with our project: the first editor (LWB) is grateful to his home institution of Indiana University–Purdue University Fort Wayne and also to Purdue University for an award of sabbatical leave during which he was a guest of the Mathematical Institute at Oxford University, while the second editor (RJW) has had the cooperation of the Open University as well as Keble College and Pembroke College, Oxford.

LOWELL W. BEINEKE
ROBIN J. WILSON

Preliminaries

LOWELL W. BEINEKE and ROBIN J. WILSON

1. Graph theory

This section presents the basic definitions, terminology and notation of graph theory, along with some fundamental results. Further information can be found in the many standard books on the subject – for example, Bondy and Murty [1], Chartrand, Lesniak and Zhang [2], Gross and Yellen [3] or West [5], or, for a simpler treatment, Marcus [4] or Wilson [6].

Graphs

A *graph* G is a pair of sets (V, E), where V is a finite non-empty set of elements called *vertices*, and E is a finite set of elements called *edges*, each of which has two associated vertices. The sets V and E are the *vertex-set* and *edge-set* of G, and are sometimes denoted by $V(G)$ and $E(G)$. The number of vertices in G is called the *order* of G and is usually denoted by n (but sometimes by $|G|$); the number of edges is denoted by m. A graph with only one vertex is called *trivial*.

An edge whose vertices coincide is a *loop*, and if two edges have the same pair of associated vertices, they are called *multiple edges*. In this book, unless otherwise specified, graphs are assumed to have neither loops nor multiple edges; that is, they are taken to be *simple*. Hence, an edge e can be considered as its associated pair of vertices, $e = \{v, w\}$, usually shortened to vw. An example of a graph of order 5 is shown in Fig. 1(a).

The *complement* $\overline{G}$ of a graph G has the same vertices as G, but two vertices are adjacent in $\overline{G}$ if and only if they are not adjacent in G. Fig. 1(b) shows the complement of the graph in Fig. 1(a).

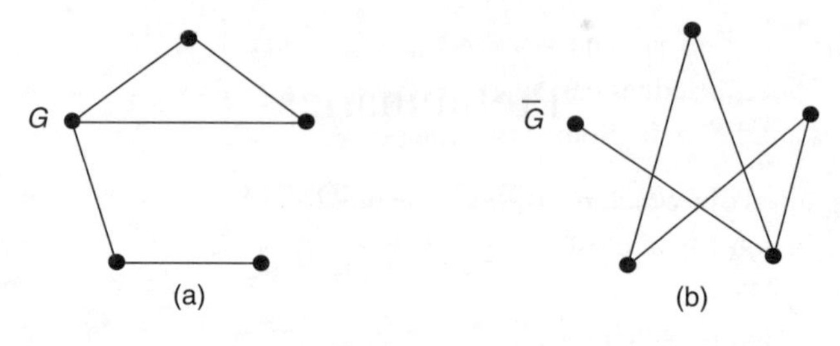

Fig. 1.

Adjacency and degrees

The vertices of an edge are its *endpoints* and the edge is said to *join* these vertices. An endpoint of an edge and the edge are *incident* with each other. Two vertices that are joined by an edge are called *neighbours* and are said to be *adjacent*; if v and w are adjacent vertices we sometimes write $v \sim w$, and if they are not adjacent we write $v \nsim w$. Two edges are *adjacent* if they have a vertex in common.

The set $N(v)$ of neighbours of a vertex v is called its *neighbourhood*. If $X \subseteq V$, then $N(X)$ denotes the set of vertices that are adjacent to some vertex of X.

The *degree* $\deg v$, or $d(v)$, of a vertex v is the number of its neighbours; in a non-simple graph, it is the number of occurrences of the vertex as an endpoint of an edge, with loops counted twice. A vertex of degree 0 is an *isolated vertex* and one of degree 1 is an *end-vertex* or *leaf*. A graph is *regular* if all of its vertices have the same degree, and is *k-regular* if that degree is k; a 3-regular graph is sometimes called *cubic*. The maximum degree in a graph G is denoted by $\Delta(G)$, or just Δ, and the minimum degree by $\delta(G)$ or δ.

Isomorphisms and automorphisms

An *isomorphism* between two graphs G and H is a bijection between their vertex-sets that preserves both adjacency and non-adjacency. The graphs G and H are *isomorphic*, written $G \cong H$, if there exists an isomorphism between them.

An *automorphism* of a graph G is an isomorphism of G with itself. The set of all automorphisms of a graph G forms a group, called the *automorphism group* of G and denoted by $\mathrm{Aut}(G)$. A graph G is *vertex-transitive* if, for any vertices v and w, there is an automorphism taking v to w. It is *edge-transitive* if, for any edges e and f, there is an automorphism taking the vertices of e to those of f. It is *arc-transitive* if, given two ordered pairs of adjacent vertices (v, w) and (v', w'), there is an automorphism taking v to v' and w to w'. This is stronger than edge-transitivity, since it implies that for each edge there is an automorphism that interchanges its vertices.

Walks, paths and cycles

A *walk* in a graph is a sequence of vertices and edges $v_0, e_1, v_1, \ldots, e_k, v_k$, in which the edge e_i joins the vertices v_{i-1} and v_i. This walk is said to *go from v_0 to v_k*

or to *connect* v_0 *and* v_k, and is called a v_0–v_k walk. It is frequently shortened to $v_0 v_1 \cdots v_k$, since the edges can be inferred from this. A walk is *closed* if the first and last vertices are the same. Some important types of walk are the following:

- a *path* is a walk in which no vertex is repeated;
- a *cycle* is a non-trivial closed walk in which no vertex is repeated, except the first and last;
- a *trail* is a walk in which no edge is repeated;
- a *circuit* is a non-trivial closed trail.

Connectedness and distance

A graph is *connected* if it has a path connecting each pair of vertices, and *disconnected* otherwise. A *(connected) component* of a graph is a maximal connected subgraph.

The number of occurrences of edges in a walk is called its *length*, and in a connected graph the *distance* $d(v, w)$ from v to w is the length of a shortest v–w path. It is easy to check that distance satisfies the properties of a metric. The *diameter* of a connected graph G is the greatest distance between any pair of vertices in G. If G has a cycle, the *girth* of G is the length of a shortest cycle.

A connected graph is *Eulerian* if it has a closed trail containing all of its edges; such a trail is an *Eulerian trail*. The following statements are equivalent for a connected graph G:

- G is Eulerian;
- every vertex of G has even degree;
- the edge-set of G can be partitioned into cycles.

A graph of order n is *Hamiltonian* if it has a spanning cycle, and is *pancyclic* if it has a cycle of every length from 3 to n. It is *traceable* if it has a spanning path. No 'good' characterizations of these properties are known.

Bipartite graphs and trees

If the set of vertices of a graph G can be partitioned into two non-empty subsets so that no edge joins two vertices in the same subset, then G is *bipartite*. The two subsets are called *partite sets*, and if they have orders r and s, G is said to be an $r \times s$ *bipartite graph*. (For convenience, the graph with one vertex and no edges is also called bipartite.) Bipartite graphs are characterized by having no cycles of odd length.

Among the bipartite graphs are *trees*, those connected graphs with no cycles. Any graph without cycles is a *forest*; thus, each component of a forest is a tree. Trees have been characterized in many ways, some of which we give here. For a graph G of order n, the following statements are equivalent:

- G is connected and has no cycles;
- G is connected and has $n - 1$ edges;

- G has no cycles and has $n - 1$ edges;
- G has exactly one path between any two vertices.

The set of trees can also be defined inductively: a single vertex is a tree; and for $n \geq 1$, the trees with $n + 1$ vertices are those graphs obtainable from some tree with n vertices by adding a new vertex adjacent to precisely one of its vertices.

This definition has a natural extension to higher dimensions. The *k-dimensional trees*, or *k-trees* for short, are defined as follows: the complete graph on k vertices is a k-tree, and for $n \geq k$, the k-trees with $n+1$ vertices are those graphs obtainable from some k-tree with n vertices by adding a new vertex adjacent to k mutually adjacent vertices in the k-tree. Fig. 2 shows a tree and a 2-tree.

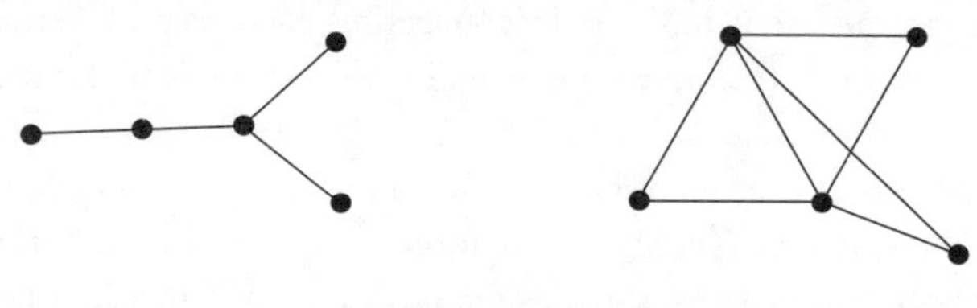

Fig. 2.

An important concept in the study of graph minors (introduced later) is the *tree-width* of a graph G, the minimum dimension of any k-tree that contains G as a subgraph.

Special graphs

We now introduce some individual types of graph:

- the *complete graph* K_n has n vertices, each adjacent to all the others;
- the *null graph* $\overline{K}_n$ has n vertices and no edges;
- the *path graph* P_n consists of the vertices and edges of a path of length $n - 1$;
- the *cycle graph* C_n consists of the vertices and edges of a cycle of length n;
- the *complete bipartite graph* $K_{r,s}$ is the $r \times s$ bipartite graph in which each vertex is adjacent to all of the vertices in the other partite set;
- the *complete k-partite graph* $K_{r_1,r_2,\ldots,r_k}$ has its vertices in k sets with orders $r_1, r_2, \ldots, r_k$, and every vertex is adjacent to all of the vertices in the other sets; if the k sets all have order r, the graph is denoted by $K_{k(r)}$.

Examples of these graphs are given in Fig. 3.

Operations on graphs

Let G and H be graphs with disjoint vertex-sets $V(G) = \{v_1, v_2, \ldots, v_r\}$ and $V(H) = \{w_1, w_2, \ldots, w_s\}$.

- The *union* $G \cup H$ has vertex-set $V(G) \cup V(H)$ and edge-set $E(G) \cup E(H)$. The union of k graphs isomorphic to G is denoted by kG.

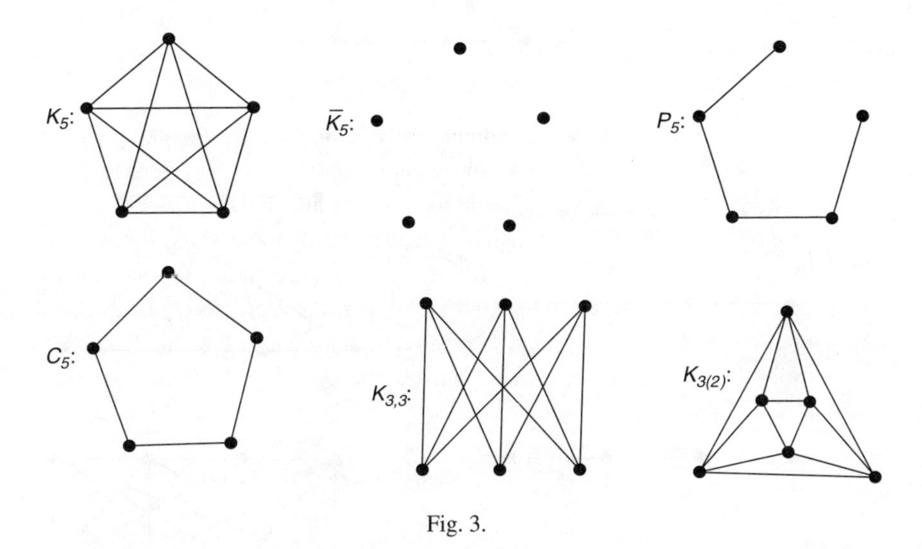

Fig. 3.

- The *join* $G + H$ is obtained from $G \cup H$ by adding an edge from each vertex in G to each vertex in H.
- The *Cartesian product* $G \times H$ (or $G \square H$) has vertex-set $V(G) \times V(H)$, with (v_i, w_j) adjacent to (v_h, w_k) if either v_i is adjacent to v_h in G and $w_j = w_k$, or $v_i = v_h$ and w_j is adjacent to w_k in H; in less formal terms, $G \times H$ can be obtained by taking n copies of H and joining corresponding vertices in different copies whenever there is an edge in G.
- The *lexicographic product* (or *composition*) $G[H]$ also has vertex-set $V(G) \times V(H)$, but with (v_i, w_j) adjacent to (v_h, w_k) if either v_i is adjacent to v_h in G or $v_i = v_h$ and w_j is adjacent to w_k in H.

Examples of these binary operations are given in Fig. 4.

Subgraphs and minors

If G and H are graphs with $V(H) \subseteq V(G)$ and $E(H) \subseteq E(G)$, then H is a *subgraph* of G, and is a *spanning subgraph* if $V(H) = V(G)$. The subgraph $\langle S \rangle$ (or $G[S]$) *induced* by a non-empty set of S of vertices of G is the subgraph H whose vertex-set is S and whose edge-set consists of those edges of G that join two vertices in S. A subgraph H of G is called an *induced subgraph* if $H = \langle V(H) \rangle$. In Fig. 5, H_1 is a spanning subgraph of G, and H_2 is an induced subgraph.

The *deletion of a vertex* v from a graph G results in the subgraph obtained by removing v and all of its incident edges; it is denoted by $G - v$ and is the subgraph induced by $V - \{v\}$. More generally, if S is any set of vertices in G, then $G - S$ is the graph obtained from G by deleting all of the vertices in S and their incident edges; that is, $G - S = \langle V(G) - S \rangle$. Similarly, the *deletion of an edge* e results in

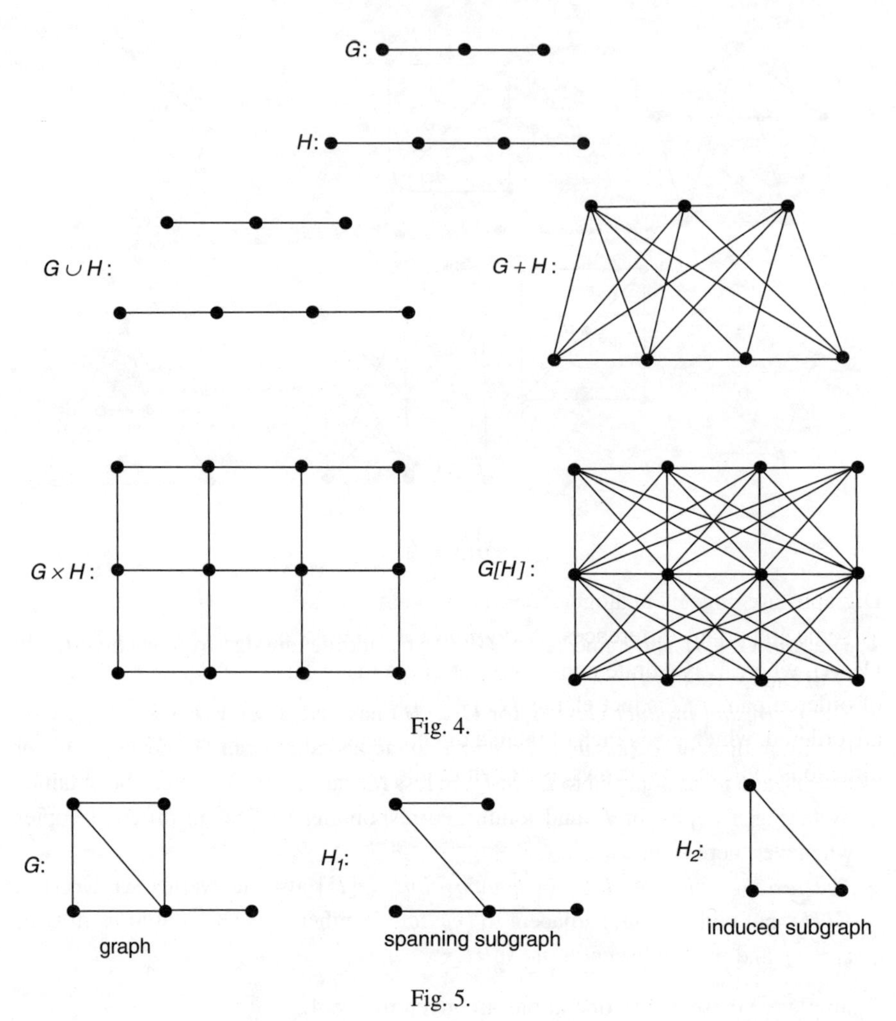

Fig. 4.

Fig. 5.

the subgraph $G - e$ and, for any set X of edges, $G - X$ is the graph obtained from G by deleting all the edges in X.

If the edge e joins vertices v and w, then the *subdivision* of e replaces e by a new vertex u and two new edges vu and uw. Two graphs are *homeomorphic* if there is some graph from which each can be obtained by a sequence of subdivisions. The *contraction* of e replaces its vertices v and w by a new vertex u, with an edge ux if v or w is adjacent to x in G. The operations of subdivision and contraction are illustrated in Fig. 6.

If H can be obtained from G by a sequence of edge-contractions and the removal of isolated vertices, then G is *contractible* to H. A *minor* of G is any graph that can be obtained from G by a sequence of edge-deletions and edge-contractions, along with deletions of isolated vertices. Note that if G has a subgraph homeomorphic to H, then H is a minor of G.

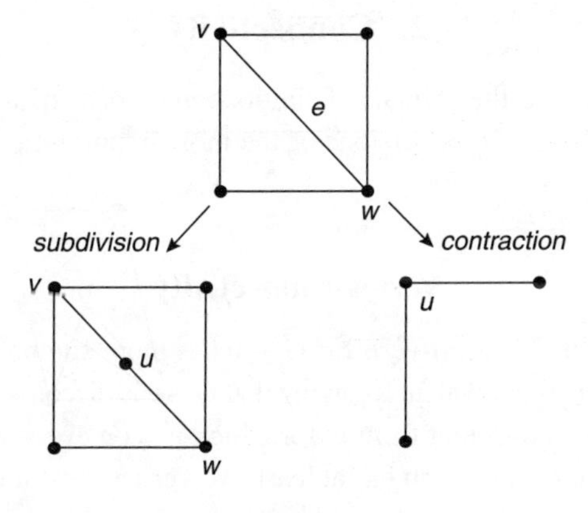

Fig. 6.

Digraphs

Digraphs are directed analogues of graphs, and thus have many similarities, as well as some important differences. A *digraph* (or *directed graph*) D is a pair of sets (V, A), where V is a finite non-empty set of elements called *vertices*, and A is a set of ordered pairs of distinct elements of V called *arcs*. Note that the elements of A are ordered, which gives each of them a direction. An example of a digraph, with the directions indicated by arrows, is shown in Fig. 7.

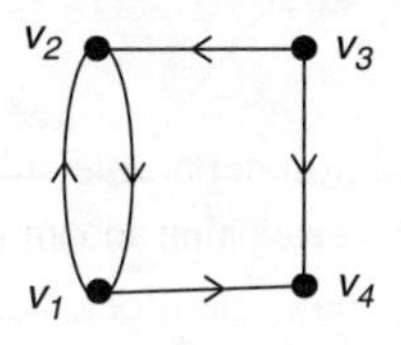

Fig. 7.

Because of the similarities between graphs and digraphs, we mention only the main differences here and do not redefine those concepts that carry over easily. An arc (v, w) in a digraph may be written as vw, and is said to *go from v to w*, or to *go out of* v and *into* w. In the context of digraphs, walks, paths, cycles, trails and circuits are all understood to be directed, unless otherwise indicated. A digraph D is *strongly connected* or *strong* if there is a path from each vertex to each of the others; note that the digraph in Figure 7 is itself strong. A *strong component* is a maximal strongly connected subgraph. Every vertex is in at least one strong component, and an edge is in a strong component if and only if it is on a directed cycle.

The *out-degree* $d^+(v)$ of a vertex v in a digraph D is the number of arcs out of v, and the *in-degree* $d^-(v)$ is the number of arcs into v. The minimum out-degree in a digraph is denoted by δ^+, the minimum in-degree δ^-, and the minimum of the two is denoted by δ^0, or sometimes just δ.

2. Connectivity

In this section, we give the primary definitions and some of the basic results on connectivity, including several versions of the most important one of all, Menger's theorem.

Vertex-connectivity

A vertex v in a graph G is a *cut-vertex* if $G - v$ has more components than G. For a connected graph, this is equivalent to saying that $G - v$ is disconnected, and that there exist vertices u and w, different from v, for which v is on every u–w path. It is easy to see that every non-trivial graph has at least two vertices that are not cut-vertices.

A non-trivial graph is *non-separable* if it is connected and has no cut-vertices. Note that under this definition the graph K_2 is non-separable. There are many characterizations of the other non-separable graphs, as the following statements are all equivalent for a connected graph G with at least three vertices:

- G is non-separable;
- every two vertices of G share a cycle;
- every vertex and edge of G share a cycle;
- every two edges of G share a cycle;
- for any three vertices u, v and w in G, there is a v–w path that contains u;
- for any three vertices u, v and w in G, there is a v–w path that does not contain u;
- for any two vertices v and w and any edge e in G, there is a v–w path that contains e.

A *block* in a graph is a maximal non-separable subgraph. Each edge of a graph lies in exactly one block, while each vertex that is not an isolated vertex lies in at least one block, those that are in more than one block being cut-vertices. An *end-block* is a block with only one cut-vertex; every connected separable graph has at least two end-blocks. The graph in Fig. 8, which has four blocks, illustrates these concepts.

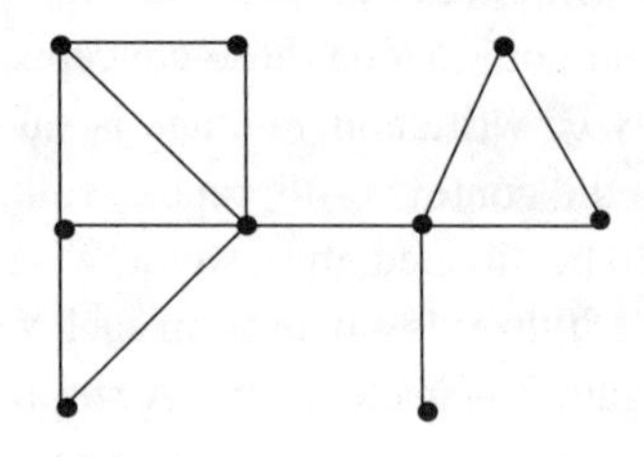

Fig. 8.

The basic idea of non-separability has a natural generalization: a graph G is *k-connected* if the removal of fewer than k vertices always leaves a non-trivial connected graph. The main result on graph connectivity – indeed, it might well be called the Fundamental theorem of connectivity – is Menger's theorem, first published in 1927. It has many equivalent forms, and the first that we give here is the global vertex version. Paths joining the same pair of vertices are called *internally disjoint* if they have no other vertices in common.

Menger's theorem (Global vertex version) *A graph is k-connected if and only if every pair of vertices are joined by k internally disjoint paths.*

The *connectivity* $\kappa(G)$ of a graph G is the maximum non-negative integer k for which G is k-connected; for example, the connectivity of the complete graph K_n is $n - 1$, and a graph has connectivity 0 if and only if it is trivial or disconnected.

For non-adjacent vertices v and w in a graph G, a *v–w separating set* is a set of vertices whose removal leaves v and w in different components, and the *v–w connectivity* $\kappa(v, w)$ is the minimum order of a v–w separating set. Here is another version of Menger's theorem. A minimal separating set is called a *cutset*.

Menger's theorem (Local vertex version) *If v and w are non-adjacent vertices in a graph G, then the maximum number of internally disjoint v–w paths is $\kappa(v, w)$.*

Edge-connectivity

There is an analogous body of material that involves edges rather than vertices, and because of the similarities, we treat it in less detail.

An edge e is a *cut-edge* (or *bridge*) of a graph G if $G - e$ has more components than G. (In contrast to the situation with vertices, the removal of an edge cannot increase the number of components by more than 1.) An edge e is a cut-edge if and only if there exist vertices v and w for which e is on every v–w path. The cut-edges in a graph are also characterized by the property of not lying on a cycle; thus, a graph is a forest if and only if every edge is a cut-edge. Graphs with no cut-edges can be characterized in a variety of ways similar to those having no cut-vertices – that is, non-separable graphs. The concepts corresponding to cycles and paths for vertices are circuits and trails for edges.

Moving beyond cut-edges, we have the following definitions. A graph G is *l-edge-connected* if the removal of fewer than l edges always leaves a connected graph. Here is a third version of Menger's theorem.

Menger's theorem (Global edge version) *A graph is l-edge-connected if and only if each pair of its vertices is joined by l edge-disjoint paths.*

The *edge-connectivity* $\lambda(G)$ of a graph G is the greatest non-negative integer l for which G is l-edge-connected. Obviously, $\lambda(G)$ cannot exceed the minimum degree of a vertex of G; furthermore, it is at least as large as the connectivity – that is,

$$\kappa(G) \leq \lambda(G) \leq \delta(G).$$

For non-adjacent vertices v and w in a graph G, a v–w *cut set* is a set of edges whose removal leaves v and w in different components, and the v–w *edge-connectivity* $\lambda(v, w)$ is the minimum number of edges in a v–w cut set.

Menger's theorem (Local edge version) *If v and w are vertices in a graph G, then the maximum number of edge-disjoint v–w paths is $\lambda(v, w)$.*

When working with Menger's theorem, it can be useful to have a name for the collections of paths whose existence is guaranteed by the theorem. For vertices v and w, a v–w *k-skein* is a collection of k internally disjoint v–w paths, and a v–w *edge-k-skein* is a collection of k edge-disjoint v–w paths. Thus, with this terminology, the two local versions of Menger's theorem state that v and w are k-connected if and only if there is a v–w k-skein, and they are l-edge-connected if and only if there is a v–w edge-l-skein.

Along with the four undirected versions of Menger's theorem, there are corresponding directed versions (with directed paths and strong connectivity). Furthermore, there are weighted versions, as we describe briefly in the next section. In Chapter 1, Menger's theorem is explored more fully, with two proofs and many of its consequences.

3. Flows in networks

The classic result known as the 'max flow – min cut theorem' is the form that Menger's theorem takes when the arcs in a digraph have weights. This result, equivalent to Menger's theorem, was first stated by L. R. Ford and D. R. Fulkerson in 1956, and within their proof there lies a good algorithm for finding the maximum flow in a network. In order for us to present this material, we need a number of definitions.

- A *network* N is a directed graph with vertex-set V and arc-set A, with two vertices s (the 'start') and t (the 'terminus') specified, and with each arc a assigned a positive integer weight $c(a)$, called its *capacity*. An example of a network is shown in Fig. 9.

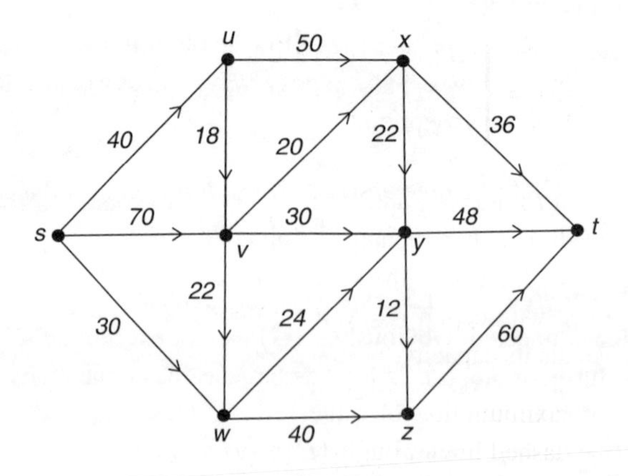

Fig. 9.

- A *flow* in N is a function f from the arc-set A to the set of non-negative integers such that:
  (i) $0 \leq f(a) \leq c(a)$ for each arc a (the *capacity constraint*);
  (ii) $f^+(v) = f^-(v)$, for each vertex v other than s and t (the *conservation constraint*). (Here, $f^+(v)$ denotes the sum of the flows $f(a)$ on the arcs a that are directed into v, and $f^-(v)$ denotes the sum of the flows $f(a)$ on the arcs a that are directed out of v.)
- The *value* $\text{val}(f)$ of a flow f is the net flow out of s and into t
- A *cut* K in N is the set of arcs directed away from some set of vertices S that contains s but not t.
- The *capacity* $\text{cap}(K)$ of a cut K is the sum of the capacities of its arcs.

Max flow – min cut theorem *In any network, the maximum value of a flow equals the minimum capacity of a cut.*

As noted above, this result is equivalent to Menger's theorem. It can be proved algorithmically by an augmentation procedure that increases the value of any flow that is not maximal. The following procedure achieves this.

Given a flow f in a network N, an *f-augmenting path* is an s–t 'path' P in which arcs can be forwardly or reversely oriented, but if a is a forward arc of P then $f(a) < c(a)$, while if a is a reverse arc of P then $f(a) > 0$. The algorithm starts with a flow (such as the flow with all values 0) and then grows a tree from the vertex s by a process that ends either when the vertex t is reached or when no more arcs can be added (under the given rules). As with a path, the edges of a tree are oriented (in a way that depends on the rules). The tree is grown by adding a new vertex v if it is joined by an arc a to a vertex w already in the tree if either a is a forward arc and $f(a) < c(a)$, or a is a reverse arc and $f(a) > 0$.

- If t can be reached, then there exists an f-augmenting path P obtained as follows: Let $i(P) = \min(\{c(a) - f(a) : a \text{ is a forward arc in } P\} \cup \{f(a) : a \text{ is a reverse arc in } P\})$, and define a new flow g on the network N with

$$g(a) = \begin{cases} f(a) + i(P) & \text{if } a \text{ is a forward arc,} \\ f(a) - i(P) & \text{if } a \text{ is a reverse arc,} \\ f(a) & \text{otherwise.} \end{cases}$$

  Then $\text{val}(g) = \text{val}(f) + i(P)$, and since $i(P) > 0$, g is a flow of greater value than f. The procedure is then repeated using the new flow.
- If t cannot be reached, let S be the set of vertices in the tree obtained. The set S determines a cut K with $\text{cap}(K) = \text{val}(f)$; that is, the flow on each arc directed out of the tree equals its capacity.

An example of a maximum flow of value 136 (in parentheses) and a minimum cut of capacity 136 (the dashed line through xt, yt, yz and wz) in the network in Fig. 9 is shown in Fig. 10.

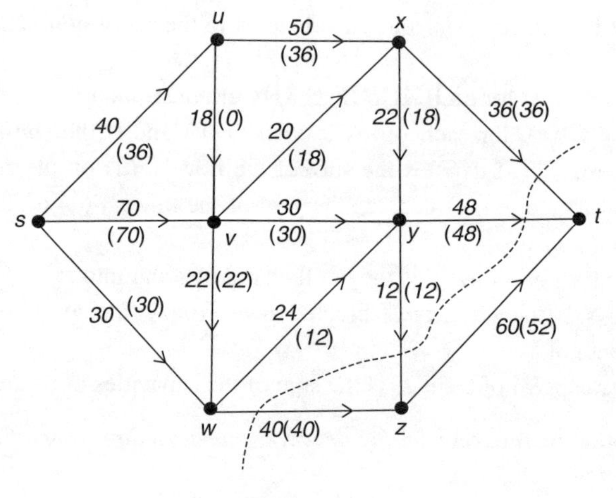

Fig. 10.

The max flow – min cut theorem of Ford and Fulkerson can be adapted to a variety of situations, such as having multiple sources and terminals: add a new source with appropriate capacities on arcs into each of the original sources, and similarly add a new terminal and arcs to it from the original terminals.

References

1. J. A. Bondy and U. S. R. Murty, *Graph Theory*, Springer, 2008.
2. G. Chartrand, L. Lesniak and P. Zhang, *Graphs and Digraphs* (5th edn.), CRC, 2011.
3. J. L. Gross and J. Yellen, *Graph Theory and its Applications* (2nd edn.), CRC, 2005.
4. D. A. Marcus, *Graph Theory*, Math. Assoc. of America, 2008.
5. D. B. West, *Introduction to Graph Theory* (2nd edn.), Pearson, 2001.
6. R. J. Wilson, *Introduction to Graph Theory* (5th edn.), Pearson, 2010.

1

Menger's theorem

ORTRUD R. OELLERMANN

In its best-known version, Menger's theorem states that the maximum number of pairwise internally disjoint paths between a given pair of non-adjacent vertices in a graph equals the minimum number of vertices whose deletion disconnects the pair. Thus, the maximum number of pairwise internally disjoint paths that connect a given pair of vertices is a local measure that indicates how well two given vertices are connected. The connectivity of a graph is the minimum number of vertices whose deletion disconnects the graph, and can be expressed in terms of the connectivity between pairs of vertices. We survey some classical results from this field and highlight some advances in closely related areas.

1. Introduction

Since a given pair of vertices in a graph may be connected by many paths, it is natural to ask how one should measure the connectedness between them. One way of doing this is to determine the largest number of such paths that are pairwise 'independent' of one another, in sharing no other vertices. Another way of measuring their connectedness is to determine the smallest number of vertices whose deletion from the graph destroys every path between this pair. Menger's elegant theorem [56] states that, for each pair of non-adjacent vertices, these two measures are equal. An equivalent formulation of this theorem states that if V and W are non-empty sets of vertices in a graph, then the maximum number of internally disjoint V–W paths equals the minimum number of vertices whose deletion destroys every such path.

In Section 2 we give two proofs of Menger's theorem, one for each of our formulations, and discuss some edge versions in Section 3. We then proceed to the global measure of the 'connectivity of a graph' and Whitney's version of Menger's theorem [73], where the connectivity of a non-complete graph is the smallest number of vertices whose deletion produces a disconnected graph. In Section 4 we discuss mixed versions of Menger's theorem. We consider disconnecting a pair of vertices in a graph by deleting both vertices and edges, and relate this to the problem of finding a largest set of edge-disjoint paths between a pair of vertices, some subcollection of which is internally disjoint.

Graphs with the same connectivity may have vastly differing degrees of 'connectedness'. Motivated by Menger's theorem we consider, in Section 5, another global measure called the 'average connectivity', defined as the average of the maximum number of internally disjoint paths connecting pairs of vertices. Section 6 is devoted to Menger-type results for paths of bounded length between a pair of vertices. In particular, we discuss relationships between the maximum number of internally disjoint paths of a constrained length connecting a pair of vertices and the minimum number of vertices whose deletion destroys all such paths between this pair of vertices.

We conclude the chapter with two sections that survey Menger-type results for sets of more than two vertices.

Throughout most of the chapter, we focus on vertex results, but usually, as with Menger's theorem itself, there is a corresponding edge result; when appropriate, we also discuss these. Unless specifically stated otherwise, we assume that all graphs are simple, finite and undirected.

2. Vertex-connectivity

Let v and w be non-adjacent vertices in a graph G. A set S of vertices is a v–w *separating set* if v and w lie in different components of $G - S$: that is, if every v–w path contains a vertex in S. The minimum order of a v–w separating set is called the v–w *connectivity* and is denoted by $\kappa(v, w)$.

For any two vertices v and w, a set of v–w paths is called *internally disjoint* if the paths are pairwise disjoint except for the vertices v and w. The maximum number of internally disjoint v–w paths is denoted by $\mu(v, w)$. Since every v–w separating set must contain an internal vertex from each path in any set of internally disjoint $v - w$ paths, $\mu(v, w) \leq \kappa(v, w)$.

As an example, consider the graph in Fig. 1. It is easy to verify that both $\kappa(v, w)$ and $\mu(v, w)$ are 3. That the two parameters are equal in this case is no coincidence, and the fact that this holds in general is the content of one version of Menger's theorem.

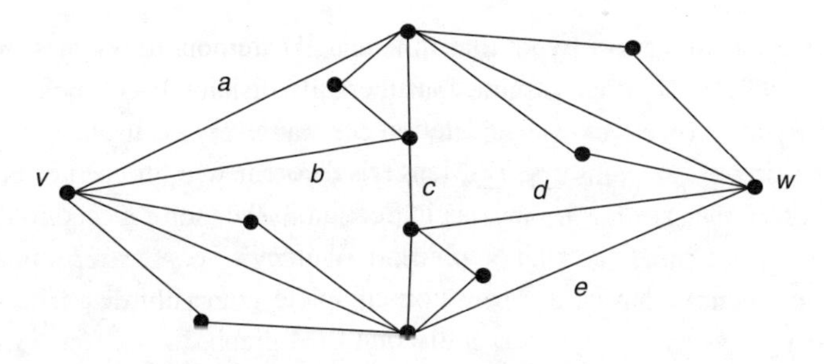

Fig. 1. A graph with $\kappa(v, w) = \mu(v, w) = 3$

Theorem 2.1 *If v and w are non-adjacent vertices in a graph G, then the maximum number of internally disjoint v–w paths equals the minimum number of vertices in a v–w separating set.*

Menger discovered his result in the course of research that he was conducting into the theory of curves in point-set topology (see [58]). However, as König [42] pointed out, Menger's original proof contained a significant gap. The first flawless proof was provided by G. Nöbeling, and appeared in another paper by Menger [57]; it uses an earlier result of König on bipartite graphs [40]. Other early proofs were given by König [41] and Hajós [33], since when numerous other proofs have been given (see [55] for an extensive list). We give two proofs here, the first of which uses induction and follows one by McCuaig [55]. Our second proof is for an equivalent version of the theorem, and uses an algorithmic approach. It is due to Ore [65], and bears a strong resemblance to the max-flow min-cut algorithm of Ford and Fulkerson [24] (see the Preliminaries chapter), from which Menger's theorem can also be deduced.

Proof Let v and w be a pair of non-adjacent vertices in a graph G. As observed earlier, $\mu(v, w) \leq \kappa(v, w)$ since any v–w separating set must contain at least one vertex from each path in any set of internally disjoint v–w paths.

We now show that $\mu(v, w) \geq \kappa(v, w)$. Let $k = \kappa(v, w)$. Then no set of fewer than k vertices separates v and w. We proceed to show, by induction on k, that if $\kappa(v, w) \geq k$, then $\mu(v, w) \geq k$. (Thus, in particular, $\kappa(v, w) = k$ implies that $\mu(v, w) \geq k$, which is the desired result.)

If $k = 1$, then there is a v–w path. Assume then that $k \geq 1$, and that if $\kappa(v, w) \geq k$, then $\mu(v, w) \geq k$. Assume further that v and w are non-adjacent vertices in G with $\kappa(v, w) \geq k + 1$.

By the induction hypothesis, there are k internally disjoint v–w paths P_1, P_2, ..., P_k. Since the set of vertices that immediately follow v on these paths (there are k of these) do not separate v and w, there is a v–w path P whose initial edge is not on any P_i. Let x be the first vertex after v on P that belongs to some P_i. Let P_{k+1} be the v–x subpath of P. Assume that $P_1, P_2, \ldots, P_{k+1}$ have been chosen in such a way that the distance from x to w in $G - v$ is a minimum. If $x = w$, then we have the desired set of $k + 1$ internally disjoint paths. We assume therefore that $x \neq w$.

Again, by the induction hypothesis, there are k internally disjoint v–w paths $Q_1, Q_2, \ldots, Q_k$ in $G - x$. Assume that these paths have been chosen so that a minimum number of edges not on any of the paths P_i are used. Let H be the graph consisting of the paths $Q_1, Q_2, \ldots, Q_k$, together with the vertex x. Let B be the edges of H that are not in any P_i. Choose some P_j for $1 \leq j \leq k+1$, whose initial edge is not in H, and let y be the first vertex on P_j after v which is in H. If $y = w$, then we have the desired set of $k+1$ internally disjoint v–w paths. So assume $y \neq w$.

If $y = x$, then let R be the shortest x–w path in $G - v$. Let z be the first vertex on R that lies on some Q_i. Then the distance in $G - v$ from z to w is less than the distance from x to w, which contradicts our choice of $P_1, P_2, \ldots, P_{k+1}$. So $y \neq x$.

If y lies on some Q_i, for $1 \leq i \leq k$, then the v–y subpath of Q_i has an edge in B. Otherwise, two paths from among $P_1, P_2, \ldots, P_{k+1}$ intersect at a vertex other than v, w or x. If we replace the v–y subpath of Q_i by the v–y subpath of P_j, we get a set of k internally disjoint v–w paths in $G - x$ that uses fewer edges from B than $Q_1, Q_2, \ldots, Q_k$ do, which is a contradiction. The result follows. ■

Before presenting a second proof of Menger's theorem, we introduce an equivalent formulation. Assume that V and W are sets of vertices in a graph G. Then a *V–W path* is a path from some $v \in V$ to some $w \in W$ that passes through no other vertex of V or W. If a vertex x belongs to both V and W, then x is itself a V–W path. We say that a set S of vertices *separates* V and W if every V–W path contains a vertex of S, and that S is a *V–W separating set*. In particular, both V and W are themselves V–W separating sets. The following theorem is equivalent to Menger's theorem.

Theorem 2.2 *Let V and W be sets of vertices in a graph G. For any positive integer k, there are k (pairwise) disjoint V–W paths in G if and only if every V–W separating set contains at least k vertices.*

To see that this theorem is equivalent to Theorem 2.1, suppose first that Theorem 2.1 holds, and let V and W be sets of vertices in G. Introduce two new vertices v and w, join v to every vertex of V and w to every vertex of W, and let H be the resulting graph. Then v and w are not adjacent in H, and every set of internally disjoint v–w paths in H corresponds to a set of disjoint V–W paths in G, and vice versa. Moreover, a set S is a v–w separating set in H if and only if it is a V–W separating set in G. Hence Theorem 2.1 implies Theorem 2.2.

Suppose now that Theorem 2.2 holds. Let v and w be two non-adjacent vertices in a graph G. Let V be the set of neighbours of v and let W be the set of neighbours of w. As before, each pair of internally disjoint v–w paths in G corresponds to a pair of disjoint V–W paths in G, and vice versa. Moreover, S is a v–w separating set if and only if it is a V–W separating set. Hence Theorem 2.2 implies Theorem 2.1.

We now outline a second proof of Menger's theorem by proving Theorem 2.2.

Proof Let V and W be sets of vertices of a graph G. As before, it is obvious that if there are k pairwise disjoint V–W paths in G, then every V–W separating set must have at least k vertices. For the converse, suppose that every V–W separating set has at least k vertices. We show that there must then exist k disjoint V–W paths. We assume that $V \cap W = \varnothing$; the result then follows when $V \cap W \neq \varnothing$, since every vertex in both V and W is itself a V–W path. Begin with a set Π_l of l disjoint V–W paths $P_1, P_2, \ldots, P_l$ (possibly $l = 0$). We now describe how this set of l paths can be used to construct a set of $l+1$ disjoint V–W paths, if there is such a set. For this purpose we define a trail Q: $u_0, e_0, u_1, e_1, \ldots, e_{k-1}, u_k$ to be an *alternating trail* with respect to Π_l if the following three conditions are satisfied.

- If e_i belongs to a path P_j in Π_l, then P_j traverses e_i in the opposite direction to Q.
- If $u_r = u_s$ with $r \neq s$, then u_r lies on some P_i – that is, if a vertex of Q appears more than once on Q, then it lies on a path in Π_l.
- If u_i lies on some P_j, then at least one of e_{i-1} and e_i belongs to P_j.

An *augmenting V–W trail with respect to* Π_l is an alternating trail from a vertex $v \in V$ to a vertex $w \in W$, with neither v nor w on any P_i.

Note that if Q is such a trail, then every vertex of Q that is not on any P_i appears exactly once on Q, and, by the third condition, every vertex of Q that lies on some P_i appears at most twice on Q. Let H be the union of the l paths P_i. If $u_r = u_s$ for some r, s with $1 \leq r < s \leq k$, then either e_{r-1} and e_s are in H and e_r and e_{s-1} are not in H, or e_r and e_{s-1} are in H and e_{r-1} and e_s are not in H.

Using this observation, one can show that if Q is an augmenting V–W trail with respect to Π_l, then G has $l+1$ disjoint V–W paths. Indeed, the subgraph consisting of the union of Q and the paths $P_1, P_2, \ldots, P_l$, with the edges on both Q and some P_i deleted, consists of $l+1$ disjoint V–W paths, which we denote by Π_{l+1} (see Fig. 2 for an illustration).

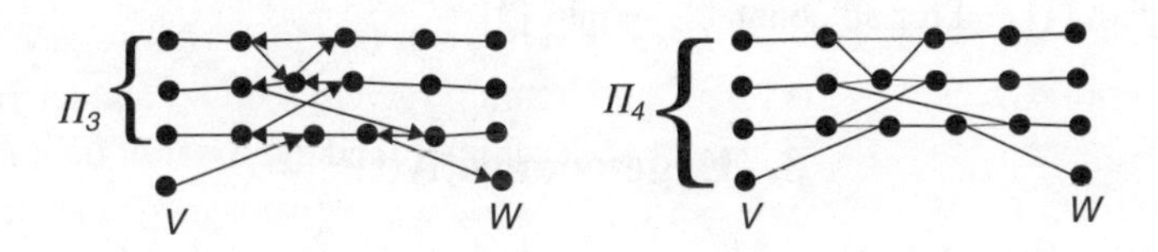

Fig. 2. Augmenting Π_3 along an augmenting V–W trail to obtain Π_4

We continue in this manner until we arrive at a set Π_L of L disjoint V–W paths for which there is no augmenting V–W trail. For each path P in Π_L, let v_P be the last vertex of P that lies on some alternating trail that begins with a vertex in V but does not lie on any path in Π_L (if such a trail exists); otherwise, let v_P be the first vertex of P. Then it can be shown that the set X consisting of the L vertices v_P for $P \in \Pi_L$ is a V–W separating set. Hence $L = |X| \geq k$. ■

Recall that the *connectivity* $\kappa(G)$ of a graph G is the smallest number of vertices whose deletion from G produces a disconnected or trivial graph. Clearly, a complete graph cannot be disconnected by deleting vertices, but all other graphs can. It is not difficult to see that, in any case, $\kappa(G) = \min\{\kappa(v, w) : v, w \in V(G)\}$.

A graph G for which $\kappa(G) \geq k$ is said to be *k-connected*. The first characterization of k-connected graphs was given by Whitney [73] in 1932. He proved this result independently of Menger's theorem, but with Menger's theorem this result can be obtained in a straightforward manner.

Theorem 2.3 *A graph G is k-connected if and only if each pair of vertices is connected by k internally disjoint paths.*

We note that Menger's theorem also holds for digraphs and multigraphs. Both of the proofs given in this section can easily be adapted to digraphs, and the existence of multiple edges does not change the proofs.

We now turn our attention briefly to infinite graphs. For sets of vertices V and W in an infinite graph G, V–W paths and separating sets are defined as for finite graphs. Erdős showed that if there are at least k vertices in any V–W separating set, then there are at least k disjoint V–W paths in G, a result published in König [41]. Menger's theorem also holds for infinite cardinals, but Erdős proposed a better way of extending the theorem to infinite graphs when he made the following classic conjecture.

Conjecture A *Let V and W be sets of vertices in an infinite graph G. Then G contains a set Π of disjoint V–W paths and a V–W separating set S that are in one-to-one correspondence, where each vertex of S lies on exactly one path in Π, and each path in Π contains exactly one vertex of S.*

Podewski and Steffens [67] made some progress on this conjecture by proving it to be true for countable graphs that contain no infinite paths. Aharoni strengthened their result when he showed the conjecture to be true for all graphs that contain no infinite path [1] and for all countable graphs [2].

3. Edge-connectivity

The vertex versions of Menger's theorem discussed in Section 2 have edge analogues that we briefly describe here. Let v and w be two vertices in a graph G. A set S of edges is a *v–w edge-separating* set if v and w lie in different components of $G - S$: that is, if every v–w path contains an edge of S. The minimum cardinality of a v–w edge-separating set is the *v–w edge-connectivity* and is denoted by $\lambda(v, w)$.

The maximum number of edge-disjoint v–w paths in G is denoted by $\nu(v, w)$. Since each such path must contain an edge from every v–w edge-separating set, $\nu(v, w) \leq \lambda(v, w)$. For example, consider again the graph of Fig. 1: it is easy to see that $\nu(v, w) = \lambda(v, w) = 5$. That the two parameters are equal in this case is again not a mere coincidence, and the fact that this holds in general is the local edge version of Menger's theorem.

Theorem 3.1 *For any vertices v and w in a graph G, $\nu(v, w) = \lambda(v, w)$.*

One may well ask whether there always exists a system of $\nu(v, w)$ edge-disjoint paths that contains a system of $\mu(v, w)$ internally disjoint v–w paths. Beineke and Harary [4] showed that this need not be the case. For the graph of Fig. 1, $\mu(v, w) = 3$ and $\nu(v, w) = 5$, but no set of three internally disjoint v–w paths is contained in a set of five edge-disjoint v–w paths. To see this, note that every set of three internally disjoint v–w paths contains all five edges a, b, c, d, e of a minimal v–w edge-separating set, and thus cannot be extended to five edge-disjoint v–w paths.

If v and w are not adjacent, then both deg v and deg w may exceed $\kappa(v, w)$ by an arbitrarily large amount. Take, for example, two complete graphs K_n and join one vertex from each copy by an edge. The resulting graph has minimum degree $n - 1$ and $\kappa(v, w) = 1$ for each pair of non-adjacent vertices v and w in the resulting graph. However, Mader [50], [51] proved the following result.

Theorem 3.2 *Every non-null graph has adjacent vertices v and w for which $\mu(v, w) = \min\{\deg v, \deg w\}$.*

An immediate consequence of the above theorem is that there exist vertices v and w for which $\mu(v, w) = \nu(v, w) = \lambda(v, w) = \min\{\deg v, \deg w\}$.

We note that this theorem is not true for multigraphs, since a multigraph formed from a cycle by doubling every edge does not satisfy the theorem. However, it is true that every multigraph has adjacent vertices v and w for which $\nu(v, w) = \min\{\deg v, \deg w\}$ (see [54]).

The *edge-connectivity* $\lambda(G)$ of a non-trivial graph G is the smallest number of edges whose deletion produces a disconnected graph, while that of the trivial graph is defined to be 0. It is not difficult to see that $\lambda(G) = \min\{\lambda(v, w) : v, w \in V(G)\}$. A graph G is *l-edge-connected* if $\lambda(G) \geq l$. The following is a global edge version of Menger's theorem.

Theorem 3.3 *A graph is l-edge-connected if and only if every two vertices are connected by at least l edge-disjoint paths.*

4. Mixed connectivity

In this section we consider the problem of disconnecting pairs of vertices by permitting the removal of both vertices and edges. We also look at optimal notions for this concept, and relate it to the existence of a combination of edge-disjoint and internally disjoint paths.

Connectivity pairs

In their 1967 paper, Beineke and Harary [4] considered the problem of disconnecting a graph by deleting both vertices and edges. In a graph G with vertices v and w, a set S of vertices and a set T of edges form a v–w *disconnecting pair* if v and w belong to different components of $G-(S \cup T)$. The vertices v and w are (k, l)-*connected* if there is no disconnecting pair of s vertices and t edges with $s<k$ and $t \leq l$ or $s \leq k$ and $t<l$. The pair of integers (k, l) is a v–w *connectivity pair* if v and w are neither $(k+1, l)$-connected nor $(k, l+1)$-connected. Beineke and Harary claimed to prove a mixed version of Menger's theorem, but Mader [54] pointed out a gap in their proof. A correction is claimed to have been given in Xu [74], but no English translation is currently available.

The graph of Fig. 3 illustrates that the Beineke–Harary conjecture is in some sense best possible. The pair (2, 2) is a connectivity pair for vertices v and w, and there are four edge-disjoint v–w paths (but no more) of which two are internally disjoint. This graph also serves to illustrate that k internally disjoint v–w paths in a system of $k+l$ edge-disjoint paths cannot always be chosen in such a way that each of these paths is internally disjoint from the remaining $k+l-1$ paths.

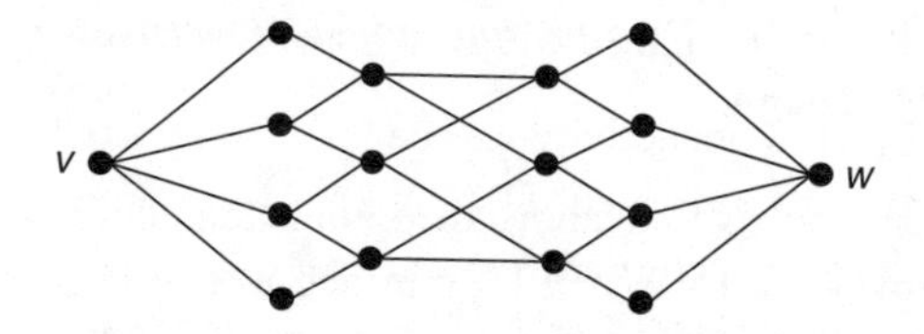

Fig. 3. Connectivity pairs and edge-disjoint paths

Even if true, the Beineke–Harary conjecture is not as strong as one might have hoped. If (k, l) is a connectivity pair for v and w, then there can be more than $k+l$ edge-disjoint v–w paths, of which more than k are internally disjoint. For example, (2, 1) is a v–w connectivity pair in the graph G of Fig. 4. So, by the conjecture, there are three edge-disjoint v–w paths, but in fact there are five such paths, with three internally disjoint.

There are some interesting open questions related to this conjecture. For example, if v and w are vertices in a graph G, and if k is an integer with $0 \leq k \leq \kappa(v, w)$, what is the maximum number of edge-disjoint v–w paths of which k are internally disjoint? The graph of Fig. 1 shows that this number need not be $\lambda(v, w)$. A follow-up problem is to determine how difficult it is to compute the second coordinate in a v–w connectivity pair (k, l).

If G is a graph and $\kappa=\kappa(G)$, then it is readily seen that, for each k with $0 \leq k \leq \kappa$, there is a unique connectivity pair (k, l_k). Thus, the connectivity pairs of a graph determine a function f from $\{0, 1, \ldots, \kappa\}$ into the set of non-negative integers for which $f(\kappa)=0$; this function is called the *connectivity function* of G. It is not

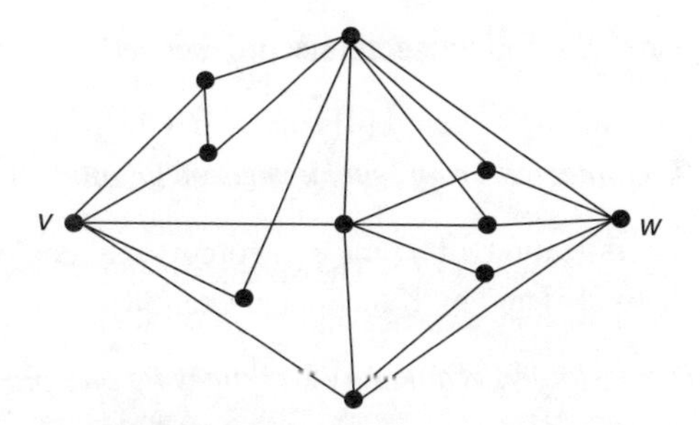

Fig. 4. Three internally disjoint paths among five edge-disjoint paths

difficult to show that f is a strictly decreasing function. Beineke and Harary [4] showed that these two conditions characterize these functions.

Theorem 4.1 *Suppose that κ is a positive integer. A function $f : \{0, 1, \ldots, \kappa\} \to \mathbb{Z}^+ \cup \{0\}$ is the connectivity function of a graph if and only if f is decreasing and* $f(\kappa) = 0$.

Edge-disjoint skeins

Another mixed version of connectivity was considered by Egawa, Kaneko and Matsumoto [16]. A *v–w k-skein* is a set of k internally disjoint v–w paths. Menger's theorem states that non-adjacent vertices v and w in a graph G are k-connected if and only if G has a v–w k-skein. Let k and r be positive integers and v and w distinct vertices of a graph G. Then v and w are *$[k, r]$-joined* if, for each set S of vertices (not containing v or w) and each set T of edges with $r|S| + |T| < kr$, the vertices v and w are in the same component of $G - (S \cup T)$. The following mixed version of Menger's theorem was established by Egawa *et al.*

Theorem 4.2 *Let v and w be vertices of a graph G and let k and l be positive integers. Then v and w are $[k, r]$-joined if and only if G contains r pairwise edge-disjoint v–w k-skeins.*

Extending these concepts, we define a graph G to be *$[k, r]$-joined* if each pair of vertices is *$[k, r]$-joined* – that is, for any set S of s vertices and any set T of t edges with $rs + t < kr$, the graph $G - (S \cup T)$ is connected. Thus a graph is $[k, 1]$-joined if and only if it is k-connected, and $[1, r]$-joined if and only if it is r-edge-connected. As Kaneko and Ota [38] pointed out, the following result is a consequence of Theorem 4.2 and extends Whitney's characterization of k-connected graphs.

Theorem 4.3 *A graph G of order at least $k+1$ is $[k, r]$-joined if and only if any two vertices are joined by r pairwise edge-disjoint k-skeins.*

Dirac [14] established the following connection between k-connected graphs and cycles.

Theorem 4.4 *In a k-connected graph, any k vertices lie on a common cycle.*

The following generalization of Dirac's theorem was conjectured by Egawa, Kaneko and Matsumoto [16] and by Kaneko and Ota [38].

Conjecture B *In a $[k, l]$-joined graph, any k vertices lie on l pairwise edge-disjoint cycles.*

Enomoto and Kaneko [17] established a link between pairs of vertices being $[k, l]$-joined and connectivity pairs. Using Theorem 4.3, they showed that the Beineke–Harary conjecture holds for certain pairs (k, l). They also made the following stronger conjecture.

Conjecture C *If (k, l) is a connectivity pair for vertices v and w of a graph G, then G contains $k + l$ edge-disjoint v–w paths, of which $k + 1$ are internally disjoint.*

5. Average connectivity

The connectivity of a graph is a worst-case measure and as such may not distinguish between graphs that obviously have different degrees of connectedness. For example, for $n \geq 4$, if G is the graph obtained from the complete graph K_{n-1} by adding an end-vertex, and if T is any tree of order n, then they have the same order and the same connectivity, but G appears much more connected than T. Several measures of reliability, including toughness, binding number and integrity, have been introduced as more sensitive measures of reliability. However, computing these parameters appears to be difficult. An average measure inspired by Menger's theorem is the topic of this section, and it can be computed in polynomial time (see Chapter 12 for good algorithms for computing the connectivity).

Average vertex-connectivity

From Menger's theorem, we know that the connectivity $\kappa(v, w)$ between two non-adjacent vertices v and w of a graph G is the maximum number of internally disjoint v–w paths. For the purposes of this section, we extend this to cover all pairs of vertices, and thus define $\kappa(v, w)$, for adjacent vertices v and w, to be the maximum number of internally disjoint v–w paths.

The *average connectivity* $\overline{\kappa}(G)$ of a graph G of order n is the average of the connectivities of all pairs of vertices of G – that is,

$$\overline{\kappa}(G) = \frac{\sum_{v,w} \kappa(v, w)}{\frac{1}{2}n(n-1)}.$$

Since $\kappa(v, w) \le \min\{\deg v, \deg w\}$, for all pairs of vertices v and w in a graph G, it is not difficult to see that if G has degrees $d_1 \ge d_2 \ge \ldots \ge d_n$, then

$$\overline{\kappa}(G) \le \frac{\sum_{v,w} (i-1)d_i}{\frac{1}{2}n(n-1)}.$$

Using this observation and the fact that the average degree $\overline{d}(G)$ is $2m/n$, Beineke, Oellermann and Pippert [5] established the upper bound in the next result. The lower bound was established by Dankelmann and Oellermann [12] by counting the number of 'short paths' that are guaranteed to connect pairs of vertices.

Theorem 5.1 *Let G be a graph with n vertices and $m (\ge n)$ edges and average degree $\overline{d}$, and let $r = 2m - n\lfloor \overline{d} \rfloor$. Then*

$$\frac{\overline{d}^2}{n-1} \le \overline{\kappa}(G) \le \overline{d} - \frac{r(n-r)}{n(n-1)}.$$

Moreover, these bounds are sharp.

Wang and Kleitman [72] considered the problem of constructing a graph of largest possible connectivity with a given degree sequence. Asano [3] investigated complexity issues related to this problem. The problem of finding the largest average connectivity of a graph with a given degree sequence is open. A related problem is to determine for which graphical degree sequences there is a graph for which $\overline{\kappa}(G) = \sum_{i=1}^{n} (i-1)d_i \big/ \frac{1}{2}n(n-1)$.

Dankelmann and Oellermann [12] found bounds on the average connectivity of several families of graphs, including planar and outerplanar graphs, and Cartesian products of graphs. For planar graphs and Cartesian products these bounds are sharp, and the established bound is asymptotically sharp for maximal outerplanar graphs.

Mader [49] posed another interesting question in asking about the greatest connectivity of a subgraph in a graph with a given number of edges. He showed that, for each positive integer t, there is a number $g(t)$ such that every graph G of sufficiently large order n and with more than $g(t)(n-t+1)$ edges contains a t-connected subgraph. Furthermore, there are infinitely many graphs G with $g(t)(n-t+1)$ edges and no t-connected subgraph.

In the same paper Mader showed that

$$\tfrac{3}{2}t - 2 \le g(t) < \tfrac{1}{2}(2+\sqrt{2})(t-1)$$

and conjectured the following.

Conjecture D *There is an integer n_0 such that, if G is a graph of order $n \ge n_0$ and at least $\frac{3}{2}t - 2$ edges, then G contains a t-connected subgraph.*

For sufficiently large orders, Mader's inequalities can be used to show that every graph with average connectivity at least k has a subgraph of connectivity at least $\frac{1}{2}(2-\sqrt{2})k + 1$, and if Mader's conjecture is true, then this can be improved to

$\frac{1}{3}(k+4)$. An upper bound is provided by the graph $2K_r + \overline{K}_r$, whose average connectivity is asymptotically equal to $\frac{4}{9}(4r-3)$, and for which the largest connectivity of a subgraph is r. For sufficiently large r, this is approximately $\frac{9}{16}(k+\frac{4}{3})$.

Uniformly connected graphs

We now turn to graphs with equal connectivity and average connectivity. A graph G is *uniformly k-connected* if $\kappa(G) = \overline{\kappa}(G) = k$. A graph G with connectivity k is *critically k-connected* if $\kappa(G-v) < k$ for each vertex v, and is *minimally k-connected* if $\kappa(G-e) < k$ for each edge e. Some necessary conditions for a graph to be uniformly k-connected were given by Beineke, Oellermann and Pippert [5].

Theorem 5.2 *Every uniformly k-connected graph is minimally k-connected if $k \geq 1$, and is critically k-connected if $k \geq 2$.*

These conditions are not sufficient, as there are graphs that are both minimally and critically k-connected, but not uniformly k-connected. For example, for $r \geq 4$, let G be the graph obtained from the $2r$-cycle $u_1u_2 \ldots u_{2r}u_1$ by adding two new vertices v and w and joining v to each u_i with an even subscript and w to each u_i with an odd subscript. Then G is both minimally and critically 3-connected, but is not uniformly 3-connected since $\kappa(v, w) = r > 3$. Using similar constructions, but with the cycle replaced by circulants of sufficiently large order and connectivity $k-1$, one can construct graphs that are both minimally and critically k-connected, but are not uniformly k-connected for $k > 3$.

It is easy to see that the uniformly 0-connected, 1-connected and 2-connected graphs are the null graphs, the trees and the cycles, and that the uniformly $(n-1)$-connected and $(n-2)$-connected graphs of order n are the complete graphs and the complete graphs minus a 1-factor. Moreover, for $k \geq 3$, a graph is uniformly k-connected if and only if it does not contain a subdivision of $K_2 + \overline{K}_k$. However, finding non-trivial characterizations of uniformly k-connected graphs remains an open problem.

Average edge-connectivity

Analogous to the connectivity between a pair of vertices, the *edge-connectivity* $\lambda(v, w)$ between v and w is the maximum number of edge-disjoint v–w paths in G. The *average edge-connectivity* of G, denoted by $\overline{\lambda}(G)$, is defined to be the average edge-connectivity between pairs of vertices of G – that is,

$$\bar{\lambda}(G) = \frac{\sum_{v,w} \lambda(v, w)}{\frac{1}{2}n(n-1)}.$$

Much less is known about the average edge-connectivity than the average connectivity. We give here a few bounds on this parameter and state some open problems and conjectures. It is well known that, for all vertices v and w in a graph G,

$\kappa(v, w) \le \lambda(v, w) \le \min\{\deg v, \deg w\}$. Thus, if G has average degree $\overline{d}$, then

$$\overline{\kappa}(G) \le \overline{\lambda}(G) \le \overline{d}.$$

It follows from Theorem 5.1 that, if G has order n, then

$$\overline{d}^2/(n-1) \le \overline{\lambda}(G) \le \overline{d}.$$

Much research has focused on conditions that guarantee equality of the edge-connectivity and the minimum degree of a graph. Some of these conditions were given by Chartrand [8], Lesniak [44], Plesník [66] and Volkmann [71]; for example, Chartrand showed that if the minimum degree of an n-vertex graph is at least $\frac{1}{2}n$, then its edge-connectivity equals its minimum degree. It turns out that these four conditions imply something even stronger – namely, that $\lambda(v, w) = \min\{\deg v, \deg w\}$ for all vertices v and w – thereby implying that the average edge-connectivity is as large as possible for these graphs.

Edmonds [15] proved that if **D** is the degree sequence of a connected graph, then there is a graph G with **D** as its degree sequence and $\lambda(G) = \delta(G)$. Thus, if **D** is the degree sequence of a connected graph, then it can be realized by a graph whose edge-connectivity is as large as possible. It is natural to ask a similar question with regard to the average edge-connectivity – namely, if **D** is the degree sequence of a connected graph, can it be realized by a graph for which $\lambda(v, w) = \min\{\deg v, \deg w\}$ for all pairs of vertices v and w? If there is a graph G with degree sequence **D** and $\lambda(v, w) = \min\{\deg v, \deg w\}$ for all v and w, then **D** is called *optimal*. Dankelmann and Oellermann [13] proved the following result about such sequences for multigraphs.

Theorem 5.3 *Let* D $= d_1 \ge d_2 \ge \ldots \ge d_n$ *be the degree sequence of a multigraph and let r be the number of terms equal to* 1. *Then* **D** *is optimal if and only if* $r \le d_1 - d_2$ *or* $r = n - 1$.

In order for such a graphical sequence to be optimal, it is necessary that $d_1 - r \ge d_2$. Indeed, **D** is optimal if and only if the sequence obtained from **D** on replacing d_1 by $d_1 - r$ and deleting all r terms equal to 1 is graphical. Fricke, Oellermann and Swart [25] conjectured the following.

Conjecture E *Every degree sequence with no* 0*s or* 1*s is optimal.*

As was done for the average connectivity, it is natural to ask for the largest edge-connectivity of a subgraph in a graph with given average edge-connectivity. We make a few observations related to this question. Mader [47] showed that if t is a positive integer and if G is a graph of order $n \ge t$ and at least $(t-1)n - \binom{t}{2}$ edges, then G contains a t-edge-connected subgraph. With a little algebra, it can be shown that if $l = 2s$ is an even integer, then a graph G with average edge-connectivity at least l satisfies the hypotheses of Mader's result with $t = s + 1$. Hence, G contains a $(\frac{1}{2}l + 1)$-edge-connected subgraph. However, we believe that this lower bound

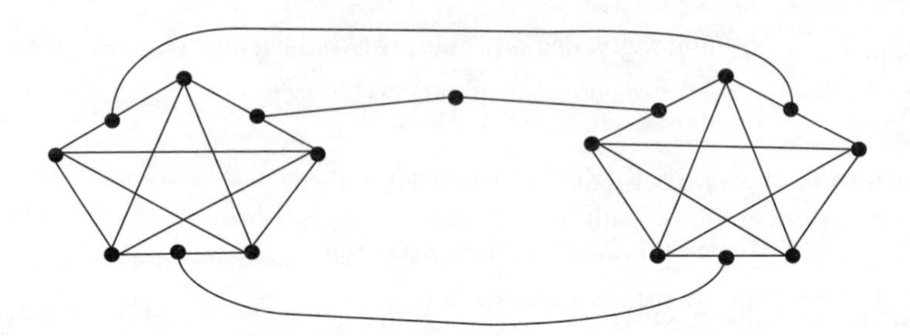

Fig. 5. A graph with average edge-connectivity greater than 3

can be improved significantly. As for upper bounds for this parameter, the graph in Fig. 5, due to Jacques Verstraëte, has average edge-connectivity $\frac{103}{34}$, but has no 3-edge-connected subgraph. Hence, in general, if $\bar{\lambda}(G) \geq l$ for some integer l, then the largest edge-connectivity of a subgraph may be less than l.

Another question to which it would be interesting to have an answer is: when do all pairs of vertices have the same edge-connectivity? That is, which graphs have equal edge-connectivity and average edge-connectivity?

Average connectivity of digraphs

For vertices v, w of a digraph D, $\kappa(v, w)$ denotes the maximum number of internally disjoint directed v–w paths. As one would expect, the *average connectivity* $\overline{\kappa}(D)$ of D is the average of $\kappa(v, w)$ over all ordered pairs v, w – that is,

$$\overline{\kappa}(D) = \frac{\sum_{v,w} \kappa(v, w)}{n(n-1)},$$

where n is the order of D. Clearly, $\kappa(v, w) \leq \min\{d^+(v), d^-(w)\}$, from which it follows that the average connectivity is bounded by the average out-degree – that is, for a digraph with n vertices and m arcs,

$$\overline{\kappa}(D) \leq \frac{m}{n(n-1)}.$$

Rather than look at other digraph results similar to those for graphs, we turn to another type of problem. An *orientation* of a graph G is the result of assigning a direction to each edge of G. A major focus of work on the average connectivity of digraphs has been to investigate the maximum average connectivity among the orientations of a given graph. Let $\overline{\kappa}_{\max}(G)$ and $\overline{\kappa}_{\min}(G)$ denote the maximum and minimum average connectivity among the orientations of G.

Our starting point for this discussion is the class of trees. For a tree T, the problem of finding $\overline{\kappa}_{\max}(T)$ is equivalent to finding an orientation of T that maximizes the number of pairs of vertices v and w for which there exists a directed v–w path. The *centroid* of a tree of order n consists of those vertices for which no component of $T - v$ has more than $\frac{1}{2}n$ vertices. A basic result in this area is that the centroid of every tree consists of either a single vertex or a pair of adjacent vertices.

Let v be a vertex of a tree T. Recall that a *branch at* v is the subtree of T induced by v and those vertices whose path from v begins with a specified edge. An orientation D of T is v*-based* if each branch at v is oriented as either an in-tree or an out-tree rooted at v. Note that D then consists of in-trees at v and out-trees at v with only v in common. The following result is due to Henning and Oellermann [36].

Theorem 5.4 *If D is an orientation of a tree T that achieves the maximum average directed connectivity, then D is a v-based orientation at a centroid vertex v, with the difference in the orders of the in-trees and out-trees at v as small as possible.*

Henning and Oellermann also established bounds on the maximum average connectivity of trees.

Theorem 5.5 *If T is a tree of order at least* 3, *then*

$$\tfrac{2}{9} < \overline{\kappa}_{\max}(T) \leq \tfrac{1}{2}.$$

The bounds in this theorem are sharp, in that the upper bound is attained by oriented paths (and only paths) and there are orientations of a class of trees whose average connectivity gets arbitrarily close to $\frac{2}{9}$. The minimum average connectivity of an orientation of a tree T of order n can readily be seen to be $1/n$, by using the fact that T is a bipartite graph and orienting the edges from one partite set to the other. Indeed, this argument shows that, for any bipartite graph G with n vertices and m edges, $\overline{\kappa}_{\min}(G) = m/n(n-1)$.

The values of $\overline{\kappa}_{\max}(G)$ and $\overline{\kappa}_{\min}(G)$ for regular complete multipartite graphs were found by Henning and Oellermann [36], [35]. These have tournaments as a special case, and bounds on the average connectivity are given in the next theorem.

Theorem 5.6 *If T is a tournament of order n, then*

$$\tfrac{1}{6}(n+1) \leq \overline{\kappa}(T) \leq \begin{cases} \tfrac{1}{2}(n-1) & \text{for } n \text{ odd,} \\[2ex] \dfrac{2n^2 - 5n + 4}{4(n-1)} & \text{for } n \text{ even.} \end{cases}$$

Moreover, the upper bounds sharp, and the lower bound is attained if and only if T is a transitive tournament.

Although $\overline{\kappa}_{\max}(G)$ can be no greater than $\overline{\kappa}(G)$, it seems unlikely that equality ever holds for non-trivial graphs. It would be interesting to have good bounds on $\overline{\kappa}_{\max}(G)$ in terms of $\overline{\kappa}(G)$.

6. Menger results for paths of bounded length

Some paths in a graph may be so long that they are of little practical value. This observation has led to the investigation of connectivity being restricted to paths no longer than some prescribed distance. This section focuses on parameters analogous to κ and μ for paths of bounded length.

Short paths

Following Lovász, Neumann-Lara and Plummer [46], for any integer $d \geq 2$ and any non-adjacent vertices v and w in a graph G, we let $\kappa_d(v, w)$ be the minimum number of vertices in G whose deletion destroys all v–w paths of length at most d. Further, for any positive integer d and any vertices v and w in G, we let $\mu_d(v, w)$ denote the maximum number of internally disjoint v–w paths of length at most d. Obviously, for $d \geq 2$, $\mu_d(v, w) \leq \kappa_d(v, w)$, and there are values of d, such as $d = n - 1$, when G has order n and equality holds. However, equality does not always hold, as the graph in Fig. 6 shows. Here, $\mu_5(v, w) = 1$ and $\kappa_5(v, w) = 2$. Lovász *et al.* [46], Entringer, Jackson and Slater [18] and Hartman and Rubin [34] independently showed that equality also holds at the low end of the range of path-lengths.

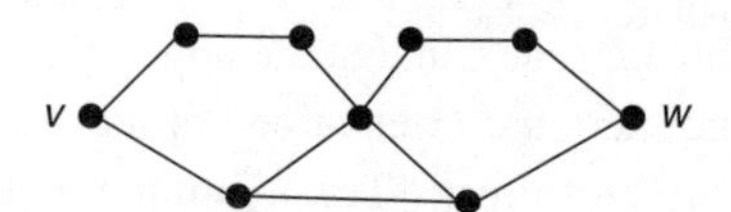

Fig. 6. Destroying paths of bounded length

Theorem 6.1 *If d is the distance between vertices v and w in graph G, then* $\kappa_d(v, w) = \mu_d(v, w)$.

We now turn to upper bounds for $\kappa_d(v, w)$ in terms of $\mu_d(v, w)$. It is convenient to look at these in the form of the ratio

$$\rho_d(v, w) = \frac{\kappa_d(v, w)}{\mu_d(v, w)}.$$

The following results were established in [46].

Theorem 6.2 *If v and w are non-adjacent vertices in a graph G, then*

(a) $\rho_2(v, w) = \rho_3(v, w) = \rho_4(v, w) = 1$;
(b) $\rho_5(v, w) = 2$;
(c) *for* $d \geq 6$, $\rho_d(v, w) \leq \lfloor \frac{1}{2} d \rfloor$.

It is not known, for general d, how close $\rho_d(v, w)$ can get to $\frac{1}{2}d$. Let $\rho_d(G) = \max\{\rho_d(v, w)\}$ over all pairs of non-adjacent vertices v and w in a graph G, and let

$$\rho^*(d) = \sup \{\rho_d(G)\},$$

where the supremum is taken over all graphs G for which $\rho_d(G)$ is defined. Chung [11] obtained bounds for this invariant.

Theorem 6.3 *For* $d \geq 6$,

$$\tfrac{1}{3}(d+1) \leq \rho^*(d) \leq \tfrac{1}{2}d.$$

Having established the lower bound in the above theorem, Chung [11] conjectured that the lower bound in the theorem is close to being exact.

Conjecture F $\rho^*(d) = \frac{1}{3}d + O(1)$.

It was pointed out earlier that $\mu_d(v, w) \leq \kappa_d(v, w)$, and that this inequality may be strict. Intuitively, one may expect though that, for a fixed d, there is a sufficiently large value of c_d for which $\kappa_d(v, w) \leq \mu_{c_d}(v, w)$. This question was answered in the affirmative in [46].

Theorem 6.4 *Let* v *and* w *be non-adjacent vertices of a graph* G*, and let* d *and* k *be positive integers. Then there is a constant* $c(d, k)$ *for which, if* $\kappa_d(v, w) \geq k$*, then* $\mu_{c(d,k)}(v, w) \geq k$.

Let $c_{\min}(d, k)$ be the smallest value for such a constant $c(d, k)$. An upper bound on $c_{\min}(d, k)$ was established in [46] and improved by Pyber and Tuza [68] to

$$c_{\min}(d, k) < \binom{d+k-2}{k}\binom{d+2k-2}{d+k-2}.$$

They also showed by construction that $c(d, k) \geq \lfloor (d/k) - 1 \rfloor^k$.

Menger path systems

So far, we have focused more on the connectivity parameter $\kappa_d(v, w)$ than on the path parameter $\mu_d(v, w)$, but now we shift our focus. For positive integers d and k, a (d, k)-*path system* (or a (d, k)-*skein*) is a set of k internally disjoint paths of length at most d joining a pair of vertices. A graph is said to be (d, k)-*Mengerian* if there is a (d, k)-path system between each pair of its vertices – that is, if $\mu_d(v, w) \geq k$ for all v and w in G.

In introducing this idea, Ordman [64] related it to computer networks and distributed processing. Faudree, Jacobson, Ordman, Schelp and Tuza [23] found a variety of conditions, involving parameters such as the minimum degree, the number of edges and the connectivity, that guarantee this Mengerian property. One example is the following result.

Theorem 6.5 *If* G *is a graph of order* n *with connectivity greater than* $k - 1 + (n-k)/d$*, then* G *is* (d, k)-*Mengerian.*

Other conditions for a graph to be (d, k)-Mengerian were given by Faudree, Gould, Lesniak and Schelp (see [21], [22]). Path systems for edge-disjoint paths were studied by Fathnezhad [19]. An extensive survey of results on Menger path systems is included in Faudree [20].

Long paths

Montejano and Neumann-Lara [59] focused on long paths, rather than short ones. For $d \geq 2$ and non-adjacent vertices v and w in a graph G, let $\kappa'_d(v, w)$ denote the minimum number of vertices in $G - \{v, w\}$ whose deletion destroys all v–w paths of length at least d, and let $\mu'_d(v, w)$ denote the maximum number of internally disjoint v–w paths with length at least d. Montejano and Neumann-Lara were interested in a lower bound for the maximum-sized skein of paths joining vertices with 'long-path connectivity' κ' at least k. More precisely, given a graph G, for $d \geq 2$ and $k \geq 0$, let $\psi_d(G, k) = \min\{\mu'_d(v, w) : \kappa'_d(v, w) \geq k\}$ and let $\psi_d(k) = \inf_G\{\psi_d(G, k)\}$. The following theorem combines results of Mader (see Hager [31]) for the case $d = 3$ with general bounds of Montejano and Neumann-Lara [59] and Hager [31].

Theorem 6.6

(a) *For* $k \geq 0$, $\psi_2(k) = k$ *and* $\psi_3(k) = \lceil \frac{1}{2}k \rceil$.
(b) *For* $k \geq 0$ *and* $d \geq 4$, $\lceil k/(3d-5) \rceil \leq \psi_d(k) \leq \lceil k/(d-1) \rceil$.

7. Connectivity of sets

The beauty of Menger's theorem lies in the way that it relates separating sets to paths joining pairs of vertices. In this section we consider extensions of these concepts to sets of more than two vertices. Our story begins with sets of paths in a graph joining (independent) sets of vertices in a connected subgraph. In Section 8 we move on to trees joining unrestricted sets of vertices.

Total separation

Let G be a graph, and let A be an independent set of vertices with at least two vertices. A subset S of $V - A$ is a *total A-separating set* if the vertices of A are all in different components of $G - S$, and the *total A-connectivity* $\kappa(A)$ is the minimum cardinality of a total A-separating set. An *A-path* is a path whose only vertices in A are its endpoints. We let $\mu(A)$ denote the maximum cardinality of a set of internally disjoint A-paths. Clearly $\mu(A) \leq \kappa(A)$ and, from Menger's theorem, equality holds when $|A| = 2$. However, this need not be the case for larger sets. In fact, $\mu(A)$ can be as small as $\frac{1}{2}\kappa(A)$, as the following example shows. For $r \geq 1$, let G be the graph obtained from K_{2r+1} by adding an end-vertex at each of its vertices, and let A be the set of end-vertices of G. Then $\mu(A) = r$ and $\kappa(A) = 2r$ (see Fig. 7 for the case $r = 1$). It was conjectured by Gallai [26] and proved by Mader [53] that this is the extreme case.

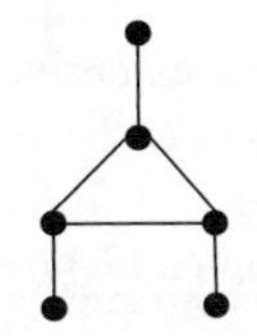

Fig. 7. Paths between pairs of vertices in a set

Theorem 7.1 *If A is an independent set of vertices in a graph, then*

$$\tfrac{1}{2}\kappa(A) \le \mu(A) \le \kappa(A),$$

and these bounds are sharp.

The upper bound is attained by the complete bipartite graph $K_{r,r}$, where A is one of the partite sets.

Analogous to the edge-connectivity of a graph G, there is also an edge version of total connectivity. For a set of vertices A with $|A| \ge 2$, a set F of edges is a *total edge-A-separating set* if the vertices of A are all in different components of $G - F$. The *total edge-A-connectivity* $\lambda(A)$ is the minimum cardinality of a total edge-A-separating set. The number $\nu(A)$ is the maximum number of paths between pairs of vertices in A that are pairwise edge-disjoint. Mader [52] established results similar to Theorem 7.1.

Theorem 7.2 *If A is any set of vertices in a graph, then*

$$\tfrac{1}{2}\lambda(A) \le \nu(A) \le \lambda(A),$$

and these bounds are sharp.

The sharpness of the lower bound follows from the complete bipartite graph $K_{2r+1,2r+1}$, with A being one of the partite sets, since then $\lambda(A) = 2r(2r+1)$ and $\nu(A) = r(2r+1)$. The upper bound holds for any complete graph with A as the full vertex-set.

Disconnecting a graph into more than two components

There are also global parameters related to A-paths. Just as the connectivity of a non-complete graph is the minimum connectivity of a pair of vertices in the graph, so the *r-connectivity* $\kappa_r(G)$ of a graph G with independence number at least r is the minimum value of $\kappa(A)$, where A is a set of r independent vertices, and that of a graph of order n and independence number less than r is $n - r$: that is, $\kappa_r(G)$ is the smallest number of vertices whose removal produces a graph with at least r components or a graph with fewer than r vertices. This concept was first introduced by Chartand, Kapoor, Lesniak and Lick [10], who also defined the *connectivity sequence* of a graph G of order n to be the sequence $(\kappa_2(G), \kappa_3(G), \ldots, \kappa_n(G))$. One of their results gives criteria for a sequence of non-negative integers to be the connectivity sequence of some graph.

Theorem 7.3 *A sequence $r_2, r_3, \ldots, r_n$ of non-negative integers is the connectivity sequence of a graph of order n if and only if there is an integer a such that $r_2 \leq r_3 \leq \ldots \leq r_a \leq r_{a+1}$ and $r_{a+i} = n - (a+i) + 1$, for $i = 1, 2, \ldots, n-a$. Moreover, a is the independence number of any graph with this as its connectivity sequence.*

There is naturally a counterpart for edge-deletion: the *r-edge-connectivity* $\lambda_r(G)$ of a graph G is the minimum number of edges whose removal results in a graph with at least r components. As in the vertex case, $\lambda_2(G) = \lambda(G)$. This concept was introduced by Boesch and Chen [6] and was studied further by Goldsmith [27], [28] and Goldsmith, Manvel and Faber [29]. The focus of these papers is on heuristics and bounds for approximating the r-edge-connectivity.

Much of the other work on r-connectivity concerns degree conditions that guarantee lower bounds for this parameter (see [62], [63], [69]). The next result was established in Kapoor, Lesniak and Lick [10].

Theorem 7.4 *Let G be a graph of order n with independence number at least r. If*

$$\delta(G) \geq \frac{n + (r-1)(k-2)}{r},$$

then $\kappa_r(G) \geq k$.

When $r = 2$, this result gives the well-known minimum degree condition guaranteeing k-connectedness in a graph (see [9]). This minimum degree condition also guarantees the existence of a large number of 'short' internally disjoint paths (see [62]).

Theorem 7.5 *Let G be a graph of order $n \geq 2$ and let $r \geq 3$ and k be an integer such that $1 \leq k \leq n - r + 1$. If*

$$\delta(G) \geq \frac{n + (r-1)(k-2)}{r},$$

then for each set A of r vertices there exist at least k internally disjoint A-paths, each of length at most 2.

Whitney [73] showed that $\kappa(G) \leq \lambda(G)$ for any graph G. It seems natural to seek extensions for Whitney's inequality – namely, that $\kappa_r(G) \leq \lambda_r(G)$ for any graph G. However, it turns out that the proposed inequality fails even for $r = 3$. Take, for example, the graph G obtained from $K_{l,l} \cup K_2$ ($l \geq 3$) by adding an edge between the two components; then $\lambda_3(G) = 2$ and $\kappa_3(G) = l$.

8. Connecting with trees

In 1961, Tutte [70] and Nash-Williams [60] independently proved a classic result on packing a graph with trees. (An analogue for covering a graph with trees was published three years later by Nash-Williams [61].) In order to state their result, it is convenient to have some additional terminology and notation. The *boundary* ∂S of a set S of vertices in a graph G is the set of edges that join a vertex in S to a vertex not in S. More generally, given a set of pairwise disjoint sets $\mathcal{P} = \{u_1, u_2, \ldots, u_t\}$ of vertices of G, $|\mathcal{P}|$ is the number of sets in $\mathcal{P}$, and the *boundary* $\partial\mathcal{P}$ of $\mathcal{P}$ is the union of the boundaries of the elements of $\mathcal{P}$. Tutte and Nash-Williams' theorem is then as follows.

Theorem 8.1 *A graph G has r edge-disjoint spanning trees if and only if*

$$|\partial\mathcal{P}| \geq r(|\mathcal{P}| - 1),$$

for each partition $\mathcal{P}$ of its vertex-set V.

The following corollary can readily be deduced from this result.

Corollary 8.2 *Every $2k$-edge-connected graph has a set of k edge-disjoint spanning trees.*

While this result is similar in appearance to Menger's theorem, there is also a deeper connection. It concerns a form of separation that is less restricted than the total version considered in Section 7, and involves connecting sets of vertices with trees rather than paths. We begin with a result of Keijsper and Schrijver [39].

Given a graph G with $S \subseteq V$, a set F of edges of G is an S–$\overline{S}$ *connector* if every component of the subgraph of G with edge-set F meets both S and $\overline{S}$. A *subpartition* $\mathcal{P}$ of a set X is a set of pairwise disjoint non-empty subsets of X. Keijsper and Schrijver proved the following extension of the theorem of Tutte and Nash-Williams.

Theorem 8.3 *Let S be a set of vertices in a graph G. Then G has r edge-disjoint S–$\overline{S}$ connectors if and only if $|\partial\mathcal{P}| \geq r|\mathcal{P}|$, for each subpartition $\mathcal{P}$ of S or $\overline{S}$.*

In the case where S is a singleton, an S-connector is a connected spanning subgraph of G, so Theorem 8.1 is a special case of Theorem 8.3. Indeed, Keijsper and Schrijver used the result of Nash-Williams and Tutte to establish their own result. Their proof also yields a polynomial algorithm for finding r edge-disjoint S-connectors when they exist.

Given a set A of vertices in a graph G, a subgraph of G is called an *A-tree* if it is a tree containing the vertices in A, and if each of its end-vertices is in A. Just as with most of the concepts we have considered, there are both vertex and edge versions, but here, primarily for historical reasons, we begin with the edge results.

Edge-disjoint trees

Kriesell [43] further considered the case where $2 < |A| < n$. A set of vertices A is *l-edge-connected* if $\lambda(v, w) \geq l$, for all vertices v and w in A. Kriesell showed that for any positive integers r and s there is a smallest number $l = f_s(r)$ such that, for each l-edge-connected set A of order at most s, there is a set of r edge-disjoint A-trees. He settled the case of $s = 3$; for larger values of s the problem of finding $f_s(r)$ remains open.

Theorem 8.4 *Let G be a graph and let A be a $\lfloor \frac{1}{3}(4k+1) \rfloor$-edge-connected set of three vertices of G. Then G has a set of k edge-disjoint A-trees. Furthermore, this bound on the edge-connectedness of the set is sharp.*

Kriesell also conjectured an analogue to Corollary 8.2.

Conjecture G *If A is a $2k$-edge-connected set of vertices in a graph G, then G has a set of k edge-disjoint A-trees.*

Note that the hypothesis of the conjecture does not include the assumption that G itself be $2k$-edge-connected. The following theorem, also due to Kriesell [43], gives the minimum edge-connectivity that guarantees that a graph has a specified number of edge-disjoint spanning trees.

Theorem 8.5 *For $n, k \geq 2$, every $(2k + 1 - \lceil (2k+2)/n \rceil)$-edge-connected graph of order n has k edge-disjoint spanning trees. Furthermore, this value is sharp.*

Internally disjoint trees

We now turn to the vertex analogue, in which the A-trees are required to be disjoint, except for the vertices in A; parallel to other usage, we call such A-trees *internally disjoint*. For a set A of independent vertices in a graph G, let $\mu_\tau(A)$ be the maximum number of trees in any set of internally disjoint A-trees in G. For $k, t \geq 2$, let $\beta(t, k)$ be the minimum value of $\mu_\tau(A)$, taken over all graphs G with order at least t and all k-connected sets A of t independent vertices in G. It follows from Menger's theorem that $\beta(2, k) = k$, and exact values are also known for $t = 3$ and 4. However, for larger values of t, only bounds are known. The results in the following theorem are due to Hind and Oellermann [37].

Theorem 8.6

(a) *For $k \geq 2$, $\beta(2, k) = k$, $\beta(3, k) = \lfloor \frac{1}{4}(3k+1) \rfloor$ and $\beta(4, k) = \lfloor \frac{2}{3}k \rfloor$.*
(b) *For $t \geq 5$ and $k \geq 2$, $\left\lceil \left(\frac{2}{3}\right)^{t-2} k \right\rceil \leq \beta(t, k) \leq \left\lfloor \left\lceil \frac{1}{2}tk \right\rceil / (t-1) \right\rfloor$.*

We note that, for $t = 2$, 3 and 4, the upper bound in part (b) equals the value in part (a), and Hind and Oellermann [37] conjectured that this is always an equality. The upper bound was established by constructing a multigraph H with t vertices and no more than $\lfloor \frac{1}{2}tk/(t-1) \rfloor$ edge-disjoint spanning trees. If G is the result of subdividing each edge of H once, and if A is the set of vertices in H, the result follows from the observation that a family of internally disjoint A-trees in G corresponds to a family of edge-disjoint A-trees in H. The lower bound can be established using induction and the fact that if a set A of vertices in a graph G is k-connected, then so is every subset of A with order at least 2.

Hager [30] studied sets of internally disjoint A-trees with additional structural properties and in a global setting. For positive integers t and k, a graph G of order at least $t+k$ is *(t,k)-pendant-tree-connected* if, for each set A of t vertices, there are k internally disjoint A-trees in each of which every vertex of A is an end-vertex (also known as a pendant vertex). For a graph G, the *pendant-tree-connectivity* $\tau_t(G)$ is the largest integer k for which G is (t,k)-pendant-tree-connected. Let $\tau(t,k) = 1 + \max\{\kappa(G)\colon \tau_t(G) < k\}$. It is not hard to see that, for all t, $\tau(t,1) = t$. The following theorem contains some of Hager's results [30].

Theorem 8.7

(a) *For* $k \geq 2$, $\tau(1,k) = \tau(2,k) = k$.
(b) *For* $k \geq 2$ *and* $t \geq 3$, $t+k+1 \leq \tau(t,k) \leq 2^{k(t+1)}$.

The lower bound was established by a recursive construction. In establishing the upper bound, it was first shown that $\tau_t(G) \geq k$ for every kt-connected graph containing a subdivision of the complete graph $K_{k(t+1)}$. This gives the upper bound when combined with the result of Mader [48] that every graph of order n with at least $2^{r-1}n$ edges contains a subdivision of K_r.

Hager [30] also established some relationships between k-linked graphs and the pendant-tree-connectivity.

Theorem 8.8

(a) *If G is a graph with minimum degree* $\delta(G) \geq 2k(t-1)+1$ *for which* $G - v$ *is* $k(t-1)$*-linked for each vertex v, then* $\tau_t(G) \geq k$.
(b) *If* $\tau_{2k}(G) \geq k$, *then G is k-linked.*

Dirac [14] showed that, in a k-connected graph, every set of $k+1$ vertices lie on a common path, a result that motivated Hager [32] to ask how many internally disjoint A-paths exist for a given set A of vertices in a graph G. This question is most interesting when G satisfies extra conditions, often involving connectivity. Similarly, one can seek conditions on G such that, for each set A of t vertices, there are at least k internally disjoint A-paths. A graph G with at least $\max\{t, k+1\}$ vertices is called *(t,k)-path-connected* if every set A of t vertices has k internally disjoint A-paths. The *t-th path-connectivity number* $\pi_t(G)$ is the largest integer k for which G is (t,k)-path-connected; thus, $\pi_2(G) = \kappa(G)$. Hager [32] established the following result on connectivity that ensures t-path-connectivity of a given size.

Theorem 8.9

(a) *If G is $3k$-connected, then* $\pi_4(G) \geq k$.
(b) *If G is* $\lceil \frac{9}{2}k \rceil$*-connected, then* $\pi_5(G) \geq k$.
(c) *If G is* $2^{t-2}k$*-connected, where* $t \geq 6$, *then* $\pi_t(G) \geq k$.

Analogous to the function $\tau(t, k)$ discussed above, we define, for $t \geq 2$ and $k \geq 1$, $\pi(t, k) = 1 + \max\{\kappa(G) : \pi_t(G) < k\}$. Hager [32] established the results in the next theorem.

Theorem 8.10

(a) *For* $k \geq 2$, $\pi(2, k) = k$, $\pi(3, k) = 2k$ *and* $\pi(4, k) = 3k$.
(b) *For* $t \geq 5$ *and* $k \geq 2$, $\pi(t, k) \geq k(t-1)$.

Note that the lower bound in (b) is the actual value for $\pi(t, k)$ for the cases $t = 2, 3$ and 4. Hager conjectured that this is always the case.

Conjecture H *For* $t \geq 5$ *and* $k \geq 2$, $\pi(t, k) = k(t-1)$.

References

1. R. Aharoni, Menger's theorem for graphs containing no infinite paths, *Europ. J. Combin.* **4** (1983), 201–204.
2. R. Aharoni, Menger's theorem for countable graphs, *J. Combin. Theory (B)* **43** (1987), 303–313.
3. T. Asano, An $O(n \log\log n)$ time algorithm for constructing a graph of maximum connectivity with prescribed degrees, *J. Comp. Sys. Sciences* **51** (1995), 503–510.
4. L. W. Beineke and F. Harary, The connectivity function of a graph, *Mathematika* **14** (1967), 197–202.
5. L. W. Beineke, O. R. Oellermann and R. E. Pippert, The average connectivity of a graph, *Discrete Math.* **252** (2002), 31–45.
6. F. T. Boesch and S. Chen, A generalization of line connectivity and optimally invulnerable graphs, *SIAM J. Appl. Math.* **34** (1978), 657–665.
7. S. M. Boyles and G. Exoo, A counterexample on a conjecture on paths of bounded lengths, *J. Graph Theory* **6** (1982), 205–209.
8. G. Chartrand, A graph theoretic approach to a communications problem, *SIAM J. Appl. Math.* **14** (1996), 778–781.
9. G. Chartrand and F. Harary, Graphs with prescribed connectivities, *Theory of Graphs, Proceedings of the Colloquium Held at Tihany, Hungary*, Akadémiai Kaidó, 1968.
10. G. Chartrand, S. F. Kapoor, L. Lesniak and D. R. Lick, Generalized connectivity in graphs, *Bull. Bombay Math. Colloq.* **2** (1984), 1–6.
11. F. R. K. Chung, Problem on short Menger path systems, *Finite and Infinite Sets* (eds. A. Hajnal, L. Lovász and V. T. Sós), Colloq. Math. Soc. János Bolyai **37** (1984), 873.
12. P. Dankelmann and O. R. Oellermann, Bounds on the average connectivity of a graph. *Discrete Appl. Math.* **129** (2003), 305–318.
13. P. Dankelmann and O. R. Oellermann, Degree sequences of optimally edge-connected multigraphs, *Ars Combin.* **77** (2005), 161–168.
14. G. A. Dirac, In abstrakten Graphen vorhandene vollständige 4-Graphen und ihre Unterleitungen, *Math. Nachr.* **22** (1960), 61–85.
15. J. Edmonds, Existence of k-edge connected ordinary graphs with prescribed degrees, *J. Res. Nat. Bur. Std.-B, Math. and Math. Phys.* **68B** (1964), 73–74.

16. Y. Egawa, A. Kaneko and M. Matsumoto, A mixed version of Menger's theorem, *Combinatorica* **11** (1991), 71–74.
17. H. Enomoto and A. Kaneko, The condition of Beineke and Harary on edge-disjoint paths some of which are openly disjoint, *Tokyo J. Math.* **17** (1994), 355–357.
18. R. Entringer, D. E. Jackson and P. J. Slater, Geodetic connectivity of graphs, *IEEE Trans. Circuits Systems* **24** (1977), 460–463.
19. F. Fathnezhad, *Generalized degree and connectivity conditions that imply edge Menger path systems*, Ph.D. thesis, Memphis State University, 1992.
20. R. J. Faudree, Some strong variations of connectivity, *Combinatorics, Paul Erdős is Eighty*, Bolyai Soc. Math. Stud. **1** (1993), 125–144.
21. R. J. Faudree, R. J. Gould and L. Lesniak, Generalized degrees and Menger path systems, *Discrete Appl. Math.* **38** (1992), 1–13.
22. R. J. Faudree, R. J. Gould and R. H. Schelp, Menger path systems, *J. Combin. Math. Combin. Comput.* **6** (1989), 9–21.
23. R. J. Faudree, M. S. Jacobson, E. T. Ordman, R. H. Schelp and Z. Tuza, Menger's theorem and short paths, *J. Combin. Math. Combin. Comput.* **2** (1987), 235–253.
24. L. R. Ford Jr. and D. R. Fulkerson, Maximal flow through a network, *Canad. J. Math.* **8** (1956), 399–404.
25. G. Fricke, O. R. Oellermann and H. C. Swart, The average edge-connectivity and degree conditions, preprint.
26. T. Gallai, Maximum-minimum Sätze and verallgemeinerte Faktoren von Graphen, *Acta Math. Acad. Sci. Hungar.* **12** (1961), 131–173.
27. D. L. Goldsmith, On the second-order edge-connectivity of a graph, *Congr. Numer.* **29** (1980), 479–484.
28. D. L. Goldsmith, On the nth order connectivity of a graph. *Congr. Numer.* **32** (1981), 375–382.
29. D. L. Goldsmith, B. Manvel and V. Faber, Separation of graphs into three components by removal of edges, *J. Graph Theory* **4** (1980), 213–218.
30. M. Hager, Pendant tree-connectivity, *J. Combin. Theory (B)* **38** (1985), 179–189.
31. M. Hager, A Mengerian theorem for paths of length at least three, *J. Graph Theory* **10** (1986), 533–540.
32. M. Hager, Path-connectivity in graphs, *Discrete Math.* **59** (1986), 53–59.
33. G. Hajós, Zum Mengerschen Graphensatz, *Acta Litt. Sci. Szeged* **7** (1934), 44–47.
34. J. Hartman and I. Rubin, On diameter stability of graphs, *Theory and Applications of Graphs* (eds. Y. Alavi and D. R. Lick), Lecture Notes in Mathematics **642**, Springer (1976), 247–254.
35. M. A. Henning and O. R. Oellermann, The average connectivity of regular multipartite tournaments, *Australas. J. Combin.* **23** (2001), 101–113.
36. M. A. Henning and O. R. Oellermann, The average connectivity of a digraph, *Discrete Appl. Math.* **140** (2004), 143–153.
37. H. R. Hind and O. R. Oellermann, Menger-type results for three or more vertices, *Congr. Numer.* **113** (1996), 179–204.
38. A. Kaneko and K. Ota, On minimally (n, λ)-connected graphs, *J. Combin. Theory (B)* **80** (2000), 156–171.
39. J. Keijsper and A. Schrijver, On packing connectors, *J. Combin. Theory (B)* **73** (1998), 184–188.
40. D. König, Graphok és matrixok, *Mat. Fiz. Lapok* **38** (1931), 116–119.
41. D. König, Über trennende Knotenpunkte in Graphen, *Acta Litt. Sci. Szeged* **6** (1933), 155–179.

42. D. König, *Theorie der Endlichen und Unendlichen Graphen*, Chelsea, 1950.
43. M. Kriesell, Edge-disjoint trees containing some given vertices in a graph, *J. Combin. Theory (B)* **88** (2003), 53–65.
44. L. Lesniak, Results on the edge-connectivity of graphs, *Discrete Math.* **8** (1974), 351–354.
45. L. Lovász, On some connectivity properties of eulerian graphs, *Acta Math. Acad. Sci. Hungar.* **28** (1976), 129–138.
46. L. Lovász, V. Neumann-Lara and M. Plummer, Mengerian theorems for paths of bounded length, *Periodica Math. Hungar.* **9** (1978), 269–276.
47. W. Mader, Minimale n-fach kantenzusammenhängende Graphen, *Math. Ann.* **191** (1971), 21–28.
48. W. Mader, Hinreichende Bedingungen für die Existenz von Teilgraphen, die zu einem vollständigen Graphen homöomorph sind, *Math. Nachr.* **53** (1972), 145–150.
49. W. Mader, Existenz n-fach zusammenhängender Teilgraphen in Graphen genügend grosser Kantendichte, *Abh. Math. Sem. Univ. Hamburg* **37** (1972), 86–97.
50. W. Mader, Grad und lokaler Zusammenhang in endlichen Graphen, *Math. Ann.* **205** (1973), 9–11.
51. W. Mader, Ecken mit starken Zusammenhangseigenschaften in endlichen Graphen, *Math. Ann.* **216** (1975), 123–126.
52. W. Mader, Über die Maximalzahl kantendisjunkter A-Wege, *Arch. Math.* **30** (1978), 325–336.
53. W. Mader, Über die Maximalzahl kreuzungsfreier H-Wege, *Arch. Math.* **31** (1978), 387–402.
54. W. Mader, Connectivity and edge-connectivity in finite graphs, *Surveys in Combinatorics* (ed. B. Bollobás), London Math. Soc. Lecture Notes **38**, Cambridge University Press (1979), 66–95.
55. W. McCuaig, A simple proof of Menger's theorem, *J. Graph Theory* **8** (1984), 427–429.
56. K. Menger, Zur allgemeinen Kurventheorie, *Fund. Math.* **10** (1927), 96–115.
57. K. Menger, *Kurventheorie*, Teubner, 1932.
58. K. Menger, On the origin of the n-arc theorem, *J. Graph Theory* **5** (1981), 341–350.
59. L. Montejano and V. Neumann-Lara, A variation on Menger's theorem for long paths, *J. Combin. Theory (B)* **36** (1984), 213–217.
60. C. St.J. A. Nash-Williams, Edge-disjoint spanning trees of finite graphs, *J. London Math. Soc.* **36** (1961), 445–450.
61. C. St.J. A. Nash-Williams, Decomposition of finite graphs into forests, *J. London Math. Soc.* **39** (1964), 12.
62. O. R. Oellermann, On the l-connectivity of a graph, *Graphs and Combin.* **3** (1987), 285–299.
63. O. R. Oellermann, A note on the l-connectivity function of a graph, *Congr. Numer.* **60** (1987), 181–188.
64. E. T. Ordman, Fault-tolerant networks and graph connectivity, *J. Combin. Math. Combin. Comput.* **1** (1987), 191–205.
65. O. Ore, *Theory of Graphs*, Amer. Math. Soc. Colloq. Publ. **XXXVIII**, American Mathematical Society, 1962.
66. J. Plesník, Critical graphs of given diameter, *Acta. Fac. Rerum Natur. Univ. Comm. Math.* **30** (1975), 71–93.
67. K. P. Podewski and K. Steffens, Über Translationen und der Satz von Menger in unendlichen Graphen, *Acta. Math. Acad. Sci. Hungar.* **30** (1977), 69–84.
68. L. Pyber and Z. Tuza, Menger-type theorems with restrictions on path lengths, *Discrete Math.* **120** (1993), 161–174.

69. E. Sampathkumar, Connectivity of a graph – a generalization, *J. Combin. Info. Syst. Sci.* **9** (1984), 71–78.
70. W. T. Tutte, On the problem of decomposing a graph into n connected factors, *J. London Math. Soc.* **36** (1961), 221–230.
71. L. Volkmann, Edge-connectivity in p-partite graphs, *J. Graph Theory* **13** (1989), 1–6.
72. D. L. Wang and D. J. Kleitman, On the existence of n-connected graphs with prescribed degrees, *Networks* **3** (1973), 225–239.
73. H. Whitney, Congruent graphs and the connectivity of graphs, *Amer. J. Math.* **54** (1932), 150–168.
74. J. M. Xu, A proof of the mixed form of Menger's theorem [in Chinese], *J. Huashong Univ. Sci. Tech.* **12** (1984), 17–18.

2

Maximally connected graphs

DIRK MEIERLING and LUTZ VOLKMANN

That the edge-connectivity of a graph does not exceed the minimum degree of a vertex in the graph follows at once from the definitions. A graph is called maximally edge-connected if equality holds, and in this chapter, we explore conditions on a graph that guarantee that it is maximally edge-connected. Similarly, a graph is called maximally connected if its connectivity equals the minimum degree. These graphs are also investigated, as are the corresponding types of digraphs. Furthermore, we study those graphs and digraphs in which analogous equalities hold for all pairs of vertices. We conclude with the discussion of results related to conditional connectivity.

1. Introduction

The starting point for this chapter is a result proved by Chartrand [9] in 1966, that if the minimum degree in a graph of order n is at least $\frac{1}{2}(n-1)$, then the edge-connectivity equals the minimum degree. Since it is well known that the edge-connectivity of any graph is bounded from above by the minimum degree, this gives a family of graphs that are optimally edge-connected. We define a graph to be *maximally edge-connected* if its edge-connectivity equals its minimum degree. In Section 2, we discuss a variety of conditions on a graph that guarantee that it is maximally edge-connected, many of which involve the degrees of vertices. Section 3 is devoted to the corresponding concept for edge-connectivity in digraphs.

In 1932, Whitney [88] proved that the (vertex-)connectivity $\kappa(G)$ of a graph G never exceeds the edge-connectivity, thereby establishing this classic set of inequalities:

$$\kappa(G) \leq \lambda(G) \leq \delta(G),$$

where $\delta(G)$ is the minimum degree in G. Section 5 is devoted to graphs and digraphs in which $\kappa(G) = \delta(G)$, as well as the universally local versions of these concepts; these are based on the following set of inequalities that hold for every pair of vertices v and w in a graph G:

$$\kappa(v, w) \leq \lambda(v, w) \leq \min\{d(v), d(w)\},$$

where $\kappa(v, w)$ is the maximum number of internally vertex-disjoint paths between v and w and $\lambda(v, w)$ is the maximum number of edge-disjoint paths between v and w.

In the last two sections we survey results related to conditional edge-connectivity and the corresponding notion of conditional vertex-connectivity. The *conditional edge-connectivity* of a graph G with respect to some property P is the cardinality of the smallest set of edges in G whose removal results in a disconnected graph in which each component has property P. In the special case where P is the property of having at least two vertices, this is referred to as *restricted edge-connectivity.*

Since every connected graph with a leaf has connectivity 1, and therefore satisfies the definition of being maximally edge-connected, in this chapter all of the graphs are connected and have order $n \geq 3$ and minimum degree $\delta \geq 2$.

2. Maximally edge-connected graphs

As we stated in Section 1, the first result on maximally edge-connected graphs is due to Chartrand [9].

Theorem 2.1 *Every graph of order n with minimum degree $\delta \geq \frac{1}{2}(n - 1)$ is maximally edge-connected.*

Degree conditions

Over the years, this theorem has been strengthened many times and in many ways. Here we focus on the most recent results, a great number of which involve the degrees of vertices in a graph. For this reason, we adopt the convention that **D** denote the degree sequence of a given graph G with the degrees in non-increasing order: $d_1 \geq d_2 \geq \cdots \geq d_n$.

One small extension, due to Boesch and Chen [4], is that the minimum degree δ in Theorem 2.1 can be replaced by d_{n-1}. Other improvements are more substantial, one of which is the following theorem of Plesník [67].

Theorem 2.2 *Every graph of diameter* 1 *or* 2 *is maximally edge-connected.*

Theorem 2.2 implies not only Chartrand's theorem but also the following result of Lesniak [47].

Corollary 2.3 *If G is a graph of order n in which $d(v)+d(w) \geq n-1$ for all pairs v and w of non-adjacent vertices, then G is maximally edge-connected.*

In contrast to Lesniak's condition on non-adjacent vertices, Hellwig and Volkmann [35] discovered a result that involves only adjacent vertices.

Theorem 2.4 *If a graph of order n is such that at least one vertex of each pair of adjacent vertices has degree at least $\frac{1}{2}(n-1)$, then it is maximally edge-connected.*

We now state some results that specifically address graphs to which Chartrand's theorem does not apply. The first is due to Dankelmann and Volkmann [14] and involves the sums of certain combinations of degrees.

Theorem 2.5 *Let G be a graph of order n with degree sequence* **D** *and minimum degree $\delta < \frac{1}{2}(n-1)$. If*

$$\sum_{i=1}^{k}(d_i + d_{n+i-\delta-1}) \geq k(n-2) + 2\delta - 1,$$

for some $k \leq \delta$, then G is maximally edge-connected.

Theorem 2.5 implies the following result of Bollobás [5].

Corollary 2.6 *Let G be a graph of order n with degree sequence* **D** *and minimum degree $\delta < \frac{1}{2}(n-1)$. If*

$$\sum_{i=1}^{k}(d_i + d_{n-i}) \geq kn - 1,$$

for each $k \leq \frac{1}{2}n - 1$, then G is maximally edge-connected.

Hellwig and Volkmann [31] gave a sufficient condition for maximal edge-connectedness that uses only the lower end of the degree sequence.

Theorem 2.7 *Let G be a graph of order n with degree sequence* **D** *and minimum degree $\delta < \frac{1}{2}(n-1)$. If*

$$\sum_{i=1}^{2k} d_{n+1-i} \geq \max\{k(n-1) - 1, (k-1)n + 2\delta - 1\},$$

for some k with $2 \leq k \leq \delta$, then G is maximally edge-connected.

In 1979, Goldsmith and Entringer [27] showed that a graph is maximally edge-connected if the sum of the degrees of the neighbours of each vertex of minimum degree is sufficiently large. This result is sharp and implies Chartrand's theorem.

Theorem 2.8 *If G is a connected graph of order n with*

$$\sum_{w \in N(v)} d(w) \geq \begin{cases} \left\lfloor \frac{1}{2}n \right\rfloor^2 - \left\lfloor \frac{1}{2}n \right\rfloor & \text{for all even } n \text{ and for odd } n \leq 15, \\ \left\lfloor \frac{1}{2}n \right\rfloor^2 - 7 & \text{for odd } n \geq 15, \end{cases}$$

for each vertex v of minimum degree, then G is maximally edge-connected.

If G is a graph without isolated vertices, then its *inverse degree* is defined as $\sum_{v \in V(G)} 1/d(v)$. Using Jensen's inequality from analysis, Dankelmann, Hellwig and Volkmann [12] proved the following result.

Theorem 2.9 *If G is a connected graph of order n, minimum degree $\delta \leq n - 3$ and inverse degree less than*

$$2 + \frac{2}{\delta(\delta+1)} + \frac{n - 2\delta}{(n - \delta - 2)(n - \delta - 1)},$$

then G is maximally edge-connected.

Dankelmann, Hellwig and Volkmann [12] also presented an infinite class of examples which show that the bound in Theorem 2.9 is sharp and gave a corresponding result for triangle-free graphs.

Diameter and girth conditions

As Theorem 2.2 says, every graph of diameter at most 2 is maximally edge-connected. Soneoka, Nakada, Imase and Peyrat [73], [74] established some related results, including one that involves both the diameter and the girth of a graph. Their condition is sharp for infinitely many values of δ when the girth is 4 or odd.

Theorem 2.10 *Let G be a graph with diameter h and girth g. If $h \leq g - 1$ when g is odd, or $h \leq g - 2$ when g is even, then G is maximally edge-connected.*

Refinements of Theorem 2.10 have been found by Fiol and Fàbrega [19], [23]. Another result of Soneoka *et al.* [74] is the following.

Theorem 2.11 *A graph with minimum degree δ, maximum degree $\Delta \geq 3$ and diameter h is maximally edge-connected if it has order greater than*

$$(\delta - 1)\frac{(\Delta - 1)^{h-1} - 1}{\Delta - 2} + \Delta + \delta - 2.$$

The next condition by Fàbrega and Fiol [20] depends on the girth of a bipartite graph.

Theorem 2.12 *Every bipartite graph of minimum degree at least 2, girth g and diameter $h \leq g-1$ is maximally edge-connected.*

Let G be a bipartite graph. In the case that the diameter is at most 3, Theorem 2.12 yields $\lambda(G) = \delta(G)$. If the diameter is at least 4, then a result of Bondy and Murty [7] implies that the diameter of the complementary graph $\overline{G}$ is at most 2, so, Plesník's Theorem 2.2 leads to $\lambda(\overline{G}) = \delta(\overline{G})$. So, if G is a bipartite graph, then $\lambda(G) = \delta(G)$ or $\lambda(\overline{G}) = \delta(\overline{G})$. Recently, Hellwig and Volkmann [37] proved that this statement is valid for all graphs.

Theorem 2.13 *For any graph G, either G or $\overline{G}$ is maximally edge-connected.*

Corollary 2.14 *Every self-complementary graph is maximally edge-connected.*

Clique number conditions

The story here begins with a theorem of Volkmann [78] on r-partite graphs.

Theorem 2.15 *Every r-partite graph with minimum degree δ and order at most $2\lfloor(\delta r/r-1)\rfloor - 1$ is maximally edge-connected.*

Results on r-partite graphs have generally been superseded by results involving the clique number $\omega(G)$, the order of a largest complete subgraph of a graph G. Clearly, the clique number of an r-partite graph cannot be greater than r. Turán's well-known theorem on K_r-free graphs implies that the maximum number of edges in a graph of order n and clique number ω is at most $\frac{1}{2}\delta\omega n^2/(\omega-1)$. This leads to the following strengthening of Theorem 2.15 due to Dankelmann and Volkmann [13].

Theorem 2.16 *Every graph with minimum degree δ, clique number ω and order at most $2\lfloor(\delta\omega/\omega-1)\rfloor - 1$ is maximally edge-connected.*

The special case $\omega = 2$ in Theorem 2.15 leads to a result of Volkmann [77] on maximally edge-connected bipartite graphs.

Corollary 2.17 *Every bipartite graph of order n and minimum degree $\delta \geq \frac{1}{4}(n+1)$ is maximally edge-connected.*

During the past quarter-century, various improvements and extensions of Volkmann's result have been obtained. For example Plesník and Znám [68] proved the following result which is also implied by Theorem 2.12.

Corollary 2.18 *Every bipartite graph with diameter 2 or 3 is maximally edge-connected.*

Other conditions combine the clique number and the degree sequence of a graph. The following result is also due to Dankelmann and Volkmann [15]; we state it only for graphs to which Theorem 2.16 does not apply.

Theorem 2.19 *Let G be a graph with degree sequence* **D**, *minimum degree δ and clique number ω, having order $n \geq 2\lfloor \delta\omega/(\omega-1) \rfloor$. If*

$$\sum_{i=1}^{k} d_i + \sum_{i=1}^{(2\omega-1)k} d_{n+1-i} \geq k(\omega-1)n + 2\delta - 1,$$

for some $k \leq \delta/(\omega-1)$, then G is maximally edge-connected.

Other results of a similar nature were found by Volkmann [79]. The following result gives conditions for graphs of even and odd order, separately. It is independent of the above results and can be thought of as a condition on the average of a tail-end set of degrees. We state it only for graphs to which Chartrand's theorem does not apply.

Theorem 2.20 *Let G be a graph of order $n \geq 6$, clique number ω, minimum degree $\delta \leq \lfloor \frac{1}{2}n \rfloor - 1$ and degree sequence* **D**. *Then G is maximally edge-connected if*

$$\sum_{i=1}^{\delta+1} d_{n+1-i} \geq \frac{\delta+1}{\omega}\left(\frac{(\omega-1)(n+1)}{2} - \frac{2}{n-3}\right), \text{ for } n \text{ odd},$$

or

$$\sum_{i=1}^{2\delta+2} d_{n+1-i} \geq \frac{\delta+1}{\omega}\left((\omega-1)(n+2) - \frac{4}{n-2}\right), \text{ for } n \text{ even}.$$

Furthermore, these bounds are sharp.

Cages

A (δ, g)-cage is a δ-regular graph with $\delta \geq 2$ and girth $g \geq 3$ having the least possible number of vertices. In 2003, Wang, Xu and Wang [85] showed that every cage with odd girth is maximally edge-connected. For even girth the correspoding result was proved by Lin, Miller and Rodger [51] in 2005. Combining these results, we obtain the next theorem, which was found independently by Moriarty and Christopher [59].

Theorem 2.21 *Every cage is maximally edge-connected.*

A graph G is called *super-edge-connected* if every minimum edge-cut is trivial – that is, if every minimum edge-cut consists of the edges incident with a vertex of minimum degree. (This concept was introduced by Bauer, Suffel, Boesch and Tindell [3].) Thus every super-edge-connected graph is also maximally edge-connected, and the next result by Marcote and Balbuena [53] and Lin, Miller, Balbuena and Marcote [50] is an extension of Theorem 2.21.

Theorem 2.22 *Every cage with $\delta \geq 3$ is super-edge-connected.*

Cycles of length 4 or more show that in Theorem 2.22 the condition $\delta \geq 3$ is necessary.

3. Maximally edge-connected digraphs

Many of the results in Section 2 have analogues for digraphs, and these analogues lead in a natural way to the corresponding statements for graphs as follows. The *associated digraph* D_G of a graph G is obtained by replacing each edge of G by a pair of mutually opposite oriented edges. If G is a graph and D_G is its associated digraph, then each $v-w$ path in G corresponds to a $v-w$-path and a $w-v$-path in D_G. This immediately leads to $\kappa(G) = \kappa(D_G)$ and $\lambda(G) = \lambda(D_G)$. (We define the *degree* of a vertex in a digraph to be the minimum of its in- and out-degrees; furthermore, the *minimum degree* $\delta(D)$ of the digraph D is the minimum of all the vertex degrees.)

In 1971, Geller and Harary [25] wrote: 'Connectivity of graphs has been extensively investigated. On the other hand, connectivity in digraphs has until recently been almost completely neglected. In this article, we begin with an expository review of connectivity concepts and results concerning both graphs and digraphs.' Among other things, Geller and Harary obtained an analogue to Whitney's theorem for digraphs.

Theorem 3.1 *If D is a digraph, then*

$$\kappa(D) \leq \lambda(D) \leq \delta(D).$$

In view of this result, we call a digraph D *maximally connected* if $\kappa(D) = \delta(D)$ and *maximally edge-connected* if $\lambda(D) = \delta(D)$.

Diameter conditions

A pair of sets X and Y in a digraph D with distance $d(X, Y) = k$ is called *k-distance maximal* if there do not exist $X_1 \supseteq X$ and $Y_1 \supseteq Y$, with $X_1 \neq X$ or $Y_1 \neq Y$, such that $d(X_1, Y_1) = k$. The first result is due to Hellwig and Volkmann [31].

Theorem 3.2 *Let D be a strong digraph. If, for all 3-distance maximal pairs of vertex sets X and Y, there exists an isolated vertex in $\langle X \cup Y \rangle$, then D is maximally edge-connected.*

It is easy to see that Theorem 3.2 includes Theorem 2.2, as well as the digraph version of this result.

Corollary 3.3 *Every strong digraph of diameter 1 or 2 is maximally edge-connected.*

In 1985, Imase, Soneoka and Okada [44] presented a result concerning digraphs that is related to Theorem 2.11.

Theorem 3.4 *A digraph with maximum degree $\Delta \geq 2$, minimum degree δ and diameter h is maximally edge-connected if it has more than*

$$(\delta - 1)\left(\frac{\Delta^{h-1} - 1}{\Delta - 1} + \Delta + 1\right)$$

vertices.

Degree conditions

The first result on maximally edge-connected digraphs was proved by Geller and Harary [25] and is the digraph analogue of Chartrand's theorem. This result, as well as the digraph version of Lesniak's result, is included in Theorem 3.2.

Corollary 3.5 *Every digraph of order n with minimum degree $\delta \geq \frac{1}{2}(n-1)$ is maximally edge-connected.*

Corollary 3.6 *If D is a digraph of order n in which $d^+(v) + d^-(w) \geq n-1$ for all pairs v and w of non-adjacent vertices, then D is maximally edge-connected.*

As a further generalization of Volkmann's result concerning bipartite graphs, Balbuena and Carmona [1] gave a degree condition for bipartite digraphs.

Theorem 3.7 *If D is a bipartite digraph of order n for which*

$$d^+(v) + d^-(w) \geq \tfrac{1}{2}(n+1)$$

for all pairs v and w of vertices with $d(v, w) \geq 4$, then D is maximally edge-connected.

If the minimum degree in a bipartite digraph of order n is at least $\frac{1}{4}(n+1)$, then the requirements of Theorem 3.7 are trivially fulfilled.

Corollary 3.8 *Every bipartite digraph with minimum degree at least $\frac{1}{4}(n+1)$ is maximally edge-connected.*

Hellwig and Volkmann [34] also proved the digraph version of Theorem 2.4.

Theorem 3.9 *A strong digraph of order n is maximally edge-connected if for each edge vw, the in-degree of v or w is at least $\frac{1}{2}(n-1)$ and the out-degree of v or w is at least $\frac{1}{2}(n-1)$.*

As in Section 2 we now state some results that address digraphs to which Geller and Harary's theorem does not apply. The first is due to Dankelmann and Volkmann [14] and involves the sums of certain combinations of degrees.

Theorem 3.10 *Let D be a digraph of order n with degree sequence* $\mathbf{D}$ *and minimum degree $\delta < \frac{1}{2}(n-1)$. If*

$$\sum_{i=1}^{k}(d_i + d_{n+i-\delta-1}) \geq k(n-2) + 2\delta - 1,$$

for some $k \leq \delta$, then D is maximally edge-connected.

Hellwig and Volkmann [31] gave a sufficient condition for maximal edge-connectedness that uses only the lower end of the degree sequence.

Theorem 3.11 *Let D be a digraph of order n with degree sequence* $\mathbf{D}$ *and minimum degree $\delta < \frac{1}{2}(n-1)$. If*

$$\sum_{i=1}^{2k} d_{n+1-i} \geq \max\{k(n-1) - 1, (k-1)n + 2\delta - 1\}$$

for some k with $2 \leq k \leq \delta$, then D is maximally edge-connected.

The last degree sequence condition addresses bipartite digraphs with low minimum degree. It was proved by Dankelmann and Volkmann [14] in 1997.

Theorem 3.12 *Let D be a bipartite digraph of order n with degree sequence* $\mathbf{D}$ *and minimum degree $\delta < \frac{1}{4}(n+1)$. If*

$$\sum_{i=1}^{k}(d_i + d_{n+1-2\delta+k-i}) \geq k(n-2\delta) + 2\delta - 1,$$

for some $k \leq 2\delta$, then D is maximally edge-connected.

4. Maximally locally edge-connected graphs and digraphs

Recall that the local edge-connectivity $\lambda(v, w)$ of two vertices v and w in a graph is the maximum number of edge-disjoint $v{-}w$ paths. Clearly, $\lambda(v, w) \leq \min\{d(v), d(w)\}$, for all pairs v and w of vertices in the graph. A graph G is called *maximally locally edge-connected* when $\lambda(v, w) = \min\{d(v), d(w)\}$ for all pairs v and w of vertices in G.

Similarly, the local edge-connectivity $\lambda(v, w)$ of two vertices v and w in a digraph D is the maximum number of directed edge-disjoint $v{-}w$ paths. Again, it is obvious that $\lambda(v, w) \leq \min\{d^+(v), d^-(w)\}$ for all pairs v and w of vertices in the digraph, and a digraph D is called *maximally locally edge-connected* when $\lambda(v, w) = \min\{d^+(v), d^-(w)\}$ for all pairs v and w of vertices in D.

In an unpublished manuscript, Fricke, Oellermann and Swart showed that some known conditions that guarantee maximally edge-connectedness for a graph also guarantee that the graph is maximally *locally* edge-connected. Since maximally locally edge-connectedness implies maximally edge-connectedness, such results generalize the corresponding known ones.

Theorem 4.1 *Every graph of diameter* 1 *or* 2 *is maximally locally edge-connected.*

Theorem 4.2 *Every r-partite graph with minimum degree δ and order at most $2\lfloor \delta r/(r-1) \rfloor - 1$ is maximally locally edge-connected.*

Note that Theorems 2.2 and 2.15 are immediate consequences of Theorems 4.1 and 4.2, respectively. For $r = 2$ (bipartite graphs), the upper bound on the order in Theorem 4.2 is $4\delta - 1$. Recently, Holtkamp [38] showed that relaxing the condition for a graph from being bipartite to being diamond-free (that is, no two 3-cycles share a common edge) preserves the maximally locally edge-connectedness. Recall that a diamond is obtained by removing an edge from k_4.

Theorem 4.3 *Every diamond-free graph with minimum degree δ and order at most $4\delta - 1$ is maximally locally edge-connected.*

Hellwig and Volkmann [32] extended Theorem 4.1 to digraphs.

Theorem 4.4 *Every digraph of diameter* 1 *or* 2 *is maximally locally edge-connected.*

Using Turán's inequality, Hellwig and Volkmann [32] also obtained an improvement of Theorem 4.2.

Theorem 4.5 *Every graph with of minimum degree δ clique number ω and order at most $2\lfloor \delta\omega/(\omega-1) \rfloor - 1$ is maximally locally edge-connected.*

An analogue of Theorem 4.2 for r-partite digraphs is also valid, as was proved by Hellwig and Volkmann [32].

Theorem 4.6 *Every r-partite digraph with minimum degree δ and of order at most $2\lfloor \delta r/(r-1) \rfloor - 1$ is maximally locally edge-connected.*

As a strengthening of a result by Dankelmann and Volkmann [13], Hellwig and Volkmann [32] presented a sharp sufficient condition for bipartite graphs and digraphs to be maximally locally edge-connected. We state the result only for digraphs; there is a corresponding result for graphs.

Theorem 4.7 *Let D be a bipartite digraph of order n with minimum degree $\delta \geq 2$. If $d(v) + d(w) \geq \frac{1}{2}(n+1)$ for each pair v and w of vertices in the same partite set, then D is maximally locally edge-connected.*

As a generalization of Theorem 2.18, Fàbrega and Fiol [20] proved that every bipartite digraph with diameter at most 3 is maximally locally edge-connected. In [81], Volkmann constructed examples show that a bipartite digraph of diameter at most 3 is not maximally locally edge-connected in general. However, in the same paper Volkmann proved the following theorem that includes the above result of Fàbrega and Fiol.

Theorem 4.8 *If D is a bipartite digraph with diameter at most 3, then $\lambda(v, w) = \min\{d^+(v), d^-(w)\}$ for all pairs v and w of vertices in the same partite set.*

5. Maximally connected and maximally locally connected graphs and digraphs

In 1967, Watkins [87] determined the first relationship between order, diameter and connectivity. With the aid of Menger's theorem [58], the following result can be established rather easily.

Theorem 5.1 *Every connected non-trivial graph with diameter h and connectivity k has at least $k(h-1)+2$ vertices.*

Given positive integers k and h, Watkins [87] constructed graphs of order n with diameter h and connectivity k for which $n = k(h-1)+2$. Later, Kane and Mohanty [46] improved this result by $2(\delta - k)$ when the diameter of the graph in question is at least 3. Furthermore, they presented examples which demonstrate that their bound is sharp.

Theorem 5.2 *If G is a k-connected graph with diameter $h \geq 2$ and minimum degree δ, then*

$$n \geq \begin{cases} \delta + 2 & \text{if } h = 2, \\ k(h-3) + 2(\delta+1) & \text{if } h \geq 3. \end{cases}$$

Kane and Mohanty proved Theorem 5.2 without using Menger's theorem. This is somewhat surprising, since Menger's theorem yields Theorem 5.2 as easily as Watkins obtained his Theorem 5.1. When the diameter h is at least 3, this method even leads to the slightly better bound $n \geq k(h-3) + d_n + d_{n-1} + 2$, where $d_1 \geq d_2 \geq \cdots \geq d_n$ is the degree sequence of the graph in question. Using Menger's theorem, one also can prove Theorem 5.2 for digraphs (see Volkmann [80]).

As would be expected, the higher the degrees of the vertices of a graph, the more likely it is that the graph has large connectivity. There are several sufficient conditions of this type. We start with one of the simplest, originally presented by Chartrand and Harary [10].

Theorem 5.3 *Every connected non-complete graph of order n with minimum degree $\delta \geq \frac{1}{2}n$ is $(2\delta + 2 - n)$-connected.*

This implies that $\kappa(G) = \delta(G)$ when $n \leq \delta(G) + 2$. One year later, Bondy [6] generalized Theorem 5.3 by the following degree-sequence condition.

Theorem 5.4 *Let G be a connected graph of order n with degree sequence* **D**. *If $d_{n-i+1} \geq i + k - 1$, for $i = 1, 2, \ldots, n-1-d_k$ and some $k \leq n-1$, then G is k-connected.*

In 1971, Geller and Harary [25] showed that Theorem 5.3 is valid for digraphs. This also follows from the next more general result of Hellwig and Volkmann [36].

Theorem 5.5 *Let D be a digraph of order* $n \geq 4$ *with degree sequence* **D** *and minimum degree* δ. *If* $\kappa(D) \leq \delta - k$, *then*

$$\kappa(D) \geq \frac{1}{k+1} \sum_{i=0}^{2k+1} d_{n-i} + 2 - n.$$

Maximal connectedness

Soneoka, Nakada, Imase and Peyrat [74] established a sufficient condition for maximally connected graphs, depending on the diameter and the girth g, and showed that their condition is sharp for girth 4 and for odd girth.

Theorem 5.6 *Let G be a graph with girth g and diameter h. If* $h \leq g - 2$ *when g is odd, or* $h \leq g - 3$ *when g is even, then G is maximally connected.*

A refinement of Theorem 5.6 can be found in a paper by Fàbrega and Fiol [19]. Soneoka *et al.* [74] also presented the following result, which is sharp for diameters 2 and 3.

Theorem 5.7 *Let G be a graph with maximum degree* $\Delta \geq 3$, *minimum degree* δ *and diameter h. If*

$$n > (\delta - 1)(\Delta - 1)^{h-1} + 2,$$

then G is maximally connected.

Balbuena and Carmona [1] gave a degree condition for bipartite digraphs to be maximally connected.

Theorem 5.8 *Let D be a bipartite digraph of order n with minimum degree* δ. *If*

$$d^{+}(v) + d^{-}(w) \geq \tfrac{1}{2}(n + \delta)$$

for all pairs v and w of vertices with $d(v, w) \geq 3$, *then G is maximally connected.*

The above theorem implies a result by Topp and Volkmann [75].

Corollary 5.9 *If G is a bipartite graph with minimum degree* δ *and order* $n \leq 3\delta$, *then G is maximally connected.*

A graph is *diamond-free* if it contains no diamond as an induced subgraph, and C_4*-free* if it contains no cycle of length 4 as an induced subgraph. Dankelmann, Hellwig and Volkmann [11] presented an extension of the above corollary for graphs with minimum degree at least 3.

Theorem 5.10 *If G is a connected diamond-free graph with minimum degree $\delta \geq 3$ and order $n \leq 3\delta$, then G is maximally connected.*

Dankelmann, Hellwig and Volkmann [11] also considered C_4-free graphs.

Theorem 5.11 *Let G be a connected C_4-free graph of order n and minimum degree $\delta \geq 2$. If*

$$n \leq \begin{cases} 2\delta^2 - 3\delta + 2 & \text{if } \delta \text{ is even,} \\ 2\delta^2 - 3\delta + 4 & \text{if } \delta \text{ is odd,} \end{cases}$$

then G is maximally connected.

For $\delta = 2$ and $\delta = 3$, example are known with equality in Theorem 5.11. However, following a well-known construction of C_4-free graphs (see for example [26]), the authors showed in [11] that Theorem 5.11 is at least asymptotically sharp. It seems to be very difficult to find sharp bounds in general.

Fàbrega and Fiol [20] considered bipartite graphs with large girth.

Theorem 5.12 *Every bipartite graph with minimum degree $\delta \geq 2$, girth g and diameter $h \leq g - 2$ is maximally connected.*

Using Turán's inequality, Hellwig and Volkmann [34] obtained a generalization of a theorem by Topp and Volkmann [75] on multipartite graphs.

Theorem 5.13 *Let G be a connected graph of order n with clique number ω and minimum degree δ. If $n \leq (2\omega - 1)\delta/(2\omega - 3)$, then G is maximally connected.*

Examples in [75] show that Theorem 5.13 is sharp for multipartite graphs.

Maximally locally connected graphs

Recall that the local connectivity $\kappa(v, w)$ between two distinct vertices v and w of a graph G is the maximum number of internally vertex-disjoint v–w paths in G. Obviously, $\kappa(v, w) \leq \min\{d(v), d(w)\}$, and it is a well-known consequence of Menger's theorem [58] that $\kappa(G) = \min\{\kappa(v, w) \colon v, w \in V(G)\}$. A graph is called *maximally locally connected* when $\kappa(v, w) = \min\{d(v), d(w)\}$, for all pairs v and w of vertices in G. Note that being maximally locally connected implies being maximally connected.

Using Turán's inequality, Holtkamp and Volkmann [42] recently presented a new short proof of the following common generalization of Theorem 5.13 and a result of Volkmann [83].

Theorem 5.14 *Let G be a connected graph of order n with clique number ω and minimum degree δ. If $n \leq (2\omega - 1)/(2\omega - 3)$, then G is maximally locally connected.*

Holtkamp and Volkmann [41] constructed a family of graphs which show that the condition $n \leq 3\delta$ in Theorem 5.10 does not guarantee that the graph is maximally locally connected. However, the following is valid.

Theorem 5.15 *If G is a diamond-free graph with minimum degree $\delta \geq 3$ and order $n \leq 3\delta - 1$, then G is maximally locally connected.*

Using Theorem 5.11, Holtkamp and Volkmann [43] proved a similar result for C_4-free graphs to be maximally locally connected.

Hellwig and Volkmann [37] gave examples which show that Theorem 2.13 is not valid in general for the connectivity of a graph and its complement. However, they proved the following bound and showed that it is sharp.

Theorem 5.16 *If G and $\overline{G}$ are both connected graphs, then*

$$\kappa(G) + \kappa(\overline{G}) \geq \min\{\delta(G), \delta(\overline{G})\} + 1.$$

Cages

Some fundamental properties of cages were established by Fu, Huang and Rodger [24]. They proved that all cages are 2-connected, and they proposed the following conjecture.

Conjecture A *Every (δ, g)-cage is maximally connected.*

The next four results support this conjecture.

Theorem 5.17

- *Every (δ, g)-cage with $\delta \geq 3$ is 3-connected.*
- *Every $(4, g)$-cage is maximally connected.*
- *Every (δ, g)-cage with $\delta \geq 5$ and $g \geq 10$ is 4-connected.*
- *Every $(\delta, 6)$-cage and every $(\delta, 8)$-cage are maximally connected.*

The first part was been proved independently by Jiang and Mubayi [45] and Daven and Rodger [16], the second part is due to Xu, Wang and Wang [89], the third part was shown by Marcote *et al.* [55] and the last part was proved by Marcote, Balbuena and Pelayo [54].

Recently, Lin, Miller and Balbuena [49] and Lin *et al.* [48] established improved lower bounds for the connectivity of cages.

Theorem 5.18

(a) *Every (δ, g)-cage with $\delta \geq 3$ and odd girth $g \geq 7$ is $\lceil\sqrt{\delta + 1}\rceil$-connected.*
(b) *Every (δ, g)-cage with $\delta \geq 3$ and even girth g is $(r + 1)$-connected, where r is the largest integer with $r^3 + 2r^2 \leq \delta$.*

6. Restricted edge-connectivity

The *restricted edge-connectivity* $\lambda'(G)$, introduced and studied first by Esfahanian and Hakimi [18] in 1988, is the minimum cardinality of an edge-cut S in a graph G with the property that $G - S$ contains no isolated vertices. The definition of the restricted edge-connectivity is a special case of a quite general concept of *conditional edge-connectivity*, proposed by Harary [29] in 1983 (see also Section 7). A restricted edge-cut S is called a *λ'-cut* if $|S| = \lambda'(G)$. Obviously, for any λ'-cut S, the graph $G - S$ has exactly two components. A connected graph G is called *λ'-connected* if G has a restricted edge-cut. For a graph G, let $\xi(e) = d(v) + d(w) - 2$ be the *edge-degree* of the edge $e = vw$ and $\xi(G) = \min\{\xi(e) : e \in E(G)\}$ be the *minimum edge-degree* of G.

Besides the classical edge-connectivity, restricted edge-connectivity recently received much attention as a measure of fault-tolerance in networks. Obviously, λ' does not exist for any star and any graph with fewer than four vertices. In fact, Esfahanian and Hakimi [18] observed that these are the only such graphs.

Theorem 6.1 *A connected graph of order at least* 4 *is λ'-connected if and only if it is not a star $K_{1,p}$. Furthermore, every λ'-connected graph G satisfies*

$$\lambda(G) \leq \lambda'(G) \leq \xi(G).$$

Recently, Volkmann [82] proved an analogous result for strongly connected digraphs.

A λ'-connected graph G is called *λ'-optimal* or *λ'-maximal* if $\lambda'(G) = \xi(G)$. It should be noted that the above bound $\lambda'(G) \leq \xi(G)$ is sharp, in the sense that there are infinitely many graphs for which equality holds; examples are complete graphs and the class of n-cubes (see Esfahanian [17]).

Next we note some simple properties of λ'-optimal graphs.

Theorem 6.2 *If G is a λ'-optimal graph with minimum degree δ and maximum degree Δ, then $\lambda'(G) \leq \Delta + \delta - 2$ and $\lambda'(G) \geq \max\{\delta, 2\delta - 2\}$.*

We remark that the inequality $\lambda'(G) > \lambda(G)$ implies that G is super-edge-connected. Hellwig and Volkmann [35] noted further simple but interesting connections between λ'-optimality, super-edge-connectivity and maximal edge-connectivity.

Theorem 6.3 *Every λ'-optimal graph with minimum degree at least* 3 *is super-edge-connected.*

Theorem 6.4 *Every λ'-optimal graph is maximally edge-connected.*

Degree conditions

We start with a result by Ou [60] that states that regular graphs of high degree are λ'-optimal.

Theorem 6.5 *Let G be an r-regular graph of order $n \geq 4$. If $r > \frac{1}{2}n$, then G is λ'-optimal.*

Ou [60] also gave examples to show that the lower bound for the minimum degree is sharp. Wang and Li [86] gave the following sufficient condition for a graph to be λ'-optimal. It includes Theorem 6.5 and, with respect to Theorem 6.3, is an improvement of a result by Lesniak [47] for all graphs with minimum degree at least 3.

Theorem 6.6 *Let G be a connected graph of order n. If $d(v) + d(w) \geq n + 1$, for all pairs v and w of non-adjacent vertices, then G is λ'-optimal.*

Shang and Zhang [72] were able to relax the degree condition in Theorem 6.6 under an additional assumption.

Theorem 6.7 *If G is a λ'-connected graph of order n that satisfies the following two conditions, then G is λ'-optimal.*

(a) *$d(v) + d(w) \geq 2\lfloor \frac{1}{2}n \rfloor - 3$ for each pair v and w of vertices at distance 2;*
(b) *Each triangle contains at least one vertex v with $d(v) \geq \lfloor \frac{1}{2}n \rfloor + 1$.*

Yuan and Liu [91] presented a similar condition for triangle-free graphs.

Theorem 6.8 *Let G be a connected triangle-free graph of order n. If*

$$d(v) + d(w) \geq 2 \left\lfloor \tfrac{1}{4}(n+2) \right\rfloor + 1$$

for each pair v and w of vertices of distance 2, then G is λ'-optimal.

The next result involves the sum of vertex degrees of the tail end of the degree sequence of a graph. It is due to Hellwig and Volkmann [35].

Theorem 6.9 *Let G be a λ'-connected triangle-free graph of order n with degree sequence $\mathbf{D}$ and minimum degree δ. If*

$$\sum_{i=1}^{\max\{1,\delta-1\}} d_{n-i} \geq \tfrac{1}{2} \max\{1, \delta - 1\} \left(\left\lfloor \tfrac{1}{2}n \right\rfloor + 2 - \frac{4}{n-3} \right),$$

then G is λ'-optimal.

Neighbourhood conditions

Hellwig and Volkmann [33], [35] obtained several sufficient conditions for graphs to be λ'-optimal, involving the number of common neighbours of non-adjacent vertices. The first one generalizes Theorem 6.6.

Theorem 6.10 *Let G be a graph of order at least 4. If $|N(v) \cap N(w)| \geq 3$, for all pairs v and w of non-adjacent vertices, then G is λ'-optimal.*

The above bound of 3 can be lowered to 2 for graphs with bounded edge-degree.

Theorem 6.11 *Let G be a graph of order $n \geq 10$. If $|N(v) \cap N(w)| \geq 2$, for all pairs v and w of non-adjacent vertice, and if $\xi(G) \leq \lfloor \frac{1}{2}n \rfloor + 2$, then G is λ'-optimal.*

A family of examples in [33] shows that the condition for $\xi(G)$ in Theorem 6.11 is sharp. If the graph G is required to be triangle-free, then we can drop the requirement on the minimum edge-degree.

Theorem 6.12 *Let G be a triangle-free graph. If $|N(v) \cap N(w)| \geq 2$ for all pairs v and w of non-adjacent vertices, then G is λ'-optimal.*

Another graph class of λ'-optimal graphs is as follows.

Theorem 6.13 *Let G be a λ'-connected graph. If G contains an independent set I of vertices of minimum degree for which every vertex in I is adjacent to every vertex not in I, then G is λ'-optimal.*

The class of graphs satisfying the condition in Theorem 6.13 includes all complete multipartite graphs.

Diameter and girth conditions

Inspired by Hellwig and Volkmann's article [33], Balbuena, Garcia-Vázquez and Marcote [2] were able to relax the conditions in Theorem 6.10. The first of their three main theorems is the following.

Theorem 6.14 *Every graph of order $n \geq 4$ with minimum degree $\delta \geq 2$, girth g and diameter $h \leq g - 2$, is λ'-optimal.*

The second and third main theorem in [2] state sufficient conditions for graphs with odd girth g and diameter $h = g - 1$ to be λ'-optimal. For two vertices v, w at distance $d(v, w) \geq g - 1$, let $X_{v,w}$ be the set of vertices at distance at most $\frac{1}{2}(g - 1)$ from both v and w.

Theorem 6.15 *Let G be a λ'-connected graph with odd girth g, minimum degree $\delta \geq 2$ and diameter $h = g - 1$. Then G is λ'-optimal if one of the following assertions holds.*

- *All pairs v and w of vertices at distance $d(v, w) = h$ are such that neither v nor w lies on a cycle of length g.*
- *For all pairs v and w of vertices at distance $d(v, w) = h$, the set $X_{v,w}$ contains at least three vertices.*

Theorem 6.15 and Theorem 6.14 together generalize Theorem 6.10. In order to relax the order restriction of Theorem 6.15, the same authors proved the following result.

Theorem 6.16 *Let G be a λ'-connected graph with odd girth g and minimum degree $\delta \geq 2$. If $G[X_{v,w}]$ contains an edge for every pair v, w of vertices at distance $d(v, w) \geq g - 1$, then G is λ'-optimal.*

Atom size conditions

When edge-cut S is a λ'-cut of G, the vertex-sets of the two components of $G - S$ are called *λ'-fragments*. Let $r(G)$ be the order of a smallest λ'-fragment. We call a λ'-fragment X a *λ'-atom* of a graph G if $|X| = r(G)$.

Obviously, $2 \leq r(G) \leq \frac{1}{2}n(G)$, and if X is a λ'-atom, then $G[X]$ and $G[\overline{X}]$ are both connected. The next theorem, due to Xu and Xu [90], yields a necessary and sufficient condition for a λ'-connected graph to be λ'-optimal.

Theorem 6.17 *A λ'-connected graph G is λ'-optimal if and only if $r(G) = 2$.*

Zhang [93] proved a lower bound for $r(G)$ in terms of the minimum edge-degree when G is not λ'-optimal.

Theorem 6.18 *Let G be a λ'-connected graph with minimum degree $\delta \geq 2$ and minimum edge-degree ξ. If G is not λ'-optimal, then $r(G) \geq \lceil \frac{1}{2}\xi \rceil + 1$.*

Zhang [93] pointed out that the requirement $\delta(G) \geq 2$ in Theorem 6.18 is necessary, as can be seen by the following example. Let v be a vertex of the complete graph $H = K_{t+1}$ with $t \geq 4$, and let w and w' be two further vertices. Now let G consist of H, w, w' and the two edges vw and vw'. Then $\lambda'(G) = t < t + 1 = \xi(G)$ and $r(G) = 3$, and

$$r(G) = 3 < 4 \leq \left\lceil \tfrac{1}{2}(t + 1) \right\rceil + 1 = \left\lceil \tfrac{1}{2}\xi(G) \right\rceil + 1.$$

Holtkamp, Meierling and Montejano [40] showed that for triangle-free graphs the number $r(G)$ can also be bounded from below by an expression involving the minimum degree and minimum edge-degree. In the following we give an overview of results that give lower bounds for the order $r(G)$ of atoms in terms of the minimum degree or minimum edge-degree when G is not λ'-optimal. Note that each such result implies a sufficient criterion for a graph to be λ'-optimal, since a smallest 2-fragment of a graph contains at most half of the graph's vertices.

Theorem 6.19 *Let G be a λ'-connected triangle-free graph with minimum degree δ and minimum edge-degree ξ. If $r(G) < \xi + 2 - \xi/\delta$, then G is λ'-optimal.*

The following sufficient criterion for the λ'-optimality of a graph follows directly from the above theorem.

Corollary 6.20 *Let G be a connected triangle-free graph of order n with minimum degree $\delta \geq 2$ and minimum edge-degree ξ. If*

$$2(\delta - 1)\xi \geq \begin{cases} \delta(n-4) - 1 & \text{for } n \text{ odd,} \\ \delta(n-3) - 1 & \text{for even,} \end{cases}$$

then G is λ'-optimal.

The next result, by Meierling and Volkmann [56], shows that the lower bound for the minimum edge-degree in the above corollary that guarantees λ'-optimality of a graph is sufficient for the graph to be super-λ', if its minimum degree is at least 3 and its order is at least 18.

Theorem 6.21 *Let G be a connected triangle-free graph of order $n \geq 18$ with minimum degree $\delta \geq 3$ and minimum edge-degree ξ. If*

$$2(\delta - 1)\xi \geq \begin{cases} \delta(n-4) - 1 & \text{for } n \text{ odd,} \\ \delta(n-3) - 1 & \text{for } n \text{ even,} \end{cases}$$

then G is super-λ'.

Since $\xi(G) \geq 2\delta(G) - 2$, Theorems 6.17, 6.18 and 6.19 immediately imply the next results, which are due to Ueffing and Volkmann [76].

Corollary 6.22 *Let G be a λ'-connected graph with minimum degree $\delta \geq 2$. If G is not λ'-optimal, then $r(G) \geq \max\{3, \delta\}$.*

Corollary 6.23 *Let G be a λ'-connected triangle-free graph with minimum degree δ. If G is not λ'-optimal, then $r(G) \geq \max\{3, 2\delta - 1\}$.*

Examples show that Theorem 6.21 and both of those corollaries are sharp.

7. Conditional vertex-connectivity and edge-connectivity

The study of conditional connectivity has received much attention in recent years since it provides a new and interesting measure for fault-tolerance in networks.

In 1983, Harary [29] proposed the concept of conditional connectivity. The *P-connectivity* $\kappa(G, P)$ of a connected graph G is the minimum cardinality of a set S of vertices for which $G - S$ is disconnected and every component of $G - S$ has the given property P. The corresponding edge-connectivity parameter is denoted by $\lambda(G, P)$.

For a positive integer p, Fàbrega and Fiol [21] defined the *p-restricted edge-connectivity* $\lambda_p(G)$ of a connected graph G to be the cardinality of a minimum edge-cut S of G for which each component of $G-S$ contains at least p vertices. Note that $G-S$ has exactly two components for each minimum p-restricted edge-cut.

In order to find a suitable definition for the restricted vertex-connectivity, we notice that, for a minimum vertex-cut S, the remaining graph $G-S$ can have more than two components. Since components (except two of them) that do not satisfy a given property can be removed by adding them to the vertex-cut, it makes sense to require only two components to have specified properties. Following this idea, Hellwig, Rautenbach and Volkmann [30] generalized the conditional connectivity of Harary as follows. Let P_1 and P_2 be two graphical properties. The parameter $\kappa(G, P_1, P_2)$ of a connected graph G equals the minimum cardinality of a set S of vertices for which one component of $G-S$ has property P_1 and another component has property P_2. The corresponding edge parameter is denoted by $\lambda(G, P_1, P_2)$.

In this section, we consider the special case in which one component has at least p vertices and another one has at least q vertices. This leads to the following definition by Hellwig, Rautenbach and Volkmann [30]. Let G be a connected graph, and let p and q be positive integers. A vertex-cut S of G is a *p-q-vertex-cut* if one component of $G-S$ has at least p vertices and another component has at least q vertices; a graph is *$\kappa_{p,q}$-connected* if a p-q-vertex-cut exists. The *p-q-restricted connectivity* $\kappa_{p,q}(G)$ of a $\kappa_{p,q}$-connected graph G is the minimum cardinality of a p-q-vertex-cut of G. A p-q-vertex-cut S of G is called a *minimum p-q-vertex-cut*, or a *$\kappa_{p,q}$-cut*, if $|S|=\kappa_{p,q}(G)$. (The corresponding edge parameter is denoted by $\lambda_{p,q}(G)$.) Note that $\lambda_{1,1}(G)=\lambda(G)$, $\kappa_{1,1}(G)=\kappa(G)$ and $\lambda_{2,2}(G)=\lambda'(G)$. In addition, if S is a $\lambda_{p,q}$-cut of G, then $G-S$ has exactly two components.

Hellwig, Rautenbach and Volkmann [30] observed that, for fixed p and q the values $\lambda_{p,q}$ and $\kappa_{p,q}$ can be computed in polynomial time by contracting all choices of disjoint vertex-sets of cardinalities p and q that induce connected subgraphs of G, and determining minimum sets of edges (or vertices) that separate the two vertices created by the contractions, which can clearly be done using max-flow algorithms.

Existence of conditional edge-cuts and vertex-cuts

First we present a sufficient and necessary conditions for graphs to be $\lambda_{p,q}$-connected and $\kappa_{p,q}$-connected. They are due to Hellwig, Rautenbach and Volkmann [30].

Theorem 7.1 *Let G be a connected graph and let $p, q \geq 1$ be integers. Then*

- *the graph G is $\lambda_{p,q}$-connected if and only if there exist two disjoint vertex-sets X, Y, with $|X| \geq p$ and $|Y| \geq q$, for which $G[X]$ and $G[Y]$ are connected;*
- *the graph G is $\kappa_{p,q}$-connected if and only if there exist two disjoint vertex-sets X, Y, with $|X| \geq p$ and $|Y| \geq q$, for which $G[X]$ and $G[Y]$ are connected and $(X, Y)=\varnothing$.*

For some following special cases, more transparent characterizations of $\lambda_{p,q}$-connected and $\kappa_{p,q}$-connected graphs are known.

Theorem 7.2 *Let G be a connected graph of order n and maximum degree Δ. Then*

- *G is $\lambda_{1,q}$-connected if and only if $n \geq q+1$ and $\lambda_{1,q}(G) \leq \Delta$;*
- *G is $\lambda_{2,2}$-connected if and only if $n \geq 4$ and G is not a star;*
- *G is $\lambda_{3,3}$-connected if and only if $n \geq 6$ and G is not isomorphic to the graph N or to any graph of the family F in Fig. 1.*

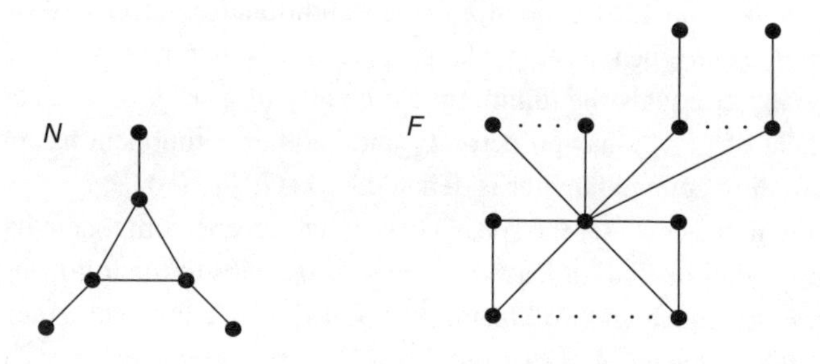

Fig. 1.

The last part of Theorem 7.2 is due to Bonsma, Ueffing and Volkmann [8]. In 2004, Ou [62] characterized the class of $\lambda_{4,4}$-connected graphs.

Concerning the general case, Volkmann observed the following: if p and q are positive integers, and G is a connected graph of order at leat $p+q+1$, then G is $\lambda_{p,q}$-connected if and only if G is $\lambda_{p,q+1}$-connected or $\lambda_{p+1,q}$-connected. Using this observation and Theorem 7.2, we conclude that a connected graph G of order at least 5 is $\lambda_{2,3}$-connected if and only if G is not a star, and a connected graph G of order at least 7 is $\lambda_{3,4}$-connected if and only if G is not isomorphic to a member of the family F in Fig. 1. In an unpublished manuscript, Volkmann also characterized all $\lambda_{2,4}$-connected graphs of order at least 6, all $\lambda_{2,5}$-connected graphs of order at least 7 and all $\lambda_{4,5}$-connected graphs of order at least 9.

Hellwig, Rautenbach and Volkmann [30] characterized the class of $\kappa_{1,2}$-connected and $\kappa_{1,3}$-connected graphs. Their first result is stated below.

Theorem 7.3 *Let G be a connected graph of order $n \geq 4$. Then G is $\kappa_{1,2}$-connected if and only if G is not a complete multipartite graph.*

The following result is a generalization of Whitney's classical inequality, $\kappa \leq \lambda$.

Theorem 7.4 *If G is both $\kappa_{1,q}$-connected and $\lambda_{q,q}$-connected, then*

$$\kappa_{1,q}(G) \leq \lambda_{q,q}(G).$$

Examples in [30] show that $\kappa_{1,q} \leq \lambda_{q-1,q}$ is not true in general. The following example demonstrates that $\kappa_{2,q} \leq \lambda_{r,s}$, with $2 \leq q \leq r, s$ and $n \geq r+s$, is not always true.

Example Let r and s be integers, with $r > 8$ and $s \geq 2$. Let B be a complete bipartite graph $K_{3,r-5}$ with partite sets $\{x_i : i = 2, 3, \ldots, r-4\}$ and $\{a, b, c\}$, and let

B' be a complete bipartite graph $K_{2,s-2}$ with partite sets $\{y_i : i = 1, 2, \ldots, s-2\}$ and $\{v, w\}$.

We define the vertex-set of the graph H as the disjoint union of the vertex-sets of B and B', together with two additional vertices x_1 and x_{r-3}. The edge-set of H contains the edge-sets of B and B' and the edges x_1a, x_1b, $x_{r-3}b$, $x_{r-3}c$, ba, bc and vw. Furthermore, we join the vertices a, b and c to v and w.

It is easy to see that $\lambda_{r,s}(G) \le |(\{a, b, c\}, \{v, w\})| = 6$. The unique possibility for a 2-2-vertex-cut is to disconnect the edges x_1a and $x_{r-3}c$. This requires the removal of the vertices v, w, b and x_i, for $i = 2, 3, \ldots, r-4$, and implies that $\kappa_{2,2}(H) = r-2 > 6$, and hence that $\kappa_{2,2}(H) > \lambda_{r,s}(H)$.

Inspired by Whitney's inequality $\kappa \le \lambda$, Harary [29] asked in 1983 whether the inequality $\kappa(G, P) \le \lambda(G, P)$ is true for any graphical property P. In 2005, Hellwig, Rautenbach and Volkmann [30] used the above example to give a negative answer: if P is the property that a graph contains at least two vertices, then $\kappa_{2,2}(G) \le \kappa(G, P)$ and $\lambda_{2,2}(G) = \lambda(G, P)$ for every graph G. It follows, for the graph H, that

$$\kappa(H, P) \ge \kappa_{2,2}(H) > \lambda_{r,s}(H) \ge \lambda_{2,2}(H) = \lambda(H, P).$$

If G is $\kappa_{1,2}$-connected, then the inequality $\kappa_{1,2}(G) \le n(G) - 3$ is immediate. Using an explicit characterization of 'claw-free and paw-free' graphs given by Faudree, Flandrin and Ryjáček [22], Hellwig, Rautenbach and Volkmann [30] were able to determine all graphs with $\kappa_{1,2}(G) = n(G) - 3$.

Theorem 7.5 *Let G be a $\kappa_{1,2}$-connected graph of order n. Then G satisfies $\kappa_{1,2}(G) = n - 3$ if and only if its complement is claw-free, paw-free, diamond-free and C_4-free.*

The following strong result is due to Győri [28] and Lovász [52].

Theorem 7.6 *For every k-connected graph G of order n, k positive integers $n_1, n_2, \ldots, n_k$ such that $n_1 + n_2 + \cdots + n_k = n$, and k vertices $v_1, v_2, \ldots, v_k$, there exists a partition $\{V_1, V_2, \ldots, V_k\}$ of $V(G)$ for which $v_i \in V_i$, $|V_i| = n_i$ and $G[V_i]$ is connected for $1 \le i \le k$.*

Using this result, Rautenbach and Volkmann [30] proved some sufficient conditions for graphs to be $\lambda_{p,q}$-connected.

Theorem 7.7 *Let p and q be integers with $q \ge p \ge 1$. A connected graph G of order $n \ge p+q$ and minimum degree δ is $\lambda_{p,q}$-connected if one of the following conditions is satisfied:*

- *G is 2-connected;*
- *G has a block of order at least $p+1$ containing at most one cut-vertex;*
- *$p = q \le \delta + 1$, and G contains a block with at least two cut-vertices;*
- *$n \ge 2q - 1$, and G contains a cut-vertex v for which all components of $G - v$ are of order at least p.*

In [70], the authors studied $\lambda_{2,q}$-connected graphs in detail. Among other results, they characterized the class of $\lambda_{2,q}$-connected trees, as follows.

Theorem 7.8 *Let $q \geq 2$ be an integer. A tree T of order $n \geq q+2$ is $\lambda_{2,q}$-connected if and only if it contains a vertex of degree k, where $2 \leq k \leq n-q$, which is adjacent to at most one vertex that is not a leaf.*

Using results about cyclic sums, Rautenbach and Volkmann [69] derived sufficient conditions for a graph of large enough order containing a cycle long enough to be $\lambda_{p,q}$-connected.

Theorem 7.9 *Let p, q and r be positive integers with $r \geq 3$ and $p+q \leq 2r-1$. If G is a connected graph of order $n \geq p+q$ which contains a cycle of length r, then G is $\lambda_{p,q}$-connected.*

The next theorem, due to Rautenbach and Volkmann [71], generalizes a result of Ou [63]. Its proof works along the same lines as Ou's proof, but [71] presents a considerably shorter argument.

Theorem 7.10 *Let p, q be integers with $2 \leq p \leq q$, and let G be a connected graph of order $n \geq \max\{2q-1, 3p-2\}$. Then G is $\lambda_{p,q}$-connected if and only if G contains no cut-vertex v with the property that each component of $G-v$ has at most $p-1$ vertices.*

Choosing $p=q$ in Theorem 7.10, we immediately obtain the above-mentioned result of Ou [63]. In [71], one can find further applications of Theorem 7.6. In [69], [71], Rautenbach and Volkmann also studied $\lambda_{p_1,p_2,\ldots,p_k}$-connected graphs: for positive integers $p_1, p_2, \ldots, p_k$ a connected graph G is *$\lambda_{p_1,p_2,\ldots,p_k}$-connected* if it has an edge-cut S with the property that $G-S$ has k components with vertex-sets $V_1, V_2, \ldots, V_k$ such that $|V_i| \geq p_i$ for $1 \leq i \leq k$.

In the following we discuss the case $p=q$ in detail. As an abbreviation, we write $\lambda_p(G)$ instead of $\lambda_{p,p}(G)$. Ou [63] gave the following sufficient and necessary conditions for graphs to be λ_p-connected.

Theorem 7.11 *Let G be a connected graph of order $n \geq 2p$. The graph G is λ_p-connected if and only if G has a spanning tree T for which $T-v$ has a component of order at least p, for any vertex $v \in V(T)$.*

The inequality $\lambda_p \leq \xi_p$

Following [8], [57], [61], a generalization of the minimum degree and minimum edge-degree of a graph is the *minimum p-edge-degree*, defined by

$$\xi_p(G) = \min\{|(X, \overline{X})| : X \subset V(G), |X| = p, G[X] \text{ is connected}\}.$$

Note that $\xi_1(G) = \delta(G)$ and $\xi_2(G) = \xi(G)$. For $p=3$, Bonsma, Ueffing and Volkmann [8] gave an inequality which is an analogue of Whitney's inequality $\lambda(G) \le \delta(G)$ and the inequality $\lambda_2(G) \le \xi_2(G)$ in Theorem 6.1.

Theorem 7.12 *If G is a λ_3-connected graph, then $\lambda_3(G) \le \xi_3(G)$.*

For $p \ge 4$, the inequality $\lambda_p(G) \le \xi_p(G)$ is not true in general. This can be seen in the following example by Bonsma, Ueffing and Volkmann [8] (see Fig. 2).

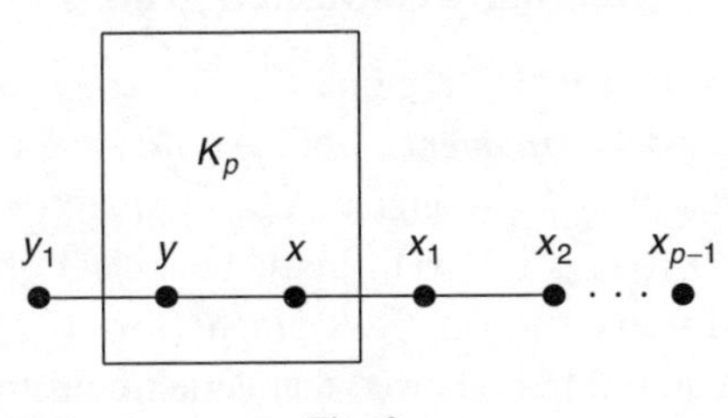

Fig. 2.

Example Let G be the disjoint union of a complete graph K_p and the vertices $y_1, x_1, x_2, \ldots, x_{p-1}$, together with the edges yy_1, xx_1 and $x_i x_{i+1}$, for $1 \le i \le p-2$, where $x, y \in V(K_p)$. Then $\xi_p(G) = |(V(K_p), \overline{V(K_p)})| = 2$ and $\lambda_p(G) = p - 1 > \xi_p(G)$.

In special cases, the inequality $\lambda_p \le \xi_p$ is true, even if $p \ge 4$, as the next results demonstrate. The first is due to Ou [64], and gives a lower bound for the order of a graph.

Theorem 7.13 *Every λ_4-connected graph G of order at least* 11 *satisfies the inequality $\lambda_4(G) \le \xi_4(G)$.*

Let $H_1, H_2, \ldots, H_r$ be r copies of the complete graph K_t and let v be a further vertex. We join the vertex v to every vertex of $V(H_i)$, for $i = 1, 2, \ldots, r$, and denote the resulting graph by $H_{r,t}$. It is easy to see that no $H_{r,t}$ is λ_{t+1}-connected. In fact, Zhang and Yuan [94] showed that they are the only such graphs with high minimum degree.

Theorem 7.14 *Let G be a connected graph of minimum degree δ and order $n \ge 2(\delta+1)$, which is not isomorphic to any $H_{r,\delta}$. Then G is λ_p-connected for every $p \le \delta + 1$, and $\lambda_p(G) \le \xi_p(G)$.*

Regular graphs satisfy the inequality $\lambda_p \le \xi_p$ if their girth is large enough, as was shown by Ou and Zhang [66].

Theorem 7.15 *Let G be an r-regular λ_p-connected graph of girth g. If $g \ge \frac{1}{2}p + 2$, then $\lambda_p(G) \le \xi_p(G)$.*

For r-regular graphs, the parameter ξ_p can be computed to be exactly $pr-2p+2$ if the girth is large enough. Thus, Wang and Li [84] were able to give an upper bound for λ_p in terms of δ.

Theorem 7.16 *Let G be a connected r-regular graph of order $n \geq 6$, with girth g, minimum degree δ and $r \geq 3$. Then G is λ_p-connected for $1 \leq p \leq \min\{g-1, \frac{1}{2}n\}$ and $\lambda_p(G) \leq p\delta - 2p + 2 = \xi_p(G)$.*

Maximally connected graphs

When an edge-cut S is a λ_p-cut of a graph G, the vertex-sets of the two components of $G - S$ are called *λ_p-fragments*. Let $r_p(G)$ denote the order of a smallest λ_p-fragment of G. We call a λ_p-fragment X a *λ_p-atom* of a graph G if $|X| = r_p(G)$. Obviously, $p \leq r_p(G) \leq \frac{1}{2}|V(G)|$. Let $\mathcal{G}$ be a class of λ_p-connected graphs with $\lambda_p(G) \leq \xi_p(G)$, for every $G \in \mathcal{G}$. A graph $G \in \mathcal{G}$ is called *λ_p-optimal* if $\lambda_p(G) = \xi_p(G)$. Wang and Li [84] showed that vertex-transitive and edge-transitive regular graphs are λ_p-optimal if their girth is large enough.

Theorem 7.17 *Let G be a regular connected edge-transitive graph of order $n \geq 6$, with girth $g \geq 4$ and minimum degree $\delta \geq 4$. Then G is λ_p-optimal for every $p \leq \min\{g-1, \frac{1}{2}n\}$.*

Theorem 7.18 *Let G be a regular connected vertex-transitive graph of order $n \geq 6$, with girth $g \geq 5$ and minimum degree $\delta \geq 4$. Then G is λ_p-optimal for every $p \leq \min\{g-1, \frac{1}{2}n\}$.*

If G is a λ_3-connected graph, then $\lambda_3(G) \leq \xi_3(G)$, by Theorem 7.12. The following characterization of λ_3-optimal graphs by Bonsma, Ueffing and Volkmann [8] was inspired by Xu and Xu's characterization of λ'-optimal graphs (see Theorem 6.17).

Theorem 7.19 *A λ_3-connected graph G is λ_3-optimal if and only if $r_3(G) = 3$.*

In the following, we survey results stating lower bounds for the order $r_p(G)$ of atoms in terms of the minimum degree or minimum edge-degree when the graph in question satisfies the inequality $\lambda_p < \xi_p$. Note that if we consider graph classes with $\lambda_p \leq \xi_p$, then each such result implies a sufficient criterion for a graph to be λ_p-optimal, since a smallest p-fragment of a graph contains at most half of the vertices. The first result in this category is an analogue of Theorem 6.18 for λ_3-connected graphs, and was proved by Zhang [93].

Theorem 7.20 *Let G be a λ_3-connected graph with minimum 3-edge-degree ξ_3. If G is not λ_3-optimal, then $r_3(G) \geq \frac{1}{3}\xi_3 + 1$.*

For triangle-free graphs this bound can be raised, as was shown by Meierling and Volkmann [56].

Theorem 7.21 *Let G be a λ_3-connected triangle-free graph with minimum degree at least 3 and minimum 3-edge-degree ξ_3. If G is not λ_3-optimal, then $r_3(G) \geq \frac{1}{2}\xi_3$.*

The bounds in both theorems are sharp, and examples demonstrate that the conditions on the minimum degree are necessary.

Since $\xi_3(G) \geq 3\delta(G) - 6$, Theorems 7.19 and 7.20 immediately imply the next result, which was originally proved by Bonsma, Ueffing and Volkmann [8].

Corollary 7.22 *Let G be a λ_3-connected graph with minimum degree δ. If G is not λ_3-optimal, then $r_3(G) \geq \max\{4, \delta - 1\}$.*

The following result, due to Zhang and Yuan [95], generalizes Corollary 7.22.

Theorem 7.23 *Let G be a λ_p-connected graph with minimum degree δ. If $\lambda_p(G) < \xi_p(G)$, then $r_p(G) \geq \max\{p+1, \delta+2-p\}$.*

For triangle-free graphs, Holtkamp, Meierling and Montejano [40] sharpened the lower bound on $r_p(G)$.

Theorem 7.24 *Let G be a λ_p-connected triangle-free graph with minimum degree δ. If $\lambda_p(G) < \xi_p(G)$, then $r_p(G) \geq \max\{p+1, 2\delta+1-p\}$.*

For graphs with bounded chromatic number or clique number, a general lower bound for $r_p(G)$ is valid. The following result is due to Holtkamp and Meierling [39].

Theorem 7.25 *Let G be a λ_p-connected graph with minimum degree δ, clique number ω and chromatic number χ. If $\lambda_p(G) < \xi_p(G)$, then $r_p(G) \geq p+1$ and*

$$(\chi - 1)(r_p(G) + p) \geq \chi\delta - \chi + 2\sqrt{\chi}.$$

Furthermore, if q divides p or is $p+1$, where $q \neq \omega$ or χ, then

$$(q-1)(r_p(G) + p) \geq q\delta + 1.$$

The following example in [39] shows that when ω or χ divides p, the lower bound for $r_p(G)$ is sharp.

Example For $q \geq 3$ let $p = q - 1$. We consider the complete q-partite graph with partite sets $V_i = \{x_i, y_i\}$ for $1 \leq i \leq q$, and remove the edges $x_i y_{i+1}$, for $1 \leq i \leq q-1$, and the edge $x_q y_1$. The resulting graph G has minimum degree $\delta = 2q - 3$, chromatic number $\chi = q$ and clique number $\omega = q$. Thus, $\xi_p(G) = p(\delta - p + 1) = (q-1)^2$. Moreover, $\lambda_k(G) \leq q(q-2) < \xi_p(G)$ (take $G[\{x_1, x_2, \ldots, x_q\}]$), which means that G is not λ_p-optimal. Furthermore, $r_p(G) = q = (q\delta - pq + p + 1)/(q-1)$.

Zhang and Yuan [95] recently presented an Ore-type condition for a graphs to be λ_p-optimal, which generalizes Theorem 6.6 for λ'-connected graphs.

Theorem 7.26 *If G is a connected graph of order $n \geq 2p$, and if*

$$d(v) + d(w) \geq n + 2p - 3$$

for every pair v and w of non-adjacent vertices, then G is λ_p-optimal.

Examples by Zhang and Yuan [95] show that the condition in Theorem 7.26 is sharp. Using similar methods, Ou [65] was able to relax it for triangle-free graphs.

Theorem 7.27 *If G is a triangle-free graph of order $n \geq 6$, and if $d(v)+d(w) \geq n-1$ for every pair v and w of non-adjacent vertices, then G is λ_3-optimal.*

Furthermore, Ou conjectured the following: if G is a connected triangle-free graph of large enough order n, for which $d(v) + d(w) \geq \frac{1}{2}n + 2$ for every pair v, w of non-adjacent vertices, then G is λ_3-optimal. Holtkamp, Meierling and Montejano [40] constructed classes of triangle-free graphs of order n with minimum degree $\delta = \frac{1}{4}n + 1$ that are not λ_3-optimal; thus, Ou's conjecture is not true if $d(v) + d(w) = \frac{1}{2}n + 2$. However, they also showed that, under a slightly stronger degree assumption, the conclusion of Ou's conjecture is valid.

Theorem 7.28 *If G is a connected triangle-free graph of order $n \geq 6$, for which $d(v) + d(w) \geq 2\lfloor \frac{1}{4}n \rfloor + 3$ for every pair v and w of non-adjacent vertices, then G is λ_3-optimal.*

The same authors also presented examples that show that their bound is sharp. The next result, due to Yuan and Liu [91], includes Theorem 6.13 as a special case.

Theorem 7.29 *For a positive integer p, let G be a triangle-free graph of order $n \geq 2p$. If $|N(v) \cap N(w)| \geq p$ for all pairs v and w of non-adjacent vertices, then G is λ_p-optimal.*

References

1. C. Balbuena and A. Carmona, On the connectivity and superconnectivity of bipartite digraphs and graphs, *Ars Combin.* **61** (2001), 3–21.
2. C. Balbuena, P. Garcia-Vázquez and X. Marcote, Sufficient conditions for λ'-optimality in graphs with girth g, *J. Graph Theory* **52** (2006), 73–86.
3. D. Bauer, C. Suffel, F. Boesch and R. Tindell, Connectivity extremal problems and the design of reliable probabilistic networks, *The Theory and Applications of Graphs* (Kalamazoo, 1980), Wiley (1981), 45–54.
4. F. Boesch and S. Chen, A generalization of line connectivity and optimally invulnerable graphs, *SIAM J. Appl. Math.* **34** (1978), 657–665.
5. B. Bollobás, On graphs with equal edge-connectivity and minimum degree, *Discrete Math.* **28** (1979), 321–323.
6. J. A. Bondy, Properties of graphs with constraints on the degrees, *Studia Sci. Math. Hungar.* **4** (1969), 473–475.
7. J. A. Bondy and U. S. R. Murty, *Graph Theory with Applications*, Macmillan, 1976.
8. P. Bonsma, N. Ueffing and L. Volkmann, Edge-cuts leaving components of order at least three, *Discrete Math.* **256** (2002), 431–439.
9. G. Chartrand, A graph-theoretic approach to a communications problem, *SIAM J. Appl. Math.* **14** (1966), 778–781.
10. G. Chartrand and F. Harary, Graphs with prescribed connectivities, *Theory of Graphs (Proc. Colloq., Tihany, 1966)*, Academic Press (1968), 61–63.

11. P. Dankelmann, A. Hellwig and L. Volkmann, On the connectivity of diamond-free graphs, *Discrete Appl. Math.* **155** (2007), 2111–2117.
12. P Dankelmann, A. Hellwig and L. Volkmann, Inverse degree and edge-connectivity, *Discrete Math.* **309** (2009), 2943–2947.
13. P. Dankelmann and L. Volkmann, New sufficient conditions for equality of minimum degree and edge-connectivity, *Ars Combin.* **40** (1995), 270–278.
14. P. Dankelmann and L. Volkmann, Degree sequence conditions for maximally edge-connected graphs and digraphs, *J. Graph Theory* **26** (1997), 27–34.
15. P. Dankelmann and L. Volkmann, Degree sequence conditions for maximally edge-connected graphs depending on the clique number, *Discrete Math.* **211** (2000), 217–223.
16. M. Daven and C. A. Rodger, (k, g)-cages are 3-connected, *Discrete Math.* **199** (1999), 207–215.
17. A. H. Esfahanian, Generalized measures of fault tolerance with application to n-cube networks, *IEEE Trans. Computers* **38** (1989), 1586–1591.
18. A. H. Esfahanian and S. L. Hakimi, On computing a conditional edge-connectivity of a graph, *Inform. Process. Lett.* **27** (1988), 195–199.
19. J. Fàbrega and M. A. Fiol, Maximally connected digraphs, *J. Graph Theory* **13** (1989), 657–668.
20. J. Fàbrega and M. A. Fiol, Bipartite graphs and digraphs with maximum connectivity, *Discrete Appl. Math.* **69** (1996), 271–279.
21. J. Fàbrega and M. A. Fiol, On the extraconnectivity of graphs, *Discrete Math.* **155** (1996), 49–57.
22. R. Faudree, E. Flandrin, and Z. Ryjáček, Claw-free graphs – A survey, *Discrete Math.* **164** (1997), 87–147.
23. M. A. Fiol and J. Fàbrega, On the distance connectivity of graphs and digraphs, *Discrete Math.* **125** (1994), 169–176.
24. H. L. Fu, K. C. Huang and C. A. Rodger, Connectivity of cages, *J. Graph Theory* **24** (1997), 187–191.
25. D. Geller and F. Harary, Connectivity in digraphs, *Recent Trends in Graph Theory (Proc. Conf., New York, 1970)*, Lecture Notes in Mathematics **186**, Springer (1971), 105–115.
26. C. Godsil and G. Royle, *Algebraic Graph Theory*, Springer, 2001.
27. D. L. Goldsmith and R. C. Entringer, A sufficient condition for equality of edge-connectivity and minimum degree of a graph, *J. Graph Theory* **3** (1979), 251–255.
28. E. Győri, On division of graphs to connected subgraphs, *Combinatorics (Proc. Fifth Hungarian Colloq., Keszthely, 1976)*, I, Colloq. Math. Soc. János Bolyai **18**, North-Holland (1978), 485–494.
29. F. Harary, Conditional connectivity, *Networks* **13** (1983), 347–357.
30. A. Hellwig, D. Rautenbach and L. Volkmann, Cuts leaving components of given minimal order, *Discrete Math.* **292** (2005), 55–65.
31. A. Hellwig and L. Volkmann, Maximally edge-connected digraphs, *Australas. J. Combin.* **27** (2003), 23–32.
32. A. Hellwig and L. Volkmann, Maximally local-edge-connected graphs and digraphs, *Ars Combin.* **72** (2004), 295–306.
33. A. Hellwig and L. Volkmann, Sufficient conditions for λ'-optimality in graphs of diameter 2, *Discrete Math.* **283** (2004), 113–120.
34. A. Hellwig and L. Volkmann, Neighborhood conditions for graphs and digraphs to be maximally edge-connected, *Australas. J. Combin.* **33** (2005), 265–277.
35. A. Hellwig and L. Volkmann, Sufficient conditions for graphs to be λ'-optimal, super-edge-connected and maximally edge-connected, *J. Graph Theory* **48** (2005), 228–246.

36. A. Hellwig and L. Volkmann, Lower bounds on the vertex-connectivity of digraphs and graphs, *Inform. Process. Lett.* **99** (2006), 41–46.
37. A. Hellwig and L. Volkmann, The connectivity of a graph and its complement, *Discrete Appl. Math.* **156** (2008), 3325–3328.
38. A. Holtkamp, Maximally local edge-connectivity of diamond-free graphs, *Australas. J. Combin.* **49** (2011), 153–158.
39. A. Holtkamp and D. Meierling, Restricted edge-connectivity in graphs with bounded clique number or bounded chromatic number, submitted.
40. A. Holtkamp, D. Meierling and L. P. Montejano, k-restricted edge-connectivity in triangle-free graphs, *Discrete Appl. Math.* (to appear).
41. A. Holtkamp and L. Volkmann, On the connectivity of p-diamond-free graphs, *Discrete Math.* **309** (2009), 6065–6069.
42. A. Holtkamp and L. Volkmann, On local connectivity of graphs with given clique number, *J. Graph Theory* **63** (2010), 192–197.
43. A. Holtkamp and L. Volkmann, On local connectivity of $K_{2,p}$-free graphs, *Australas. J. Combin.* **51** (2011), 29–40.
44. M. Imase, T. Soneoka and K. Okada, Connectivity of regular directed graphs with small diameters, *IEEE Trans. Comput.* **34** (1985), 267–273.
45. T. Jiang and D. Mubayi, Connectivity and separating sets of cages, *J. Graph Theory* **29** (1998), 35–44.
46. V. G. Kane and S. P. Mohanty, A lower bound on the number of vertices of a graph, *Proc. Amer. Math. Soc.* **72** (1978), 211–212.
47. L. Lesniak, Results on the edge-connectivity of graphs, *Discrete Math.* **8** (1974), 351–354.
48. Y. Lin, C. Balbuena, X. Marcote and M. Miller, On the connectivity of (k, g)-cages of even girth, *Discrete Math.* **308** (2008), 3249–3256.
49. Y. Lin, M. Miller and C. Balbuena, Improved lower bounds on the connectivity of $(\delta; g)$-cages, *Discrete Math.* **299** (2005), 162–171.
50. Y. Lin, M. Miller, C. Balbuena and X. Marcote, All $(k; g)$-cages are edge-superconnected, *Networks* **47** (2006), 102–110.
51. Y. Lin, M. Miller and C. Rodger, All $(k; g)$-cages are k-edge-connected, *J. Graph Theory* **48** (2005), 219–227.
52. L. Lovász, A homology theory for spanning trees of a graph, *Acta Math. Acad. Sci. Hungar.* **30** (1977), 241–251.
53. X. Marcote and C. Balbuena, Edge-superconnectivity of cages, *Networks* **43** (2004), 54–59.
54. X. Marcote, C. Balbuena and I. Pelayo, On the connectivity of cages with girth five, six and eight, *Discrete Math.* **307** (2007), 1441–1446.
55. X. Marcote, C. Balbuena, I. Pelayo and J. Fàbrega, (δ, g)-cages with $g \geq 10$ are 4-connected, *Discrete Math.* **301** (2005), 124–136.
56. D. Meierling and L. Volkmann, Sufficient conditions for triangle-free graphs to be optimally restricted edge-connected, submitted.
57. J. Meng and Y. Ji, On a kind of restricted edge connectivity of graphs, *Discrete Appl. Math.* **117** (2002), 183–193.
58. K. Menger, Zur allgemeinen Kurventheorie, *Fund. Math.* **10** (1927), 96–115.
59. M. H. Moriarty and P. R. Christopher, Cages of degree k are k-edge-connected, *Proc. 36th Southeastern International Conference on Combinatorics, Graph Theory, and Computing, Congr. Numer.* **173** (2005), 161–167.
60. J. P. Ou, Restricted edge connectivity of regular graphs, *J. Math. Study* **34** (2001), 345–350.

61. J. P. Ou, *Restricted edge connectivity and network reliability*, Ph.D. thesis, Xiamen University, 2003.
62. J. P. Ou, 4-restricted edge cuts of graphs, *Australas. J. Combin.* **30** (2004), 103–112.
63. J. P. Ou, Edge cuts leaving components of order at least m, *Discrete Math.* **305** (2005), 365–371.
64. J. P. Ou, A bound on 4-restricted edge connectivity of graphs, *Discrete Math.* **307** (2007), 2429–2437.
65. J. P. Ou, Ore-type conditions for maximizing 3-restricted edge connectivity of graphs, submitted.
66. J. P. Ou and F. J. Zhang, Bound on m-restricted edge-connectivity, *Acta Math. Appl. Sin. (English Series)* **19** (2003), 505–510.
67. J. Plesník, Critical graphs of given diameter, *Acta Fac. Rerum Natur. Univ. Commenian Math.* **30** (1975), 71–93.
68. J. Plesník and S. Znám, On equality of edge-connectivity and minimum degree of a graph, *Arch. Match. (Brno)* **25** (1989), 19–25.
69. D. Rautenbach and L. Volkmann, Cyclic sums, network sharing and restricted edge cuts in graphs with long cycles, *Networks* **52** (2008), 252–255.
70. D. Rautenbach and L. Volkmann, Some remarks on $\lambda_{p,q}$-connectedness, *Discrete Math.* **308** (2008), 5562–5569.
71. D. Rautenbach and L. Volkmann, On the existence of edge cuts leaving several large components, *Discrete Math.* **309** (2009), 1703–1707.
72. L. Shang and H. P. Zhang, Sufficient conditions for graphs to be λ'-optimal and super-λ', *Networks* **49** (2007), 234–242.
73. T. Soneoka, H. Nakada and M. Imase, Sufficient conditions for dense graphs to be maximally connected, *Proc. ISCAS* **85** (1985), 811–814.
74. T. Soneoka, H. Nakada, M. Imase and C. Peyrat, Sufficient conditions for maximally connected dense graphs, *Discrete Math.* **63** (1987), 53–66.
75. J. Topp and L. Volkmann, Sufficient conditions for equality of connectivity and minimum degree of a graph, *J. Graph Theory* **17** (1993), 695–700.
76. N. Ueffing and L. Volkmann, Restricted edge-connectivity and minimum edge-degree, *Ars Combin.* **66** (2003), 193–203.
77. L. Volkmann, Bemerkungen zum p-fachen Zusammenhang von Graphen, *An. Univ. Bucuresti Mat.* **37** (1988), 75–79.
78. L. Volkmann, Edge-connectivity in p-partite graphs, *J. Graph Theory* **13** (1989), 1–6.
79. L. Volkmann, Degree sequence conditions for equal edge-connectivity and minimum degree, depending on the clique number, *J. Graph Theory* **42** (2003), 234–245.
80. L. Volkmann, *Graphen an allen Ecken und Kanten*, RWTH Aachen University, 2006, *http://www.math2.rwth-aachen.de/volkmann*.
81. L. Volkmann, Local-edge-connectivity in digraphs and oriented graphs, *Discrete Math.* **307** (2007), 3207–3212.
82. L. Volkmann, Restricted arc-connectivity of digraphs, *Inform. Process. Lett.* **103** (2007), 234–239.
83. L. Volkmann, On local connectivity of graphs, *Appl. Math. Letters* **21** (2008), 63–66.
84. M. Wang and W. Li, Conditional edge connectivity properties, reliability comparisons and transitivity of graphs, *Discrete Math.* **258** (2002), 205–214.
85. P. Wang, B. Xu and J. Wang, A note on the edge-connectivity of cages, *Electron. J. Combin.* **10** (2003), Note 2, 4 pp. (electronic).
86. Y. Q. Wang and Q. Li, Super-edge-connectivity properties of graphs with diameter 2, *J. Shanghai Jiaotong Univ. (Chinese edn.)* **33** (1999), 646–649.
87. M. E. Watkins, A lower bound for the number of vertices of a graph, *Amer. Math. Monthly* **74** (1967), 297.

88. H. Whitney, Congruent graphs and the connectivity of graphs, *Amer. J. Math.* **54** (1932), 150–168.
89. B. Xu, P. Wang, and J. Wang, On the connectivity of (4, g)-cages, *Ars Combin.* **64** (2002), 181–192.
90. J. M. Xu and K. L. Xu, On restricted edge-connectivity of graphs, *Discrete Math.* **243** (2002), 291–298.
91. J. Yuan and A. Liu, Sufficient conditions for λ_k-optimality in triangle-free graphs, *Discrete Math.* **310** (2010), 981–987.
92. J. Yuan, A. Liu and S. Wang, Sufficient conditions for bipartite graphs to be super-k-restricted edge connected, *Discrete Math.* **309** (2009), 2886–2896.
93. Z. Zhang, Sufficient conditions for restricted-edge-connectivity to be optimal, *Discrete Math.* **307** (2007), 2891–2899.
94. Z. Zhang and J. Yuan, A proof of an inequality concerning k-restricted edge connectivity, *Discrete Math.* **304** (2005), 128–134.
95. Z. Zhang and J. Yuan, Degree conditions for restricted-edge-connectivity and isoperimetric-edge-connectivity to be optimal, *Discrete Math.* **307** (2007), 293–298.

3
Minimal connectivity

MATTHIAS KRIESELL

Three families of graphs of connectivity k that play a prominent role in the theory of graph connectivity are those for which the deletion of any edge, the deletion of any vertex, or the contraction of any edge results in a graph that is no longer k-connected. Here, we give a brief introduction to these families, with a light emphasis on reduction and construction theorems for some classes of k-connected graphs.

1. Introduction

One of the main concerns of graph connectivity theory is to find reduction methods for classes of k-connected graphs. Such methods can be used for induction proofs, but have also a constructive counterpart which might be helpful for generating the respective classes. They can also be employed for setting up problem-solving strategies for graphs in general. Roughly speaking, if a graph has a small cutset, then we can split it into two smaller parts, solve the problem, and combine the solutions; and if not, then we might be able to reduce it and use structural properties of k-connected graphs. An instructive example of this procedure is Carsten Thomassen's brilliant proof [72] of Kuratowski's theorem.

Tutte was the first to study such methods systematically, and he did it for 2-connected and 3-connected graphs. As an initial example, let us begin with the following version of his celebrated *wheel theorem* [77]. (A *wheel* is the join of a cycle and a single vertex.)

Theorem 1.1 *Every* 3*-connected graph, other than a wheel, can be reduced to a smaller* 3*-connected graph by either deleting or contracting an edge.*

This is a reduction theorem for the class of 3-connected graphs. It tells us that unless such a graph belongs to a simple subclass of *basic graphs* (the wheels), we can reduce it to a smaller one by performing a *short sequence of elementary reductions* (deleting or contracting edges). In Section 6, we discuss a number of similar methods for other classes of graphs. Not all of the reductions are forms of deletion or contraction, but there is a good reason to concentrate on these types of operations. Neither deleting nor contracting an edge, nor deleting a vertex, can create a minor in the reduced graph that was not there before. In other words, we stay inside any given class described by forbidden minors; the only reason for which the reduction might fail is that the resulting graph might not be k-connected.

The methods used to actually find reducible objects (for example, a single contractible or deletable edge, as in the wheel theorem) are developed in the proofs. They often have a potential for generalization, which typically can answer questions such as:

- are there many reducible objects?
- are they common?
- can they be reduced in such a way that some additional property survives?

In very basic terms, these are questions on the number and distribution of reducible objects. As an example, we note Negami's splitter theorem for 3-connected graphs [57]:

Theorem 1.2 *If G is a* 3*-connected graph, other than a wheel, then every* 3*-connected graph that properly contains G as a minor can be reduced to a smaller* 3*-connected graph containing G as a minor, by deleting or contracting a single edge.*

The results of Section 6 can be considered as high-end outcomes of a flourishing study of the distribution of reducible objects in a graph. We start, in Section 2, with a brief study of *minimally k-connected* graphs – that is, those k-connected graphs with the property that deleting any edge produces a graph that is no longer k-connected. These classes are far too big to become reasonably primitive base classes for a reduction theorem, even for $k=2$. The same applies to *critically k-connected graphs*, which are studied in Section 3 – that is, those k-connected graphs for which deleting any vertex produces a graph that is no longer k-connected. Indeed, the same can be said for those graphs that are both minimally and critically

k-connected. (An example of such a graph is a k-connected graph in which every edge is incident with a vertex of degree k and every vertex is adjacent to such a vertex.)

It is therefore necessary to look at yet another elementary operation, and see how far that takes us. Based on our earlier remarks, edge-contraction would seem to be a good choice, and so we define a k-connected graph to be *contraction-critical* if contracting any edge produces a graph that is no longer k-connected. Such graphs are considered in Section 4. They constitute a proper subclass of the class of critically k-connected graphs (which in turn can be treated as a special case of a more general concept explored in Section 5). For $k \leq 3$, the only contraction-critical graph is K_{k+1}, which is certainly a basic graph. However, for $k \geq 4$, there are infinitely many contraction-critically k-connected graphs, so they are rich in some sense. Still, for $k = 4$, we can describe most of them as line graphs of a class of cubic graphs, and this in turn admits a constructive characterization. Alternatively, they can be reduced by contracting *two* edges to obtain a smaller 4-connected graph [31]. This suggests an idea of what we think might be true for 5-connected graphs.

Conjecture A *There exist integers n and h for which every 5-connected graph of order at least n can be reduced to a smaller one by contracting a set of at most h edges.*

As we have pointed out, the corresponding result for k-connected graphs with $k \leq 4$ is true. But the reason that Conjecture A is an interesting question is not so much that it would settle the *next* open case, but that it would settle the *last* open case: for $k \geq 6$, the corresponding statements for k-connected graphs *are not true* – that is, there are minimally k-connected graphs for which we have to contract arbitrarily many edges to obtain a smaller k-connected graph; equivalently, for $k \geq 6$, the gap between the order of such a graph and the maximum order of its lower neighbours in the minor relation, restricted to k-connected graphs, is arbitrarily large [31].

This indicates that it may be worthwhile to look at larger substructures of a graph than just vertices or edges, and to study possible ways to employ them for reduction; this is the topic of Sections 7 and 8.

Many problems involving vertex-connectivity have a literal counterpart in terms of edge-connectivity. Although these mostly turn out to be easy, if not trivial, the restriction of the original vertex-connectivity problems to line graphs is a bit more demanding and may serve as a touchstone, as is illustrated in Section 9.

2. Edge-deletion

We continue with a brief section on graphs for which the deletion of any edge decreases the connectivity. An edge e in a k-connected graph G is called *k-essential* if $G - e$ is not k-connected, and G is called *minimally k-edge-connected* if every edge is k-essential. The connectivity of such a graph is equal to k. Minimally k-edge-connected graphs are very well understood, mostly due to the following fundamental result of Mader [42], relating vertices of degree k and k-essential edges.

Theorem 2.1 *If G is a k-connected graph and C is a cycle of k-essential edges in G, then there is a vertex of degree k in C.*

As an immediate consequence, we see that in a minimally k-connected graph G, the vertices of degree greater than k induce a forest, and it is easy to show that such a graph of order n must have more than $n(k-1)/(2k-1)$ vertices of degree k (see [42]). Moreover, Theorem 2.1 implies that every minimally k-connected graph G is the edge-disjoint union of a tree and $k-1$ other forests. (The proof is by induction on k, with the result being obvious for $k=1$. If G is minimally k-connected for $k>1$, it must contain a minimally $(k-1)$-connected subgraph H, for which the statement is true; if G has a cycle C consisting of edges not in H then, by Theorem 2.1, C contains a vertex v of degree k in G. But then v has degree $k-2$ in H, which is a contradiction.) We note that this implies that the average degree of a minimally k-connected graph is less than $2k$.

Another consequence of Theorem 2.1 is that every minimally k-connected graph has at least $k+1$ vertices of degree k (see [42]). This implies that – unlike in the case of vertex-deletion (see the next section) – deleting an edge from a k-connected graph (for $k \geq 2$) almost never produces a minimally $(k-1)$-connected graph; the only exceptions are the cycles. The argument runs as follows. If $G-e$ is minimally $(k-1)$-connected, then it has k vertices of degree $k-1$. But at most two vertices are of degree $k-1$, the ends of e. Hence $k=2$, and $G-e$ is minimally 1-connected with two vertices of degree 1, so G must be a cycle.

We note that, prior to Mader's discovery of Theorem 2.1, Halin [17] proved that every *triangle* of k-essential edges in a k-connected graph contains at least *two* vertices of degree k.

For a comprehensive survey on minimally k-connected graphs and digraphs, see Mader [49].

3. Vertex-deletion

A vertex v in a k-connected graph G is called *k-essential* if $G-v$ is not k-connected, and G is called *critically k-connected* if every vertex is k-essential. As was the case with edges, in such a graph k must equal $\kappa(G)$. Any k-connected graph in which every vertex is adjacent to a vertex of degree k is critically k-connected. Clearly, for $k>1$, such graphs might have vertices of arbitrarily large degree. One might ask if, as for minimally k-connected graphs, a critically k-connected graph must have vertices of degree k. This turns out to be true for $k=2$ and $k=3$, by the following theorem, but it is not true in general. To see this, observe that if G is a critically k-connected graph then

- the lexicographic product $G[K_l]$ of G and the complete graph of order l is critically kl-connected with $\delta(G[K_l]) = l\delta(G) + l - 1$;
- the join $G + K_1$ is critically $(k+1)$-connected, with $\delta(G + K_1) = \delta(G) + 1$.

Hence, given $k \geq 2$ and any critically 2-connected graph G, the graphs $G[K_{k/2}]$ for even k and $G[K_{(k-1)/2}] + K_1$ for odd k are critically k-connected with minimum degree $\lfloor \frac{3}{2}k - 1 \rfloor$. At the same time, they are sharpness examples of the following theorem of Lick [40].

Theorem 3.1 *If G is a critically k-connected graph, then $\delta(G) \leq \frac{3}{2}k - 1$.*

This is a consequence of a more general result due to Mader [41]. To describe it, we need to introduce some terminology. Given a graph G, we denote by $\mathcal{T}(G)$ the set of all cutsets of cardinality $\kappa(G)$. A *fragment* of G is the union of the vertex-sets of at least one, but not all, components of $G - T$, for some cutset $T \in \mathcal{T}(G)$. For a non-complete graph, the fragments of minimum cardinality are called *atoms*, and their cardinality is denoted by $a(G)$. It is easy to see that a vertex of a non-complete graph G of connectivity k is k-essential if and only if it is contained in some member of $\mathcal{T}(G)$; consequently, G is critically k-connected if and only if $\bigcup \mathcal{T}(G) = V(G)$. If A is an atom, then the neighbourhood $N(v)$ of any $v \in A$ is contained in the union $(A - \{v\}) \cup N(A)$, and $|N(A)| = \kappa(G)$ because $N(A) \in \mathcal{T}(G)$. We deduce that, $\delta(G) \leq \kappa(G) + a(G) - 1$.

Theorem 3.2 *If A is an atom of a graph G, and if $T \in \mathcal{T}(G)$ contains at least one vertex from A, then $A \subseteq T$ and $|A| \leq \frac{1}{2}|T - N(A)|$.*

We note that this is a good justification for the term *atom*, from the Greek, ἄτομος, meaning uncuttable or indivisible, as suggested by the first conclusion of the theorem, and the connotation of being small, as suggested by the second conclusion. The following is a consequence of Theorem 3.2, and it in turn implies Theorem 3.1. (In Section 5, we will come to a yet more general version of Theorem 3.2.)

Corollary 3.3 *If G is a critically k-connected graph that is not complete, then $a(G) \leq \frac{1}{2}k$.*

Unlike the case for minimally k-connected graphs, we have established the fact that the average degree of critically k-connected graphs cannot be bounded by a function of k (see [32]).

Theorem 3.2 implies that the atoms of a k-connected graph are disjoint. This was first proved by Watkins [79], and he used it to relate the degree and the connectivity of vertex-transitive graphs. (He also provided an independent proof of Theorem 3.1, for vertex-transitive graphs.)

The question of relating degree and connectivity of a vertex-transitive graph goes back to Vizing [78] (who called vertex-transitive graphs regular). For even k, the graphs $C_l[K_{k/2}]$ show that the bound of Theorem 3.1 is sharp for vertex-transitive graphs, but for odd k our examples $G[K_{(k-1)/2}] + K_1$ are not vertex-transitive.

Indeed, in this case we can improve the bound as follows. The first conclusion of Theorem 3.2 implies that distinct atoms of a critically k-connected graph are disjoint. Moreover, if G is vertex-transitive, then the atoms form a system of imprimitivity, and so every $T \in \mathcal{T}(G)$ is the disjoint union of atoms. Therefore, $a(G)$ divides $|T| = \kappa(G)$, and $a(G) \neq |T|$ by the last part of Theorem 3.2. Hence $a(G) \leq k/p$, where p is the smallest prime divisor of k, and this improves the bound from Theorem 3.1 to $\delta(G) \leq (k(p+1)/p) - 1$. This is sharp because of the vertex-transitive graphs $G[K_{k/p}]$, where G is any vertex-transitive p-regular graph of connectivity p (see also [79]).

Later, Jung [20] used similar methods to analyse graphs in which every vertex is contained in the same number of *k-atoms* – that is, fragments F with $|F| \geq k$ and $|V(G) - (F \cup N(F))| \geq k$ of minimum cardinality. In particular, this led to an interesting structural classification of vertex-transitive graphs in terms of fragment clusters.

It was observed by Watkins [79] that, for an edge-transitive graph G, the disjointness of atoms implies that $a(G) = 1$, and hence that $\delta(G) = \kappa(G)$. Here is his argument. Assume, to the contrary, that $a(G) > 1$, and let A be an atom. Then there exist a path uvw with $v, w \in A$ and $u \in N(A)$, and an automorphism of G that maps vw to uv. Clearly, uv is contained in an atom which is distinct but not disjoint from A, which is a contradiction.

Maurer and Slater [55] suggested generalizing the concept of critically k-connected graphs. They defined a graph G to be *l-critically k-connected* if $\kappa(G - X) = k - |X|$ for all $X \subseteq V(G)$ with $|X| \leq l$; equivalently, $\kappa(G) = k$ and either $G = K_{k+1}$ or every $X \subseteq V(G)$ with $|X| \leq l$ is a subset of some $T \in \mathcal{T}(G)$. Any such graph G is called *l-critical*. Obviously, G is critically k-connected if and only if G is 1-critically k-connected, and every l-critically k-connected graph is l'-critically k-connected whenever $1 \leq l' \leq l$. Trivially, $l \leq k$, but Maurer and Slater conjectured that even $l \leq \frac{1}{2}k$ holds unless $G = K_{k+1}$. This was first proved by Su [66] in 1988; for a more accessible description, see Su [67].

Theorem 3.4 *If there exists an l-critically k-connected graph other than K_{k+1}, then $l \leq \frac{1}{2}k$.*

Note that the graph $S_l = K_{l+1}[\overline{K_2}]$ is l-critically $2l$-connected and that $S_l - v$ is $(l-1)$-critically $(2l-1)$-connected, showing that the bound in this theorem is sharp.

Later, in 1998, Jordán found a very elegant argument for Theorem 3.4, which is one of the pearls of this part of the theory [18]; it depends on Theorem 2.1. Let G be an l-critically k-connected graph, and let A be an atom of G. By Theorem 3.2 it follows easily that $G - A$ is $(l-1)$-critically $(k - |A|)$-connected, and that every fragment F of $G - A$ is a fragment of G, where $N_G(F) = N_{G-A}(F) \cup A$. Each fragment of $G - A$ contains a vertex of $S = N_G(A)$. Therefore, we may add some set N of new edges between vertices of S such that $(G - A) + N$ is $(k - |A| + 1)$-connected. If we take N minimal with respect to this property, then every new edge is $(k - |A| + 1)$-essential for $(G - A) + N$, so that N forms a forest on S, by

Theorem 2.1. The edges of a forest (or of *any* bipartite graph) can be covered by at most half of its vertices, so there exists a set X of at most $\frac{1}{2}|S|$ vertices in S meeting all of the vertices in N. Since, for every smallest separating set T of $G - A$, there must be an edge from N connecting two components of $(G - A) - T$ (for otherwise $\kappa((G-A)+N) = k - |A|$), we deduce that X is not contained in a smallest separating set of $G - A$. Consequently, $|X| \geq l$, since $G - A$ is $(l - 1)$-critically $(k - |A|)$-connected. Therefore, $l \leq |X| \leq \frac{1}{2}|S| = \frac{1}{2}\kappa(G)$, as desired.

In a survey paper in 1984, Mader [47] conjectured several properties of l-critically k-connected graphs that are stronger than the statement of Theorem 3.4, and all of these have been proved. Possibly the most difficult result along these lines is the following result of Su [68]:

Theorem 3.5 *Every l-critically k-connected graph other than K_{k+1} contains $2l+2$ disjoint fragments.*

Su's proof is very ingenious and very difficult. Unlike Theorem 3.4, no simpler proof has been found. Refining his method outlined above, Jordán [18] gave a (still simple) proof that in any l-critically k-connected graph there is an antichain (with respect to $\subseteq$) of $2l + 2$ fragments.

Given an atom A of an l-critically k-connected graph G, we may apply Theorem 3.5 to the $(l-1)$-critically $(k-|A|)$-connected graph $G-A$ (see above). It is easy to see that every fragment of $G - A$ contains at least $|A|$ vertices from $S = N_G(A)$, so that $(2(l - 1) + 2)|A| \leq |S| = k$. This yields the following generalization of Su [68] of Corollary 3.3 (also conjectured by Mader [47]).

Theorem 3.6 *If G is l-critically k-connected and not K_{k+1}, then $a(G) \leq k/2l$.*

The graphs $S_l[K_m]$ are l-critically $2lm$-connected and have exactly $2l + 2$ fragments, and these are pairwise disjoint and have cardinality m. It follows that the bounds in Theorem 3.6 (and Theorem 3.5) are sharp.

In addition to these theorems, there is the following result on *extremely critically connected graphs* – that is, the l-critically $2l$-connected graphs. It says that the only such graphs that are not complete are the generalized octahedra $K_{l(2)}$; this too had been conjectured by Mader [47] (see [26], [33], and [71]).

Theorem 3.7 *If G is l-critically $2l$-connected with $l \geq 3$, then G is either K_{2l+1} or $K_{l(2)}$.*

Substantial parts of the proof are concerned with the small cases $l = 3$ and 4 (see [27]) and the cases up to approximately $l = 20$ (see [33]). For larger values, the statement turns out to be less difficult, in a sense. The proof in Kriesell [33] was developed while Theorem 3.5 was still an open question, and is thus independent of that theorem. However, Su, Yuan and Zhao [71] showed that it is possible to simplify part of the work for the small cases using Theorem 3.5. A characterization of the extremely critically connected graphs for *odd* connectivity at least 5 – that is, the l-critically $(2l + 1)$-connected graphs – might be achieved in the future.

There are several open questions on l-critically k-connected graphs, for small values of l. For the first one, see [47].

Conjecture B *Is there a number l for which every l-critically k-connected graph contains K_4?*

The octahedron $K_{2,2,2}$ is 2-critically 4-connected and does not contain K_4; this shows that l must be at least 3 if the conjecture is true.

If v and w are vertices of a 3-critically k-connected graph G, and if A is an atom of the 1-critically $(k-2)$-connected graph $G - \{v, w\}$ such that $|A| > 1$, then there exist adjacent vertices x and y in A; by Corollary 3.3, $|A| \leq \frac{1}{2}(k-2)$, which implies that x and y have a common neighbour in $A \cup N(A)$. This shows that v and w are at distance at most 4, and as they were chosen arbitrarily, it follows that G has diameter at most 4 (see [45]). Mader [47] asked if there exists a 3-critically k-connected graph of diameter 3 or 4. In [34] we answered this by constructing an l-critically k-connected graph of diameter 3, for every $l \geq 3$. The smallest has 252 vertices and is 122-connected. We doubt that there are any of diameter 4.

Conjecture C *For all k, every 3-critically k-connected graph has diameter at most* 3.

In order to extend our construction to produce l-critically k-connected graphs of diameter 4, it would be necessary to find sufficiently large l-critically $\frac{1}{2}(k-1)$-connected graphs. In [34] we showed that this is impossible for $l \geq 5$, and that it is impossible also for $l = 3$ or 4 if the following conjecture of Mader [50] is true.

Conjecture D *There exists $c > 0$ for which every* 3*-critically k-connected graph has at most $ck^{3/2}$ vertices.*

Mader [50] showed that every 3-critically k-connected graph has at most $2k^2 - k$ vertices. It is not known whether there exist $c > 0$ and $\lambda < 2$ for which every 3-critically k-connected graph has at most ck^{λ} vertices. It is not even known whether this holds for l-critically k-connected graphs, for sufficiently large l (not depending on k).

Regarding our quest for reduction theorems for a class $\mathcal{C}$ of k-connected graphs, it seems difficult to exploit knowledge on l-critically k-connected graphs for $l \geq 3$. However, a k-connected graph G has two vertices v and w whose identification produces a smaller k-connected graph G', unless G is 2-critically k-connected. These vertices may or may not be adjacent; if they *are* connected by an edge e then $G' = G/e$, and this situation is discussed in Section 4. If they are not adjacent, then G' might not be in $\mathcal{C}$ (for example, if it is described by forbidden minors). However, identifying non-adjacent vertices can be used to reduce k-connected *bipartite* graphs (where contraction is useless!), as will be explained in Section 6.

4. Edge-contraction

As before, let G be a k-connected graph. An edge e of G is called *k-contractible* if G/e is k-connected, and G is called *contraction-critically k-connected* if it has no k-contractible edges. Again, $k = \kappa(G)$. There is a plethora of results on the existence, number and distribution of k-contractible edges (see our 2002 survey [31]). Here, we repeat the basics and concentrate on more recent developments.

Trivially, every edge of a 1 connected graph other than K_2 is 1-contractible, and it is easy to see that every vertex in a 2-connected graph G other than K_3 is incident with some 2-contractible edge. Tutte's wheel theorem (Theorem 1.1) implies that every 3-connected graph G other than K_4 has a 3-contractible edge, but the wheels of order at least 5 and other examples show that there may be vertices incident with no 3-contractible edge. From these observations we deduce the following result.

Theorem 4.1 *For $k \leq 3$, the only contraction-critically k-connected graph is K_{k+1}.*

In other words, for $k \leq 3$, every k-connected graph not isomorphic to K_{k+1} can be reduced to a smaller k-connected graph by contracting one edge, and K_{k+1} is the only minor-minimal k-connected graph. This was generalized to the size of atoms by Egawa [7].

Theorem 4.2 *If G is a contraction-critically k-connected graph other than K_{k+1}, then $a(G) \leq \frac{1}{4}k$.*

Since every 2-critically k-connected graph is contraction-critically k-connected, Theorem 4.2 also generalizes Theorem 3.6, restricted to the case $l = 2$. In the spirit of Maurer and Slater's generalization of criticality (see Section 3), Mader [50] called a graph G *l-con-critically k-connected* if $\kappa(G - X) = k - |X|$, for all $X \subseteq V(G)$ with $|X| \leq l$ and $\langle X \rangle$ connected; equivalently, $\kappa(G) = k$ and either $G = K_{k+1}$ or every $X \subseteq V(G)$ with $|X| \leq l$ and $\langle X \rangle$ connected is a subset of some $T \in \mathcal{T}(G)$. Clearly, every l-critically k-connected graph is l-con-critically k-connected, but the converse is not true; this is not easy to verify (see [50]). In contrast to the situation with 3-critically k-connected graphs (see Section 3), Mader [51] showed that for 3-con-critically k-connected graphs it is not possible to bound the number of vertices by a function of k. However, he also showed that Theorem 3.4 can be generalized as follows.

Theorem 4.3 *If there exists an l-con-critically k-connected graph other than K_{k+1}, then $l \leq \frac{1}{2}k$.*

There are infinitely many contraction-critically 4-connected graphs, and all are known, due to the following result of Fontet [11], [12] and Martinov [52], [53], [54] (compare [47]).

Theorem 4.4 *The contraction-critically 4-connected graphs are the squares of cycles of length at least 5 and the line graphs of cubic cyclically 4-edge-connected graphs.*

From this, it is easy to see that every 4-connected graph other than K_5 and $K_{2,2,2}$ can be reduced to a smaller 4-connected graph by contracting one or two edges (so K_5 and $K_{2,2,2}$ are the only minor-minimal 4-connected graphs). As we mentioned in our introduction, it is a burning question as to whether a similar result holds for 5-connected graphs (see Conjecture A).

Another, less straightforward, way to turn Theorem 4.4 into a reduction theorem is to reduce cubic cyclically 4-edge-connected graphs. This can be done as follows. To *suppress* a vertex v of degree at most 2 in a graph G means to delete v and add the edge uw if u and w are the neighbours of v and they are not adjacent. (In particular, if the degree of v is 0 or 1, then this means just to delete v.) The result is denoted by $G - -v$. To *homotopically delete* an edge e in a graph means to delete it and to suppress its endpoints if they have degree at most 2. The result is denoted by $G - -e$.

Observe that if G is cubic and triangle-free, then $G - -e$ is cubic. Now every cyclically 4-edge-connected cubic graph non-isomorphic to K_4 or to the cube Q_3 can be reduced to a smaller cyclically 4-edge-connected cubic graph by homotopically deleting an edge (see [47]). In the line graph this means deleting a vertex, and contracting the two (independent) edges in its former neighbourhood. (The drawback to this, compared to just contracting any triangle in the line graph, which would always produce a smaller 4-connected graph is that we cannot do it everywhere in general.) So if G is contraction-critically 4-connected, then either G can be reduced in this way, or G is the square C_n^2 of a cycle of length $n \geq 5$. If $n = 5$, then G is K_5; if $n = 6$, then G is $K_{2,2,2}$; if $n \geq 7$, then G has two edges whose contraction yields C_{n-2}^2.

As for minimally k-connected graphs, and unlike the situation for critically k-connected graphs, we showed in [32] that the average degree of contraction-critically k-connected graphs can be bounded above by a function $f(k)$ – in fact, $f(k) \leq ck^2 \log k$, for some constant c. Since there are contraction-critically k-connected graphs of average degree $\frac{1}{6}k^2$ (see [29]), the bound is sharp up to the logarithmic factor, and we believe that it can be omitted.

Conjecture E *There exists a constant c such that every k-connected graph of average degree at least ck^2 has a k-contractible edge.*

In [32] we constructed contraction-critically 5-connected graphs of average degree 10, and conjectured that this is sharp. However, Ando [personal communication] has found an example of a contraction-critically 5-connected graph with average degree $12\frac{1}{2}$.

A k-contractible edge can also be forced by degree-sum conditions. For $k \geq 4$, it has been proved that if the sum of the degrees of any pair of vertices at distance 1 or 2 in a non-complete k-connected graph is at least $2\lfloor\frac{5}{4}k\rfloor - 1$, then there is a

k-contractible edge. In [29] we proved the result for $k \neq 7$, and the missing case was shown by Su and Yuan [69]. In fact, they later showed [70] that when $k \geq 8$, one needs the condition only for pairs of adjacent vertices.

Theorem 4.5 *For $k \geq 8$, if G is such that the sum of the degrees of every pair of adjacent vertices is at least $2\lfloor \frac{5}{4}k \rfloor - 1$, then G has a k-contractible edge. Furthermore, this bound is sharp for all k.*

5. Generalized criticality

Most of the results on fragments presented in Section 3 can be generalized to fragments whose neighbourhood contains a member of a specified set $\mathcal{S}$ of vertex-sets of the graph in question. This approach was worked out by Mader in [48]. Given a graph G and $\mathcal{S} \subseteq \mathcal{P}(V(G))$, the power-set of $V(G)$, we call a fragment of G an *$\mathcal{S}$-fragment* if $S \subseteq N_G(F)$, for some $S \in \mathcal{S}$. An *$\mathcal{S}$-atom* is a minimum $\mathcal{S}$-fragment, and its cardinality is denoted by $a_{\mathcal{S}}(G)$. The following generalization of Theorem 3.2 is due to Mader [48].

Theorem 5.1 *Let G be a graph and let $\mathcal{S} \subseteq \mathcal{P}(V(G))$. If A is an $\mathcal{S}$-atom and there exist $S \in \mathcal{S}$ and $T \in \mathcal{T}(G)$ for which $S \subseteq T \cap (A \cup N(A))$ and $A \cap T \neq \varnothing$, then $A \subseteq T$ and $|A| \leq \frac{1}{2}|T - N(A)|$.*

In all of the above results on critical graphs, the set of critical objects – be they vertices, vertex-sets or edges – was 'dense', in the rough sense that they appear *everywhere* in the graph. However, in most basic situations we need only a weaker density condition, which just ensures that the preconditions of Theorem 5.1 are satisfied. Hence, we say that a graph G of connectivity k is *$\mathcal{S}$-critically k-connected*, where $\mathcal{S} \subseteq \mathcal{P}(V(G))$, if $\mathcal{S} \neq \varnothing$, every $S \in \mathcal{S}$ is a subset of some $T \in \mathcal{T}(G)$, and for every $\mathcal{S}$-fragment A there exist $S \in \mathcal{S}$ and $T \in \mathcal{T}(G)$ for which $S \subseteq T \cap (A \cup N_G(A))$ and $A \cap T \neq \varnothing$. Using this notion, Theorem 5.1 implies the following theorem from [48], by literally the same argument that led from Theorem 3.2 to Corollary 3.3.

Theorem 5.2 *If G is an $\mathcal{S}$-critically k-connected graph, then $a_{\mathcal{S}}(G) \leq \frac{1}{2}k$.*

Mader designed this concept as a common generalization of many of the previously mentioned criticality concepts discussed here, as well as others. What follows is a summary of some of the variations; as usual, we make the assumption that G is not complete. (The relevant definitions for the last two items follow the list.)

- G is critically k-connected if and only if G is $\mathcal{S}$-critically k-connected for $\mathcal{S} = \{\{v\} : v \in V(G)\}$;
- G is l-critically k-connected if and only if G is $\mathcal{S}$-critically k-connected for $\mathcal{S} = \{X : X \subseteq V(G), |X| \leq l\}$;
- G is l-con-critically k-connected if and only if G is $\mathcal{S}$-critically k-connected for $\mathcal{S} = \{X : X \subseteq V(G), |X| \leq l, \langle X \rangle \text{ connected}\}$;

- G is contraction-critically k-connected if and only if G is $\mathcal{S}$-critically k-connected for $\mathcal{S} = \{\{v, w\} : vw \in E(G)\}$;
- G is almost critically k-connected if and only if G is $\mathcal{S}$-critically k-connected for $\mathcal{S} = \{\varnothing\}$;
- G is clique-critically k-connected if and only if G is $\mathcal{S}$-critically k-connected for $\mathcal{S} = \{X : X \subseteq V(G), \langle X \rangle \text{ complete}\}$.

We now briefly consider the last two items in the above list. Accordingly, a k-connected graph G other than K_{k+1} is *almost critically k-connected* if every fragment contains a vertex in some $\mathcal{T}(G)$. These graphs are important in the study of the distribution of contractible edges. So suppose that a vertex v of some graph $G(\neq K_{k+1})$ of connectivity k is not incident with a k-contractible edge. Then it is easy to see that G is $\mathcal{S}_v$-critical, where $\mathcal{S}_v = \{V(e) : e \in E(v)\}$. Every fragment F of $G - v$ is a fragment of G, where $N_G(F) = N_{G-v}(F) \cup \{v\}$ and both F and $N_G(F)$ contain a vertex from $N_G(v)$. (In particular, $G - v$ is almost critically $(k-1)$-connected.) Following Mader's argument [48], we use this fact to prove that there is a triangle in G 'close' to v, by considering an $\mathcal{S}_v$-atom A. If A consists of a single vertex w, then v is on a triangle, formed by v, w and any neighbour of v in $N(A)$. Otherwise, A must contain a pair of adjacent vertices, and they must have a common neighbour in $A \cup N(A)$, because $|A| \le \frac{1}{2}k$, by Theorem 5.2 – in fact, $|A| \le \frac{1}{2}(k-1)$, by an appropriate application of Theorem 5.1. For a detailed discussion of this interconnection, see [48]. Note that we get one of the main results in [73] as a corollary.

Theorem 5.3 *Every non-complete triangle-free k-connected graph has a k-contractible edge.*

A graph is called *clique-critically k-connected* if it is not complete and if $\kappa(G - V(K)) = \kappa(G) - |V(K)|$ for every clique K in G: roughly, deleting a clique always exhausts its potential for decreasing κ — that is, decreases it by its order. A tantalizing question is: are there any clique-critical graphs? Mader [48] conjectured that there are none.

Conjecture F *No clique-critically k-connected graphs exist.*

Mader also proved that Conjecture F is true for all $k \le 6$. Since a clique-critically k-connected graph is contraction-critically k-connected, it would have to contain a triangle, by Theorem 5.3, but it is not known whether it would have to contain K_4 (compare Conjecture B). If Conjecture F is true, then every l-critically k-connected graph must contain a clique on $l + 1$ vertices, and this would provide an affirmative answer to Conjecture B, with $l = 3$.

6. Reduction methods

In this section, we summarize some of the reduction methods for k-connected graphs, beginning with k-contractible edges; for a more comprehensive survey, see Kriesell [31].

Contractible edges

For small values of k, we have the following result.

Theorem 6.1 *Let G be a k-connected graph other than K_{k+1}.*

(a) *For $k = 1$, every edge of G is 1-contractible.*
(b) *For $k = 2$, every vertex of G is incident with a 2-contractible edge.*
(c) *For $k = 3$, G has a 3-contractible edge.*
(d) *For $k = 4$, G can be reduced to a smaller 4-connected graph by contracting one or two edges, unless it is the octahedron $K_{2,2,2}$.*

Clearly, every edge of a 1-connected graph other than K_2 is 1-contractible, so statement (a) is trivial; however, the corresponding statement for 2-connected graphs is not true, since the two vertices of an edge could form a cutset. For statement (b), we observe that, at each vertex v in a 2-connected graph other than K_3, some edge is 2-contractible; this is an immediate consequence of Theorem 5.1 applied to $\mathcal{S}_v$. Statement (c) is an immediate consequence of either the wheel theorem (Theorem 1.1) or Theorem 4.2. Statement (d) can be deduced from Theorem 4.4 (see Kriesell [31]).

As we mentioned in Section 1, the option of contracting a constantly bounded number of edges in one step might yield a similar reduction theorem for 5-connected graphs, but not for k-connected graphs, where $k > 5$.

Theorem 6.1 also provides the minor-minimal k-connected graphs, for $k \leq 4$. It follows from a theorem of Robertson and Seymour [59] that the class of minor-minimal 5-connected graphs is finite, since there exist planar 5-connected graphs, but there is an 'exact' conjecture of Fijavž [9] on just what these graphs are. Here, the Turán graph $K_{1,2,2,2}$ is obtained from K_7 by deleting three independent edges, while $C_5 + \overline{K_3}$ is obtained from K_8 by deleting the edges of a triangle and of a 5-cycle disjoint from that triangle. Further, we let I be the icosahedron graph, $\tilde{I}$ the graph obtained from I by replacing the edges of a cycle $vwxyzv$ induced by the neighbours of some vertex with the edges of the cycle $vwxzyv$, and $\hat{I}$ the graph obtained from I by deleting a vertex u, replacing the edge vw of a cycle $vwxyzv$ induced by the neighbours of u with the two edges vx and vy, and identifying w and z.

Conjecture G *Every 5-connected graph contains as a minor one of the graphs K_6, $K_{1,2,2,2}$, $C_5 + \overline{K_3}$, I, $\tilde{I}$ or $\hat{I}$.*

Fijavž [9], [10] proved his conjecture for graphs that are embeddable on the projective plane. ($K_{1,2,2,2}$ and $C_5 + \overline{K_3}$ are not projective planar.) Apparently, to prove that there are only finitely many minor-minimal k-connected graphs, for any $k > 5$, we need the full power of Wagner's conjecture (proved by Robertson and Seymour [60]). Let us summarize the numbers k for which the minor-minimal k-connected graphs are known or predicted.

k	minor base	reference
1	$\{K_2\}$	obvious
2	$\{K_3\}$	obvious
3	$\{K_4\}$	well-known, Theorem 6.1(c), Tutte [77]
4	$\{K_5, K_{2,2,2}\}$	Theorem 6.1(d)
5	$\{K_6, K_{1,2,2,2}, C_5 + \overline{K_3}, I, \tilde{I}, \hat{I}\}$	conjectured by Fijavž [9], [10]

Dawes constructions

A different reduction method was developed by Dawes [3]. His starting point was Dirac's theorem that every minimally 2-connected graph can be obtained from a cycle by repeatedly attaching paths of length at least 2 to suitable pairs of vertices. The resulting graph is always 2-connected, and it is not too difficult to characterize the 'suitable' pairs for which the result is minimally 2-connected. The corresponding reduction theorem is thus that every minimally 2-connected graph that is not a cycle can be reduced to a smaller such graph by deleting the interior vertices of some path of length at least 2 whose interior vertices have degree 2 (not every such path will do the trick, however).

We note in passing that, as an easy consequence, we get the well-known theorem on 'ear-decompositions', that every 2-connected graph can be obtained from a cycle by attaching paths of length at least 1 to suitable pairs of vertices. If G is any 2-connected graph that is not a cycle, then the graph obtained from G by subdividing every edge once is minimally 2-connected, and Dirac's theorem gives us a path that corresponds to a path in G with the property that deleting all its edges and interior vertices produces a smaller 2-connected graph.

For each $k \geq 1$, Dawes suggested the following operations to construct a larger graph from a given one.

Operation A_k: For $1 \leq s \leq \frac{1}{2}k$, choose s edges and $k - 2s$ vertices, delete each of the chosen edges, and add a new vertex v adjacent to both endpoints of each of the deleted edges and to the chosen vertices. (This is a special Henneberg construction, discussed below.)

Operation B_k: For $2 \leq s \leq k - 1$, choose s edges and $k - s - 1$ vertices, subdivide each of the chosen edges, and then add an edge from every subdivision vertex to every other subdivision vertex and to every chosen vertex.

Operation C_k: Choose k vertices and add a new vertex v adjacent to each of the chosen vertices.

Here is Dawes's construction theorem for minimally 3-connected graphs (see [3], [4]).

Theorem 6.2 *The class of minimally 3-connected graphs is the class of graphs obtained from K_4 by sequences of the operations A_3, B_3 and C_3 on suitable sets of objects.*

We note that the graph resulting from applying A_3, B_3 or C_3 to *any* set of objects meeting the definition of the respective operation is 3-connected, but here it is much more difficult to characterize those sets that yield a *minimally* 3-connected graph (see [3] and [4] for details).

Theorem 6.2 implies the following reduction theorem for minimally 3-connected graphs.

Theorem 6.3 *Every minimally 3-connected graph, other than K_4, can be reduced to a smaller minimally 3-connected graph by homotopically deleting an edge or deleting a vertex.*

Theorem 6.2 'literally' holds for minimally 1-connected and minimally 2-connected graphs; note, however, that the preconditions of A_1, B_1 and B_2 cannot be fulfilled, so that we get the well-known facts that the minimally 1-connected graphs are the non-trivial trees, and the minimally 2-connected graphs are the graphs obtained from K_3 by finite sequences of subdivisions (operation A_2) and attaching paths of length 2 (which follows from Dirac's theorem). Dawes [3] conjectured that these facts, together with Theorem 6.2, generalize to generator theorems for minimally k-connected graphs for each $k \geq 4$, but he also noticed that the 'compatibility condition' characterizing the suitable sets of objects would become much more difficult. However, we can disprove his conjecture.

Theorem 6.4 *For each $k \geq 4$, there are infinitely many minimally k-connected graphs that cannot be obtained from a smaller minimally k-connected graph by operations of the form A_k, B_k or C_k.*

Proof First note that A_k generates a k-connected graph only when the chosen edges are independent and are not incident with any of the chosen vertices, for otherwise the new vertex would have degree less than k. Also, B_k generates a k-connected graph only when the chosen edges are not incident with the chosen vertices, for otherwise some subdivision vertex would have degree less than k. For all three operations, the degrees of the endpoints of the chosen edges do not change after application, and the degree of each chosen vertex increases by at least 1. Moreover, B_k generates graphs with triangles, unless $s + k - s - 1 \leq 2$ – that is, unless $k \leq 3$.

For $k \geq 4$, it follows that if a *k-regular* k-connected graph arises from some minimally k-connected graph G by applying A_k, B_k or C_k, then it must arise by applications of either A_k to G and $t = \frac{1}{2}k$ independent edges, where k is even, or B_k with $s = k - 1$ edges (where no vertices are chosen).

Consequently, for odd $k \geq 5$, every k-regular k-connected graph that arises from some minimally k-connected graph by applying A_k, B_k or C_k has to contain a triangle – but there are infinitely many triangle-free k-regular k-connected graphs.

For even $k \geq 4$, let the graph H arise from a multicycle of length at least 5 with edge multiplicity $\frac{1}{2}k$, by subdividing each edge once. Every edge of H is adjacent to precisely k others, two edges of H cannot be separated by removing fewer than k

edges, and H is edge-transitive. Its line graph $G = L(H)$ is, consequently, k-regular, k-connected and vertex-transitive. However, G cannot be obtained by applying A_k (with $s = \frac{1}{2}k$) to any minimally k-connected graph, since the neighbourhood of the new vertex introduced when applying A_k had to contain at least $\frac{1}{2}k$ disjoint pairs of non-adjacent vertices, and this situation does not occur at any vertex of G. Up to isomorphism, there is a unique graph G^- from which G arises by applying B_k (with $s = k - 1$), but G^- has a separating vertex-set of order $\frac{1}{2}k + 1 < k$. (G^- can be obtained from G by contracting any complete subgraph K_k to a single vertex.) Therefore, there are infinitely many minimally k-connected graphs that do not arise from a smaller one by applying A_k, B_k or C_k if $k(\geq 4)$ is even. ■

Splitting and soldering

Yet another approach to characterizing k-connected graphs constructively is Slater's concept of 'splitting and soldering' (see [62]–[64]). Let v be a vertex of a graph G, and let X and Y be a partition of $N(v)$. To *split v according to (X, Y)* means to delete v from G, add two new adjacent vertices x and y, and add edges from x to every vertex in X and from y to every vertex in Y. If $|X| \geq k$, $|Y| \geq k$ and $|X \cap Y| = 0$, we say that the new graph is the result of *k-vertex-splitting*, and if $|X| \geq k$, $|Y| \geq k$ and $|X \cap Y| = 1$, we say that the new graph is the result of *k-edge-splitting*. (The 'edge' to which this refers is the one that connects v to the vertex in $X \cap Y$ in G.) For example, in this terminology, the class of 3-connected graphs is the class of graphs obtained from K_4 by finite sequences of edge-addition, 3-vertex-splitting and 3-edge-splitting – as a constructive version of the wheel theorem (Theorem 1.1).

Let G be a graph, and let K be a complete subgraph of order k of G. To *solder v onto K* means to add a new vertex v and an edge from v to each vertex in K and to delete a certain set F of edges from K. If, in the new graph H, $d_H(x) \geq k$ for all $x \in V(K)$, and if $(V(K), F) = \overline{H[V(K)]}$ does not contain a 4-cycle, then we say that H arises from G by *k-soldering*. The reason for excluding C_4 here is to guarantee that H is k-connected if G is; this is not true in general (see [64]). The drawback of this operation is that G might contain a minor that is not a minor of H.

One of the main results of Slater [62] is that the class of 3-connected graphs is the class of graphs obtained from K_4 by a sequence of edge-additions and 3-soldering. This does not similarly extend to 4-connected graphs: 4-connected line graphs of cubic graphs (see above) are graphs that cannot be obtained from any smaller 4-connected graph by edge-addition, 4-vertex-splitting, 4-edge-splitting or 4-soldering.

Slater's method for overcoming this situation was to generalize vertex-splitting as follows. Let v be a vertex of a graph G, and let $X_1, X_2, \ldots, X_r$ be sets for which $X_1 \cup X_2 \cup \cdots \cup X_r = N(v)$. To *split v into $(x_1, x_2, \ldots, x_r)$ according to $(X_1, X_2, \ldots, X_r)$* means to delete v from G, add r new vertices $x_1, x_2, \ldots, x_r$ and all edges between them, and add an edge from x_j to every vertex in X_j, for $j = 1, 2, \ldots, r$. If $|X_j| \geq k$ for every j, and if the X_j are pairwise disjoint, then we say that the new graph arises from G by *r-fold k-vertex-splitting*. The following is the second main result from [62].

Theorem 6.5 *The class of* 4*-connected graphs is the class of graphs obtained from* K_5 *by sequences of edge addition,* 4*-soldering,* 4*-vertex-splitting,* 4*-edge-splitting and* 3*-fold* 4*-vertex-splitting.*

Miscellaneous reductions

A more recent reduction theorem for 4-connected graphs, due to Saito [61], uses a special 'Henneberg reduction'; these reductions have been successfully applied in the context of edge-connectivity and arboricity questions. Let v be a vertex of a graph G, and let σ be a partition of the edges at v into sets of cardinality 1 or 2. The graph $G \overset{\sigma}{-} v$ is the graph obtained from $G - v$ by adding a new edge between x and y for each pair of edges with $\{vx, vy\} \in \sigma$, and we say that $G \overset{\sigma}{-} v$ arises from G by a *Henneberg reduction of degree* $|\sigma|$ *at* v. One nice feature of a Henneberg reduction of degree l at a vertex of degree $2l$ is that the vertex degrees of the resulting graph are the same as in G; in particular, when the operation is applied to a k-connected graph, the resulting graph has minimum degree at least k. As with soldering, one drawback is that $G \overset{\sigma}{-} v$ might contain a minor that is not a minor of G, and conversely. Here is Saito's theorem.

Theorem 6.6 *For each vertex* v *of a* 4*-connected graph* G*, other than* K_5*, there exists a* 4*-contractible edge at distance at most* 1 *from* v*, or* v *has a neighbour* w *of degree* 4 *at which there is a Henneberg reduction of degree* 2 *with* $G \overset{\sigma}{-} w$ 4*-connected.*

In [37], we considered another way of reducing a 3-connected graph G to a smaller one, which we call *homotopically deleting a vertex* v; here, we delete v and then repeatedly suppress vertices of degree at most 2 for as long as possible. This might do severe damage to G: we could kill the entire graph, for example, if G is a wheel and v is its centre. However, the result is well defined – that is, independent of the choice of vertices to be suppressed in each step of the induction. Here is our main result.

Theorem 6.7 *Every* 3*-connected graph, other than* $K_{3,3}$*,* $K_2 \times K_3$ *or a wheel, can be reduced to a smaller* 3*-connected graph by homotopically deleting a vertex.*

Although homotopic vertex-deletion is by no means a 'bounded' operation, it still has a constructive counterpart in terms of series–parallel extensions (see [37]). Since the homotopic deletion of vertices or edges can be considered as a (not constantly bounded) sequence of edge-deletions vertex-deletions, or edge-contractions, they still keep us in any class that is closed under taking minors.

The following result is an example of a contraction/deletion result for the class of triangle-free 3-connected graphs (see [35]).

Theorem 6.8 *Every 3-connected triangle-free graph, other than $K_{3,3}$ or the cube Q_3, can be reduced to a smaller 3-connected triangle-free graph by at most six edge-deletions, vertex-deletions or edge-contractions.*

We conclude this section with a reduction theorem for bipartite graphs. Apart from trivial situations, we obtain a non-bipartite graph when *contracting* an edge in a bipartite graph. Instead of identifying a pair of non-adjacent vertices, we proposed identifying a pair of vertices in the same partite set, and this indeed led to a reduction theorem (see [25]).

Theorem 6.9 *If G is a k-connected bipartite graph of order at least $2k^2 - 2k + 3$, which is not isomorphic to $C_l[\overline{K_{k/2}}]$ for any $l \geq 4$ if k is even, then it can be reduced to a smaller k-connected bipartite graph by identifying two non-adjacent vertices.*

The bound on the number of vertices is sharp for infinitely many k, as is shown by the point-line incidence graphs of a projective geometry (see [25]). It is easy to see that the 'irreducible' bipartite graphs in Theorem 6.9 for $k=2$ are the cycles of even length, and in [25] we determined the nine irreducible graphs for $k=3$. There may well be similar results for the class of k-connected 3-partite graphs, or even generalizations to r-partite graphs for $r>2$.

7. Subgraph-deletion

Each of the local reduction problems considered above can be reformulated as a question about the existence of a subgraph H of a given k-connected graph G for which $\kappa(G - V(H)) > k - |V(H)|$, where H also satisfies some size or connectivity conditions (or, more generally, H belongs to a certain class $\mathcal{H}$ of graphs). For example, when asking for a k-contractible edge in a graph G, we ask for a subgraph $H \cong K_2$ of G for which $\kappa(G - V(H)) > k - 2$, and $\mathcal{H}$ would be the class consisting of K_2. Although we restrict the results here to deleting all of H from G, we note that there is a corresponding theory in which only the edges of H are removed.

As we have seen, such a subgraph H need not exist. Is there one that keeps the connectivity high? Let us formulate this question more precisely. Given a class $\mathcal{H}$ of graphs, does there exist a function $f(k)$ such that every $f(k)$-connected graph G admits a subgraph $H \in \mathcal{H}$ for which $G - V(H)$ is k-connected?

For every non-empty finite graph class $\mathcal{H}$, such a function f obviously exists – take $f(k) = k + \max\{|V(H)| : H \in \mathcal{H}\}$. Another obvious case in which f exists is when $\mathcal{H}$ contains a graph H that is contained in every sufficiently highly connected graph; any forest X is such a candidate – see, for example, [1].

There are also classes $\mathcal{H}$ for which the answer is not obvious. Thomassen [73] proved that every $(k+3)$-connected graph G has an induced cycle C for which $G - V(C)$ is k-connected (see the next section) – that is, $f(k) = k + 3$ gives the answer 'yes' to our question if $\mathcal{H}$ is the class of all cycles. On the other hand, not every infinite class $\mathcal{H}$ admits such a function f, simply because there may be highly connected graphs that do not contain any graph in $\mathcal{H}$ as a subgraph – for example, let $\mathcal{H}$ be the class of odd cycles. Thus, a first step towards answering our question could be the more fundamental question of *characterizing those graph classes $\mathcal{H}$ for which every sufficiently highly connected graph contains a member of $\mathcal{H}$*. However, for these classes our problem was solved by Kühn and Osthus (see p. 30 in their paper [38]), as follows.

Theorem 7.1 *Let $\mathcal{H}$ be a class of graphs for which there is a number l such that every l-connected graph contains a member from $\mathcal{H}$ as a subgraph. Then, for each k, there exists a number $f_{\mathcal{H}}(k)$ such that every $f_{\mathcal{H}}(k)$-connected graph contains a member H of $\mathcal{H}$ for which $G - V(H)$ is k-connected.*

Kühn and Osthus first showed that the vertex-set of every $2^{11}3k^2$-connected graph admits a partition into sets A and B for which both $\langle A \rangle$ and $\langle B \rangle$ are k-connected and every vertex in A has at least k neighbours in B; then any copy of $H \in \mathcal{H}$ in $\langle A \rangle$ will do. Consequently, we propose excluding those graphs without subgraphs in $\mathcal{H}$, as follows.

Problem *Determine the classes of graphs $\mathcal{H}$ for which there is a high-connectivity-keeping-$\mathcal{H}$ theorem – that is, for all k there exists a number $f(k)$ such that every $f(k)$-connected graph G that contains any subgraph from $\mathcal{H}$ has a subgraph H in $\mathcal{H}$ for which $G - V(H)$ is k-connected.*

The following table gives a summary of those graph classes $\mathcal{H}$ for which a high-connectivity-keeping-$\mathcal{H}$ result, in the sense of this problem, has been proved or conjectured.

$\mathcal{H}$	$f(k)$	references
cycles	$k+3$	Thomassen [73]
even cycles	$k+4$	Fujita and Kawarabayashi [14]
odd cycles	?	conjectured by Thomassen [76]
theta graphs	$k+4$	Fujita and Kawarabayashi [14]
t-connected graphs	$4k+4t-13$	$t \geq 3$, Hajnal [16]
	$k+t+1$?	conjectured by Thomassen [74, p. 167]
as in Theorem 7.1	$2^{11}3\max\{k^2, l^2\}$	Kühn and Osthus [38]

There are a number of other conjectures and results with the same flavour. The first one that we consider leads us back to contractibility. By Theorem 6.1(c), every 3-connected graph other than K_4 contains an edge vw (a connected subgraph H on two vertices) for which $G - \{v, w\}$ is 2-connected – or, equivalently, that the graph obtained from $G - \{v, w\}$ by adding a new vertex u and joining it to all the neighbours of v or w in G is 3-connected. McCuaig and Ota [56] conjectured the following generalization of this.

Conjecture H *For all l, there exists a number $f(l)$ such that every 3-connected graph of order at least $f(l)$ contains a connected subgraph H of order l for which $G - V(H)$ is 2-connected.*

They proved that this conjecture is true for $k \leq 4$, where the optimal values are $f(2) = 5$ (see [77]), $f(3) = 9$ (see [56]), and $f(4) = 8$ (see [27]). These values are sharp, and we note that (somewhat surprisingly) they are not monotonic in k. Moreover, every *cubic* 3-connected graph on at least 13 vertices has a contractible subgraph on 5 vertices, and it may be interesting to note that the 'local character' of the proof yields a generalization to 3-connected graphs of average degree at most $3\frac{1}{132}$ (see [36]). The conjecture is wide open in general. For example, it is not even known whether there exists a k for which its restriction to k-connected graphs is true. However, if we do not insist that the subgraph H in the conjecture be *connected*, then we can prove the following positive result (see [26]).

Theorem 7.2 *For all l, there exists a number $f(l)$ such that every 3-connected graph of order at least $f(l)$ contains a subgraph H of order l for which $G - V(H)$ is 2-connected.*

We note in passing that if edges are deleted, instead of vertices, then we have the following result (based on a theorem of Lemos and Oxley [39]): for all l there exists a number $f(l)$ such that every 4-connected graph with at least $f(l)$ vertices contains a path or a star H on l vertices for which $G - E(H)$ is 2-connected (see [36]).

Mader [51] generalized Theorem 7.2 to higher connectivities. He also showed that it is not possible to replace $k - 2$ by $k - 1$.

Theorem 7.3 *For every $k \geq 4$ and every $l \geq 2$, there exists a number $f_k(l)$ such that every k-connected graph of order at least $f_k(l)$ contains a subgraph H of order l for which $G - V(H)$ is $(k - 2)$-connected.*

Possibly the most prominent conjecture along these lines is Lovász's conjecture on high-connectivity-keeping paths (cited according to [75, p. 267]).

Conjecture I *For all k, there exists a number $f(k)$ such that, for any two vertices v and w in an $f(k)$-connected graph G, there exists an induced v–w path P for which $G - V(P)$ is k-connected.*

This conjecture has been verified for $k = 1$ and $k = 2$, where $f(1) = 3$ and $f(2) = 5$ are sharp (see [2], [30], and also [23]). Some years ago, we proposed proving an edge-deletion version first, and perhaps not insisting that the path be induced. Kawarabayashi *et al.* [22] have made progress in this direction.

Theorem 7.4 *For all k, there exists a number $f(k)$ such that, for any two vertices v and w in an $f(k)$-connected graph G, there exists a v–w path P for which $G - V(P)$ is k-connected.*

It is easy to see that an affirmative answer to Conjecture I would imply Theorem 7.4, even with the additional constraint that P be induced. It is not clear whether the 'non-induced' version of Conjecture I would imply Theorem 7.4. It may be that Theorem 7.4 can be generalized as follows.

Conjecture J *For all k, there exists a number $f(k)$ such that every $f(k)$-connected graph has a spanning tree T for which $G - E(T)$ is k-connected.*

The edge-connectivity version of this is obviously true, since by a well-known corollary of the tree packing theorem of Tutte and Nash-Williams, every $(2k + 2)$-edge-connected graph admits $k + 1$ edge-disjoint spanning trees. Removing one of them yields a supergraph of the union of k edge-disjoint spanning trees, and hence a k-edge-connected graph. In particular, Conjecture J is true for $k = 1$, where $f(1) = 4$ is best possible. Jordán [19] proved that every $6k$-connected graph G has k edge-disjoint 2-connected subgraphs, and this shows that $f(2) \leq 12$. The edge-sets of these subgraphs are actually bases of the *2-dimensional rigidity matroid* of G. It may be that there is a way of proving Conjecture J for larger k by using properties of higher-dimensional rigidity matroids of the graph in question, but these objects are far from being well understood.

8. Partitions under connectivity constraints

As an alternative to the notion of high-connectivity-keeping subgraphs, most of the previous results can be considered as partition statements. Given a sufficiently highly connected graph G with certain extra properties, we look for a partition $\{A, B\}$ of $V(G)$ for which the induced subgraph $\langle A \rangle$ meets a 'small or bounded size' condition and, in many cases, a mild connectivity condition, whereas we want $\langle B \rangle$ to be highly connected. In some cases, an additional condition on the location of A might be incorporated; for example, the presence of a contractible edge in a graph of connectivity k (incident with some specified vertex v) is equivalent to the existence of such a partition with $\langle A \rangle$ connected and of order 2 (and containing v), whereas $\langle B \rangle$ needs to be $(k - 1)$-connected.

In this section we look at a problem where the conditions on the partition sets are more balanced. We first mention Győri's classic characterization of k-connected graphs [15].

Theorem 8.1 *A graph G of order n is k-connected if and only if, for any k vertices $v_1, v_2, \ldots, v_k$ and any k positive integers $n_1, n_1, \ldots, n_k$ with sum n, there exist disjoint sets $V_1, V_2, \ldots, V_k \subseteq V(G)$ such that, for $i = 1, 2, \ldots, k$, $\langle V_i \rangle$ is a connected graph on n_i vertices containing v_i.*

There are versions for digraphs and also for edge-connectivity (see [15]). It is perhaps surprising that this theorem has not yet been employed for the types of problem we mentioned here; on the other hand, the same applies to Menger's theorem, as the vast majority of arguments run exclusively in terms of $\mathcal{T}(G)$. The reason might be that the non-trivial part is the necessity of the partition conditon – that is, we could apply the theorem to the graph G under consideration, but possibly not to certify a certain connectivity of some substructure.

The following conjecture is due to Thomassen [74] (see the table in the previous section).

Conjecture K *For every $(s + t + 1)$-connected graph G, there exists a partition $\{A, B\}$ of $V(G)$ for which $\langle A \rangle$ is s-connected and $\langle B \rangle$ is t-connected.*

The qualitative part of this question has been settled. The conclusion holds for $(4s + 4t + 1)$-connected graphs G (see [74]). The proof runs in three steps.

(a) Since G has minimum degree at least $4s+4t+1$, there exists a partition $\{A'', B''\}$ of $V(G)$ for which $\langle A'' \rangle$ has minimum degree at least $4s$ and $\langle B'' \rangle$ has minimum degree at least $4t$, by a famous theorem of Stiebitz [65]. In particular, $\langle A'' \rangle$ and $\langle B'' \rangle$ have average degree at least $4s$ and $4t$ respectively.
(b) By a result of Mader [43], the average degree bound in (a) ensures that $\langle A'' \rangle$ has an s-connected subgraph and $\langle B'' \rangle$ has a t-connected subgraph – that is, there exist disjoint subsets $A' \subseteq A''$ and $B' \subseteq B''$ for which $\langle A' \rangle$ is s-connected and $\langle B' \rangle$ is t-connected.
(c) Since G is $(s+t-1)$-connected, we can extend A' and B' to a partition of $\{A, B\}$ of $V(G)$ as desired – that is, $A \supseteq A'$, $B \supseteq B'$, and $\langle A \rangle$ is s-connected and $\langle B \rangle$ is t-connected; this is due to a beautiful argument of Thomassen [74].

Parts (a) and (c) show that the conclusion of Conjecture K holds also under the weaker assumption that G is $(s + t - 1)$-connected and has minimum degree $4s + 4t + 1$. Part (c) shows that to prove the conclusion of the conjecture, it suffices to find two disjoint subgraphs that are s-connected and t-connected, respectively. By careful consideration of the original bounds from Mader [43], which are slightly better than $4k$, Hajnal [16] improved the bound to $4s + 4t - 13$, for all $s, t \geq 3$.

There is also a version of Conjecture K in which we partition $E(G)$ instead of $V(G)$, and this was posed by Mader [46] as an open problem.

Problem *For $s, t \geq 2$, does every $(s+t)$-connected graph admit a partition $\{A, B\}$ of $E(G)$ for which the graph formed by the edges of A is s-connected and that formed by those of B is t-connected?*

We can sketch a proof of the qualitative part. Suppose that G is a $(2s + 4t)$-connected graph, or any s-connected graph of average degree at least $2s + 4t$. Then G contains a minimally s-connected spanning subgraph H. As we saw in Section 2, the average degree of H is less than $2s$, and so the average degree of $G - E(H)$ is at least $4t$. By Mader's theorem, mentioned in part (b) above, $G - E(H)$ contains a t-connected subgraph T. Since $S = G - E(T)$ is a supergraph of H, it follows that the edge-sets of S and T partition $E(G)$ in the desired way.

The first non-trivial case of Conjecture K was settled by Thomassen [73].

Theorem 8.2 *Every $(k + 3)$-connected graph G contains an induced cycle C for which $G - V(C)$ is k-connected.*

Thomassen's proof is as follows. By induction on the order of G, we can prove the stronger statement that there is an induced cycle C such that every vertex not in $V(C)$ has at most three neighbours in $V(C)$ and $G - V(C)$ is k-connected. Obviously, every triangle would serve as such a cycle, so we may assume that G is triangle-free. By Theorem 5.3, G contains a contractible edge e. From a cycle in G/e with the desired properties, it is easy to obtain one in G itself.

Along these lines, we note Mader's result that every $(k + 2)$-connected graph contains a cycle C for which $G - E(C)$ is k-connected (see [44]). In fact, C can be taken as an induced cycle here.

It seems to be extremely difficult to prove local versions of this statement. For example, the statement that 'there is a function g such that, for every edge e of every $g(k)$-connected graph G, there is an induced cycle C containing e for which $G - V(C)$ is k-connected' is equivalent to Conjecture I, as was observed by Thomassen. Even the following weaker problem is open.

Conjecture L *For all k, there exists a number $h(k)$ such that, for every vertex v of every $h(k)$-connected graph G, there exists an induced cycle C containing v for which $G - V(C)$ is k-connected.*

It may be that partition problems, as in Conjectures K and L, are easier to solve for graphs with high girth. This is supported by the following results that were conjectured by Thomassen and proved by Egawa in [6] and [8].

Theorem 8.3 *For each $k \geq 2$, every graph G that is either $(k + 1)$-connected and has girth at least 5, or $(k+2)$-connected and has girth at least 4, contains an induced cycle C for which $G - V(C)$ is k-connected.*

Consider now the case $s = t = k$ in Conjecture K. Hajnal's theorem mentioned earlier in this section implies that the vertex-set of every $(8k - 13)$-connected graph G has a partition $\{A, B\}$ with both $\langle A \rangle$ and $\langle B \rangle$ k-connected. This is improved by a factor of nearly 2 if we restrict the statement to graphs of girth greater than k: by a result of Mader [43], every graph with minimum degree $\delta(G) \geq 2k - 2$ and girth $g(G) > k$ contains a k-connected subgraph; therefore, following the three-point argument just after Conjecture K, the vertex-set of every $(4k - 3)$-connected graph with $g(G) > k$ has a partition into sets A and B, with both $\langle A \rangle$ and $\langle B \rangle$ k-connected. In fact, for $k = 3$, it is possible to apply a result of Mader [43] in part (b) of the argument to show that the vertex-set of every 9-connected graph without an induced subgraph $K_{1,1,2}$ has a partition $\{A, B\}$, with both $\langle A \rangle$ and $\langle B \rangle$ 3-connected. Also, there are improvements for part (a) of the argument for graphs of high girth. Whereas the general bound resulting from Stiebitz's theorem is $s + t + 1$, Kaneko [21] showed that the vertex-set of every *triangle-free* graph of minimum degree at least $s + t$ can be partitioned into sets A and B, with $\delta(\langle A \rangle) \geq s$ and $\delta(\langle B \rangle) \geq t$, and Diwan [5] showed that the bound $s + t$ can be further improved to $s + t - 1$ for graphs of girth at least 5. As a consequence, the vertex-set of every 8-connected triangle-free graph can be partitioned into sets A and B for which $\langle A \rangle$ and $\langle B \rangle$ are both 3-connected.

Also, for *large* graphs, the bound in Mader's theorem in part (b) of the argument can be improved: it follows from a theorem of Mader [43] that every sufficiently large graph of average degree at least $(2 + \sqrt{2})k$ contains a k-connected subgraph. As we have no control on the size of the partition classes when applying Stiebitz's result in part (a) of the argument, this statement does not immediately improve the bound of $4s + 4t + 1$ for large graphs. However, Carmesin, Fröhlich, Hàn and Schacht (personal communication) said that they used probabilistic methods to prove that, for all r, s and t, there exists a number $f(n, s, t)$ for which every graph G of order at least $f(n, s, t)$ and minimum degree $s + t + 1$ admits a partition $\{A, B\}$ of $V(G)$ with $|A|, |B| \geq r$ and such that $\langle A \rangle$ has average degree at least s and $\langle B \rangle$ has average degree at least t. From this, we get the following result.

Theorem 8.4 *For all s, t and n, there exists a number $f(s, t, n)$ such that, for every $((2 + \sqrt{2})(s + t) + 1)$-connected graph G of order at least $f(s, t, n)$, there is a partition $\{A, B\}$ of $V(G)$ for which $\langle A \rangle$ is s-connected and of order at least n and $\langle B \rangle$ is t-connected and of order at least n.*

9. Line graphs

Most of the problems mentioned in Sections 7 and 8 provide an analogous version in terms of edge-connectivity, instead of connectivity. The answers are often affirmative, mostly due to the presence of enough edge-disjoint spanning trees (see, for example, the paragraph following Conjecture J). Furthermore, it is straightforward to translate these results on graphs into the language of their line graphs. For example, the fact that, for any two vertices v and w of an $h(k)$-edge-connected graph G, there exists a v–w path P for which $G - E(P)$ is k-edge-connected immediately implies that, for any two *edges* e and f of an $(h(k) + 2)$-edge-connected graph G, there is a path with terminal edges e and f for which $G - E(P)$ is k-connected; consequently, in the line graph $L(G)$, between any two *vertices* there is an *induced* path P for which $L(G) - V(P)$ is k-connected. One is tempted to say that this proves Lovász's conjecture (Conjecture I) for line graphs, but this is not true because high edge-connectivity of a graph is a sufficient but not necessary condition for high connectivity of its line graph (think of pendant edges). However, it is easy to see that if $L(G)$ is k-connected, then the vertices of degree at least k in G are k-edge-connected in G (Lemma 1 of [30]). This observation was combined with a powerful theorem of Okamura [58] on removable paths in graphs with a given edge-connectivity function to prove that Lovász's conjecture is true for line graphs (see [30]).

We now give another example in which we employ this method, proving Conjecture J for line graphs. In the proof we use two other results. The first is Lemma

1 of [30], just mentioned. The second is Theorem 3.1 of [13] by Frank, Király and the present author, which states that if A is a $3k$-edge-connected set in a graph G, and if $G - A$ is edgeless, then there exist k edge-disjoint subtrees of G, each of which contains all vertices of A.

Theorem 9.1 *Every* $(12k + 11)$*-connected line graph* L *has a spanning tree* T *for which* $L - E(T)$ *is* k*-connected.*

Proof Let G be a graph for which $L(G)$ is $(6k + 6)$-connected and has minimum degree at least $12k + 11$. By Lemma 1 of [30], the set A of vertices of degree at least $6k + 6$ in G is $(6k + 6)$-connected and, since $L(G)$ has minimum degree at least $12k + 11$, $G - A$ is edgeless. By Theorem 3.1 of [13], G contains $2k + 2$ edge-disjoint trees, each of which covers A. Therefore, G has $k + 1$ edge-disjoint connected subgraphs $T_1, T_2, \ldots, T_{k+1}$, each covering A, such that every vertex of A has degree at least 2 in every T_j and $\{E(T_1), E(T_2), \ldots, E(T_{k+1})\}$ partitions $E(G)$.

For each vertex $v \in V(G)$, let K_v be the clique on $E(v)$ in $L(G)$, and for each $j \in \{1, 2, \ldots, k + 1\}$, let K_v^j be the subclique induced by $E_{T_j}(v)$ in K_v.

For each $v \in A$, let m_v be such that $|K_v^{m_v}| = \max\{|K_v^j| : 1 \leq j \leq k + 1\}$. Then $K_v^{m_v}$ is a clique with at least four vertices, and thus has a non-separating spanning path $M_v^{m_v}$. For each $j \neq m_v$, there exists a matching M_v^j in K_v for which each edge of M_v^j connects a vertex in $V(K_v^j)$ to a vertex in $V(K_v^{m_v})$, and each vertex of $V(K_v^j)$ is connected in this way. The graph H_v^1 formed by $\bigcup_{j=1}^{k+1} E(M_v^j) \cup E(K_v^1)$ is therefore a connected spanning subgraph of K_v such that, for each $j \neq 1$, the graphs $H_v^j = K_v^j - E(H_v^1)$ are connected and for every vertex $e \in V(K_v)$ and every j such that $e \notin V(H_v^j)$ there exists an edge in $E(K_v) - E(H_v^1)$ connecting e to some vertex in $V(H_v^j)$.

Let $H^j = \bigcup_{v \in V(G)-A} K_v^j \cup \bigcup_{v \in A} H_v^j$. The following claim shows that H^1 is a connected spanning subgraph of $L(G)$ and that $L(G) - E(H^1)$ is k-connected. This proves the theorem.

Claim: If e and f are two vertices in $L(G)$, then there exist an e–f path in H^1 and k openly disjoint e–f paths in $L(G) - E(H^1)$.

Proof of claim: There exist $v, w \in A$ for which $e \in V(K_v)$ and $f \in V(K_v)$, $e \in V(K_w)$. If $v = w$, then there exists an e–f path in H^1, since H_v^1 is a spanning connected subgraph of K_v. Let j_e and j_f be in $\{1, 2, \ldots, k+1\}$ such that $e \in V(K_v^{j_e})$ and $f \in V(K_v^{j_f})$. Take $s_{j_e} = e$ and $r_{j_f} = f$ and, for each $j \neq j_e$, take a vertex s_j in $V(K_v^j)$ that is a neighbour of e in $K_v - E(H^1)$ and, for each $j \neq j_f$, take a vertex t_j in $V(K_v^j)$ that is a neighbour of f in $K_v - E(H^1)$. For $j \neq 1$, there exists an s_j–t_j path L_j in $K_v^j - E(H^1)$, and since these paths L_j are disjoint by definition, the paths eL_jf are openly disjoint e–f paths. This proves the claim when $v = w$.

If $v \neq w$, let P_j be any v–w path in T_j, and let Q_j be the path induced by $E(P_j)$ in $L(G)$. By our construction, $E(Q_1) \subseteq E(H^1)$ and $E(Q_j) \cap E(H^1) = \varnothing$, for all

$j \neq 1$. Since the paths Q_j are edge-disjoint, the paths P_j are also disjoint. Each path P_j joins a vertex e_j in $V(K_v^j)$ and a vertex f_j in $V(K_w^j)$. By construction, there exist an e–e_j path R_j for which $R_j - e$ is in H_v^j and an f_j–f path S_j for which $S_j - f$ is in H_w^j. It follows that $R_1 Q_1 S_1$ is an e–f path in H^1, and that the paths $R_j Q_j S_j$ for $j \neq 1$ are k openly disjoint paths in $L(G) - E(H^1)$. ■

Another example of this phenomenon is that Conjecture K is also true for line graphs, because a line graph of minimum degree $s + t + 1$ contains K_{s+1} or K_{t+1} as a subgraph. Finally, we note our result that there is no 3-con-critically k-connected line graph (see [24]); this implies Slater's conjecture [55] that there is no 3-critically k-connected line graph.

References

1. B. Bollobás, *Extremal Graph Theory*, London Mathematical Society Monographs **11**, Academic Press, 1978.
2. G. Chen, R. J. Gould and X. Yu, Graph connectivity after path removal, *Combinatorica* **23** (2003), 185–203.
3. R. Dawes, Constructions of minimally k-connected graphs for $k = 1$, 2, and 3, *Congr. Numer.* **39** (1983), 273–289.
4. R. Dawes, Minimally 3-connected graphs, *J. Combin. Theory (B)* **40** (1986), 159–168.
5. A. A. Diwan, Decomposing graphs with girth at least five under degree constraints, *J. Graph Theory* **33** (2000), 237–239.
6. Y. Egawa, Cycles in k-connected graphs whose deletion results in a $(k-2)$-connected graph, *J. Combin. Theory (B)* **42** (1987), 371–377.
7. Y. Egawa, Contractible edges in n-connected graphs with minimum degree greater than or equal to $[5n/4]$, *Graphs Combin.* **7** (1991), 15–21.
8. Y. Egawa, Contractible cycles in graphs with girth at least 5, *J. Combin. Theory (B)* **74** (1998), 213–264.
9. G. Fijavž, *Graph minors and connectivity*, Ph.D. thesis, University of Ljubljana, 2001.
10. G. Fijavž, Minor-minimal 5-connected projective-planar graphs, Preprint 765, IMFM Preprint Series University of Ljubljana **39**, 2001.
11. M. Fontet, Graphes 4-essentiels, *C. R. Acad. Sc. Paris* **287** (1978), 289–290.
12. M. Fontet, *Connectivité des graphes automorphismes des cartes: propriétés et agorithmes*, Thèse d'etat, Université P. et M. Curie, Paris, 1979.
13. A. Frank, T. Király and M. Kriesell, On decomposing a hypergraph into k connected sub-hypergraphs, *Discrete Appl. Math.* **131** (2003), 373–383.
14. S. Fujita and K. Kawarabayashi, Non-separating even cycles in highly connected graphs, *Combinatorica* **30** (2010), 565–580.
15. E. Győri, On division of graphs to connected subgraphs, *Colloq. Math. Soc. János Bolyai* **18** (1978), 485–494.
16. P. Hajnal, Partition of graphs with condition on the connectivity and minimum degree, *Combinatorica* **3** (1983), 95–99.
17. R. Halin, Untersuchungen über minimale n-fach zusammenhängende Graphen, *Math. Ann.* **182** (1969), 175–188.
18. T. Jordán, On the existence of (k, l)-critical graphs, *Discrete Math.* **179** (1998), 273–275.

19. T. Jordán, On the existence of k edge-disjoint 2-connected spanning subgraphs, *J. Combin. Theory (B)* **95** (2005), 257–262.
20. H. A. Jung, Die Zusammenhangsstruktur symmetrischer Graphen, *J. Reine Angew. Math.* **283/284** (1976), 202–221.
21. A. Kaneko, On decomposition of triangle-free graphs under degree constraints, *J. Graph Theory* **27** (1998), 7–9.
22. K. Kawarabayashi, O. Lee, B. Reed and P. Wollan, A weaker version of Lovász path removal conjecture, *J. Combin. Theory (B)* **98** (2008), 972–979.
23. K. Kawarabayashi, O. Lee and X. Yu, Non-separating paths in 4-connected graphs, *Ann. Comb.* **9** (2005), 47–56.
24. M. Kriesell, On k-critical connected line graphs, *J. Combin. Theory (B)* **74** (1998), 1–7.
25. M. Kriesell, Contractible non-edges in triangle-free graphs, *Graphs Combin.* **15** (1999), 429–439.
26. M. Kriesell, Contractible subgraphs in 3-connected graphs, *J. Combin. Theory (B)* **80** (2000), 32–48.
27. M. Kriesell, The k-critical $2k$-connected graphs for $k \in \{3, 4\}$, *J. Combin. Theory (B)* **78** (2000), 69–80.
28. M. Kriesell, Almost all 3-connected graphs contain a contractible set of k vertices, *J. Combin. Theory (B)* **83** (2001), 305–319.
29. M. Kriesell, A degree sum condition for the existence of a contractible edge in a κ-connected graph, *J. Combin. Theory (B)* **82** (2001), 81–101.
30. M. Kriesell, Induced paths in 5-connected graphs, *J. Graph Theory* **36** (2001), 52–58.
31. M. Kriesell, A survey on contractible edges in graphs of a prescribed vertex connectivity, *Graphs Combin.* **18** (2002), 1–30.
32. M. Kriesell, Average degree and contractibility, *J. Graph Theory* **51** (2006), 205–224.
33. M. Kriesell, Mader's conjecture on extremely critical graphs, *Combinatorica* **26** (2006), 277–314.
34. M. Kriesell, There exist highly critically connected graphs of diameter three, *Graphs Combin.* **22** (2006), 481–485.
35. M. Kriesell, A constructive characterization of 3-connected triangle-free graphs, *J. Combin. Theory (B)* **97** (2007), 358–370.
36. M. Kriesell, On small contractible subgraphs in 3-connected graphs of small average degree, *Graphs Combin.* **23** (2007), 545–557.
37. M. Kriesell, Vertex suppression in 3-connected graphs, *J. Graph Theory* **57** (2008), 41–54.
38. D. Kühn and D. Osthus, Partitions of graphs with high minimum degree or connectivity, *J. Combin. Theory (B)* **88** (2003), 29–43.
39. M. Lemos and J. Oxley, On removable cycles through every edge, *J. Graph Theory* **42** (2003), 155–164.
40. D. R. Lick, Critically and minimally n-connected graphs, *The Many Facets of Graph Theory* (Proc. Conf., Western Mich. Univ., Kalamazoo, 1968), Springer (1969), 199–205.
41. W. Mader, Eine Eigenschaft der Atome endlicher Graphen, *Arch. Math. (Basel)* **22** (1971), 333–336.
42. W. Mader, Ecken vom Grad n in minimalen n-fach zusammenhängenden Graphen, *Arch. Math. (Basel)* **23** (1972), 219–224.
43. W. Mader, Existenz n-fach zusammenhängender Teilgraphen in Graphen genügend grosser Kantendichte, *Abh. Math. Sem. Univ. Hamburg* **37** (1972), 86–97.
44. W. Mader, Kreuzungsfreie a, b-Wege in endlichen Graphen, *Abh. Math. Sem. Univ. Hamburg* **42** (1974), 187–204.

45. W. Mader, Endlichkeitssätze für k-kritische Graphen, *Math. Ann.* **229** (1977), 143–153.
46. W. Mader, Connectivity and edge-connectivity in finite graphs, *Surveys in Combinatorics* (Proc. Seventh British Combinatorial Conf., Cambridge, 1979), London Math. Soc. Lecture Notes **39**, Cambridge University Press (1979), 66–95.
47. W. Mader, On k-critically n-connected graphs, *Progress in Graph Theory* (Waterloo, Ont., 1982), Academic Press (1984), 389–398.
48. W. Mader, Generalizations of critical connectivity of graphs, *Discrete Math.* **72** (1988), 267–283.
49. W. Mader, On vertices of degree n in minimally n-connected graphs and digraphs, *Combinatorics, Paul Erdős is Eighty* (Keszthely, 1993), Bolyai Soc. Math. Stud. **2**, János Bolyai Math. Soc. (1996), 423–449.
50. W. Mader, On k-con-critically n-connected graphs, *J. Combin. Theory (B)* **86** (2002), 296–314.
51. W. Mader, High connectivity keeping sets in n-connected graphs, *Combinatorica* **24** (2004), 441–458.
52. N. Martinov, On 4-connected graphs, *Mathematics and Education in Mathematics* (Proc. 10th Spring Conf. Union Bulg. Math., Sunny Beach, 1981), 152–157.
53. N. Martinov, Uncontractable 4-connected graphs, *J. Graph Theory* **6** (1982), 343–344.
54. N. Martinov, A recursive characterization of the 4-connected graphs, *Discrete Math.* **84** (1990), 105–108.
55. S. Maurer and P. J. Slater, On k-critical, n-connected graphs, *Discrete Math.* **20** (1977/78), 255–262.
56. W. McCuaig and K. Ota, Contractible triples in 3-connected graphs, *J. Combin. Theory (B)* **60** (1994), 308–314.
57. S. Negami, A characterization of 3-connected graphs containing a given graph, *J. Combin. Theory (B)* **32** (1982), 69–74.
58. H. Okamura, Paths and edge-connectivity in graphs, *J. Combin. Theory (B)* **37** (1984), 151–172.
59. N. Robertson and P. D. Seymour, Graph minors. IV: Tree-width and well-quasi-ordering, *J. Combin. Theory (B)* **48** (1990), 227–254.
60. N. Robertson and P. D. Seymour, Graph minors. XX: Wagner's conjecture, *J. Combin. Theory (B)* **92** (2004), 325–357.
61. A. Saito, Splitting and contractible edges in 4-connected graphs, *J. Combin. Theory (B)* **88** (2003), 227–235.
62. P. J. Slater, A classification of 4-connected graphs, *J. Combin. Theory (B)* **17** (1974), 281–298.
63. P. J. Slater, Generalized soldering, *Proc. 5th British Comb. Conf., Aberdeen, 1975* (1976), 559–567.
64. P. J. Slater, Soldering and point splitting, *J. Combin. Theory (B)* **24** (1978), 338–343.
65. M. Stiebitz, Decomposing graphs under degree constraints, *J. Graph Theory* **23** (1996), 321–324.
66. J. Su, Proof of Slater's conjecture on k-critical n-connected graphs, *Kexue Tongbao* (English edn.) **33** (1988), 1675–1678.
67. J. Su, On locally k-critically n-connected graphs, *Discrete Math.* **120** (1993), 183–190.
68. J. Su, Proof of Mader's conjecture on k-critical n-connected graphs, *J. Graph Theory* **45** (2004), 281–297.
69. J. Su and X. Yuan, Contractible edges in 7-connected graphs, *Graphs Combin.* **21** (2005), 445–457.
70. J. Su and X. Yuan, A new degree sum condition for the existence of a contractible edge in a κ-connected graph, *J. Combin. Theory (B)* **96** (2006), 276–295.

71. J. Su, X. Yuan and Q. Zhao, On k-critical $2k$-connected graphs, *Sci. China (A)* **46** (2003), 289–299.
72. C. Thomassen, Kuratowski's theorem, *J. Graph Theory* **5** (1981), 225–241.
73. C. Thomassen, Nonseparating cycles in k-connected graphs, *J. Graph Theory* **5** (1981), 351–354.
74. C. Thomassen, Graph decomposition with constraints on the connectivity and minimum degree, *J. Graph Theory* **7** (1983), 165–167.
75. C. Thomassen, Graph decomposition with applications to subdivisions and path systems modulo k, *J. Graph Theory* **7** (1983), 261–271.
76. C. Thomassen, The Erdős-Pósa property for odd cycles in graphs of large connectivity, *Combinatorica* **21** (2001), 321–333.
77. W. T. Tutte, A theory of 3-connected graphs, *Nederl. Akad. Wetensch. Proc. (A)* **64** = *Indag. Math.* **23** (1961), 441–455.
78. V. G. Vizing, Some unsolved problems in graph theory, *Russian Math. Surveys* **23** (1968), 125–141.
79. M. E. Watkins, Connectivity of transitive graphs, *J. Combin. Theory* **8** (1970), 23–29.

4

Contractions of k-connected graphs

KIYOSHI ANDO

An edge of a k-connected graph is k-contractible if the result of contracting the edge is still k-connected. Until recently, there were few results concerning contractible edges in graphs of connectivity 4 *or more, whereas there were many on* 3*-contractible edges. In this chapter, in addition to results on* 3*-contractible edges, we present many recent ones on contractible edges in graphs with connectivity greater than* 3.

1. Introduction

In the 1960s Tutte [49] pioneered work on 3-connected graphs, and since then, operations that preserve connectivity properties of graphs have played a significant role in the study of graph connectivity. One of the most important of these operations is edge contraction. An edge of a k-connected graph is said to be *k-contractible* if the result of its contraction is still a k-connected graph. An excellent survey of 20th-century results on contractible edges was given by Kriesell [34].

Other than K_4, every 3-connected graph has many 3-contractible edges and the number of such edges is the main topic of Section 2. For every $k \geq 4$, there are infinitely many k-connected graphs that have no k-contractible edges. In Section 3, we give a good characterization of such graphs for $k = 4$. A k-connected graph with no k-contractible edges is said to be *k-contraction-critical* (this is a shortening of the term 'contraction-critically k-connected'). In the last decade we have had rich results on the distribution of 4-contractible edges in a 4-connected graph; this is the main topic in Section 3. We have also new forbidden subgraph conditions for a k-connected graph to have a k-contractible edge. These conditions are given in Section 4.

As Kriesell wrote in [34], it is probably a tremendously hard problem to characterize k-contraction-critical graphs for $k \geq 5$. Although we still do not have enough knowledge of the global structure of 5-contraction-critical graphs, we have a local structure theorem of 5-contraction-critical graphs, which is given in Section 5.

In Section 6 we give some applications of contractible edges to the non-separating cycle problem.

In this chapter, we let V_r denote the set of vertices of degree r, and $V_{\geq r}$ the set of vertices with degree at least r; further, we let $n_r = |V_r|$ and $n_{\geq r} = |V_{\geq r}|$. If G is k-connected, we denote by E_c^k the set of k-contractible edges in G, and we denote its cardinality by m_c^k.

2. Contractible edges in 3-connected graphs

In this section we look at the number of 3-contractible edges that a 3-connected graph can have. Because K_4 is the only 3-connected graph of order less than 5 (and it has no 3-contractible edges), we consider only 3-connected graphs of order 5 or greater. As noted above, Tutte [49] observed that every such graph has a 3-contractible edge, a result independently found by Halin [30]. The following result on the number of such edges is due to Kriesell [34].

Theorem 2.1 *If G is a 3-connected graph of order $n \geq 5$, then it at least $\frac{1}{2}(n + n_{\geq 4})$ 3-contractible edges.*

For a bound in terms of the number of edges, we have the following result of Ota [43].

Theorem 2.2 *If G is a 3-connected graph of order $n \geq 19$ and with m edges, then it has at least $\frac{2}{7}(m + 6)$ 3-contractible edges.*

Our last result in this section, due to Egawa, Ota, Saito and Yu [25] gives a bound in terms of the numbers of both vertices and edges.

Theorem 2.3 *If G is a 3-connected graph with $n \geq 5$ vertices and m edges, then it has at least $m - 3n + \lfloor \frac{3}{2}(\sqrt{24n + 25} - 5 \rfloor$ edges that are not 3-contractible.*

They also showed that there are infinitely many graphs that attain this bound. Interestingly, this means that while the maximum number of 3-contractible edges in a graph is linear in n, the maximum number of non-3-contractible edges is not.

3. Contractible edges in 4-connected graphs

Whereas K_4 is the only 3-contraction-critical graph, there are infinitely many k-contraction-critical graphs for $k \geq 4$. Although it seems to be a very hard problem to characterize 5-contraction-critical graphs, there is a good characterization when $k = 4$.

The *square* of a graph G is the graph obtained from G by joining each pair of vertices at distance 2 in G by an edge. A λ-edge-connected graph is said to be *essentially* $(\lambda + 1)$*-edge-connected* if every edge-cut of order λ consists of the set of edges at some vertex. The following classification of 4-contraction-critical graphs was obtained independently by Fontet [28] and Martinov [42].

Theorem 3.1 *If G is a 4-connected graph with no 4-contractible edges, then G is either the square of a cycle of length at least 6, or the line graph of an essentially 4-edge-connected 3-regular graph.*

From Theorem 3.1, we know that every 4-contraction-critical graph is 4-regular, and that each of its edges is contained in a triangle. Hence, if a 4-connected graph has either a vertex of degree greater than 4 or an edge whose endpoints have no common neighbour, then it has a contractible edge. Thus, it seems natural to expect that every 4-connected graph has a contractible edge incident with each vertex of degree greater than 4. In fact, this is not true; however, there is a contractible edge near each such vertex (see [6]). For a set of vertices S in a graph, we let $N(S)$ denote the set of vertices not in S that are neighbours of some vertex in S.

Theorem 3.2 *Let G be a 4-connected graph that is not 4-regular, and let v be a vertex of degree greater than 4. Then there exists a 4-contractible edge e for which either e is incident with v, or at least one of the endpoints of e is adjacent to v. Furthermore, if $\langle N(v) - V_4 \rangle$ is not a path of order 4, then there are at least two such edges.*

There is a 4-connected graph with a vertex v of degree greater than 4 for which there is just one 4-contractible edge at a distance 0 or 1 from v, so the condition on $\langle N(v) - V_4 \rangle$ is necessary. There also exists a 4-connected graph with a vertex v of degree greater than 4 for which there are exactly two 4-contractible edges each at a distance of 0 or 1 from v, so the number 2 is optimal. As a consequence of this theorem, Ando and Egawa [6] found the following bound on the number of 4-contractible edges. They also showed that there are arbitrarily large graphs that attain this bound.

Theorem 3.3 *Every 4-connected graph G has $n_{\geq 5}$ 4-contractible edges.*

Given a graph G, we let $\tilde{E}$ be the set of edges that are not in any triangle, and let $\tilde{m}$ be its cardinality. In view of Theorem 3.1, we may expect that if G is 4-connected, then there is a contractible edge near each edge in $\tilde{E}$. The following theorem confirms this (see [7]).

Theorem 3.4 *Let G be a 4-connected graph with $\tilde{E} \neq \varnothing$, and let $vw \in \tilde{E}$. Suppose that vw is non-contractible, let S be a 4-cutset with $v, w \in S$, and let A be the vertex-set of a component of $G - S$. Then there exists $e \in E_c^4$ for which either e is incident with v, or there exists $w \in N(v) \cap (S \cup A) \cap V_4$ for which e is incident with w.*

In this theorem, the assumption of non-contractibility is needed. Precise arguments based on Theorem 3.4 led Ando and Egawa [7] to the following bound on the number of 4-contractible edges in a 4-connected graph G. They also showed that the bound is sharp.

Theorem 3.5 *If G is a 4-connected graph with $\tilde{m} \geq 15$, then $m_c^4 \geq \frac{1}{4}\tilde{m} + 2$.*

If G is 4-regular, then the following stronger statement holds (see [8]).

Theorem 3.6 *Let G be 4-regular and 4-connected, and let e be an edge of G whose endpoints have no common neighbour. Then either e is 4-contractible, or there is a 4-contractible edge adjacent to e.*

There is a 4-connected graph that has an edge e for which neither is e 4-contractible nor is there a 4-contractible edge adjacent to e. Hence the condition that G is 4-regular cannot be dropped.

From Theorem 3.6 we have the following.

Theorem 3.7 *If G is 4-regular and 4-connected, then $m_c^4 \geq \frac{1}{2}\tilde{m}$. Moreover, this bound is sharp.*

Recently we have obtained a bound on the number of contractible edges in a 4-connected graph, in terms of its average degree (see [9]).

Theorem 3.8 *Every 4-connected graph G with n vertices and m edges has at least $\frac{1}{34}(m - 2n)$ 4-contractible edges.*

It is known that the best value of the bound in Theorem 3.8 lies between $\frac{1}{34}$ and $\frac{1}{2}$.

4. Contractible edges in k-connected graphs

If $k \geq 4$, then we cannot expect the existence of a contractible edge in a k-connected graph with no condition. In this section, we consider conditions for a k-connected graph to have a k-contractible edge. In this section, we say 'k-sufficient condition' for a condition for a k-connected graph to have a k-contractible edge.

A *fragment* A is a non-empty union of components of $G - S$, where S is a k-cutset of G for which $V(G) - (A \cup S) \neq \varnothing$. Mader showed that every k-contraction-critical graph has a fragment A whose neighbourhood contains an edge for which $|A| \leq \frac{1}{2}(k - 1)$. From this fact we can see that 'triangle-free' is a k-sufficient condition; Thomassen pointed this out in [48].

Theorem 4.1 *Every k-connected triangle-free graph has a k-contractible edge.*

The following result, due to Egawa, Enomoto and Saito [23], gives a bound on the number of k-contractible edges in a k-connected triangle-free graph.

Theorem 4.2 *Every k-connected triangle-free graph with n vertices and m edges contains* $\min\{n + \frac{3}{2}k^2 - 3k, m\}$ *k-contractible edges.*

From Theorem 4.1 we know that every k-contraction-critical graph has triangles. The following result due to Kriesell [35] is a substantial improvement on a bound of $\frac{1}{3}n$ found by Mader [41].

Theorem 4.3 *Let G be a k-connected graph of order n with no contractible edges. Then G contains at least $\frac{2}{3}n$ triangles.*

By generalizing methods of Mader [40], [41], Egawa [21] proved that every k-contraction-critical graph has a fragment A such that $|A| \leq \frac{1}{4}k$, and gave the following minimum-degree k-sufficient condition. He also showed that the bound is sharp.

Theorem 4.4 *Let $k \geq 2$ be an integer, and let G be a k-connected graph with $\delta(G) \geq \lfloor \frac{5}{4}k \rfloor$. Then G has a k-contractible edge, unless $k = 2$ and G is isomorphic to K_3, or $k = 3$ and G is isomorphic to k_4.*

Kriesell [33] extended Theorem 4.4 and proved the following degree-sum k-sufficient condition.

Theorem 4.5 *Let G be a k-connected graph for which* $\deg v + \deg w \geq 2\lfloor \frac{5}{4}k \rfloor - 1$, *for any pair v, w of distinct vertices of G. Then G contains a k-contractible edge.*

Solving a conjecture in [33], Su and Yuan [47] proved the following stronger result.

Theorem 4.6 *Let G be a contraction-critical k-connected graph with $k \geq 8$. Then G has two adjacent vertices v, w for which* $\deg v + \deg w \leq 2\lfloor \frac{5}{4}k \rfloor - 2$.

The following interesting result is also due to Kriesell [36].

Theorem 4.7 *For every $k \geq 1$, there exists a number $f(k)$ for which every k-connected graph with average degree at least $f(k)$ has a k-contractible edge.*

In [36], he showed that $f(k) \leq ck^2 \log k$ for some constant c, and posed the following conjecture.

Conjecture A *There exists a constant c for which every finite k-connected graph with average degree at least ck^2 has a k-contractible edge.*

Another problem is to determine the best value of $f(k)$, for a given value of k. Up to now, we have no 5-contraction-critical graph whose average degree exceeds $\frac{25}{2}$. Hence we pose the following.

Conjecture B *The average degree of every 5-contraction-critical graph is less than* $\frac{25}{2}$.

In view of Theorem 4.2, a k-connected triangle-free graph has many k-contractible edges, indicating the possible existence of a weaker k-sufficient condition involving forbidden subgraphs. In this direction, Kawarabayashi [31] showed that, for an odd integer $k \geq 3$, every K_4^--free k-connected graph has a k-contractible edge, where K_4^- is the graph obtained from K_4 by removing one edge. Since K_4^- contains a triangle, this is an extension of Theorem 4.1 when k is odd.

We call the graph $K_1 + 2K_2$ a *bowtie*. Ando, Kaneko, Kawarabayashi and Yoshimoto [15] proved that every k-connected bowtie-free graph has a k-contractible edge; this is also an extension of Theorem 4.1.

Since $K_1 + P_4$ contains K_4^-, the following is a common extension of the above two forbidden-subgraph k-sufficient conditions (see [31]).

Theorem 4.8 *Let $k \geq 5$, and let G be a k-connected $(K_1 + P_4)$-free graph. If $\langle V_k \rangle_G$ is bowtie-free, then G has a k-contractible edge.*

The following is another extension of Theorem 4.1 (see [16]). Note that if $s = t = 1$, then Theorem 4.9 is equivalent to Theorem 4.1. Also note that $K_4^- \cong K_2 + 2K_1$ and bowtie $\cong K_1 + 2K_2$; hence, $K_2 + sK_1$ and $K_1 + tK_2$ may be regarded as a generalized K_4^- and generalized bowtie, respectively. Note that we cannot replace the condition $s(t-1) < k$ by $s(t-1) \leq k$ in Theorem 4.9.

Theorem 4.9 *For $k \geq 5$, take two positive integers s and t with $s(t-1) < k$. If a k-connected graph G contains neither $K_2 + sK_1$ nor $K_1 + tK_2$, then G contains a k-contractible edge.*

Theorems 4.1, 4.8 and 4.9 deal with forbidden-subgraph k-sufficient conditions. On the other hand, Theorem 4.4 gives a minimum-degree k-sufficient condition. However, if we restrict ourselves to a class of graphs that satisfy some forbidden-subgraph conditions, then we may relax the minimum-degree bound in Theorem 4.4. The following forbidden-subgraph condition relaxes the minimum-degree bound (see [16]): K_5^- is the graph obtained from K_5 by removing one edge.

Theorem 4.10 *For $k \geq 5$, let G be a k-connected graph which contains neither K_5^- nor $5K_1 + P_3$. If $\delta(G) \geq k + 1$, then G has a k-contractible edge.*

Note that if $k \geq 5$, then $\lfloor \frac{5}{4}k \rfloor \geq k + 1$. Since there is a k-regular k-contraction-critical graph which contains neither K_5^- nor $5K_1 + P_3$, we cannot replace

$\delta(G) \geq k+1$ by $\delta(G) \geq k$ in Theorem 4.10. In this sense, the minimum-degree bound in Theorem 4.10 is sharp.

5. Contraction-critical 5-connected graphs

Whereas each 4-contraction-critical graph has very restricted induced subgraphs, there are no such restrictions for a k-contraction-critical graph with $k \geq 5$. In [14], it is shown that, for any given graph G and any given integer $k \geq 5$, there exists a k-contraction-critical graph with G as its induced subgraph.

Let $d(v, w)$ denote the distance between vertices v and w in G. Let $B_r(v) = \{w \in V(G) : d(v, w) \leq r\}$, and let $B_r(v)$ be the r-ball with centre v. Note that $B_0(v) = \{v\}$ and $B_1(v) = \{v\} \cup N(v)$.

The following theorem of Su [46] is one of the most important results on 5-contraction-critical graphs. Unfortunately, this result has been little known to most researchers in this field.

Theorem 5.1 *Every vertex of a 5-contraction-critical graph has at least two neighbours of degree* 5.

There is a 5-contraction-critical graph that contains a vertex v for which $|N(v) \cap V_5| = 2$.

One may expect a similar statement to hold for 6-contraction-critical graphs, but in [13] a 6-contraction-critical graph with a vertex v such that $N(v) \cap V_6 = \varnothing$ is presented. In the same paper it is shown that, for every vertex v of a 6-contraction-critical graph G, $B_2(v) \cap V_6 \neq \varnothing$.

Let G be a k-connected graph. If $G - e$ is not k-connected for any edge $e \in E(G)$, then G is said to be *minimally k-connected*. For example, if m is large, then the complete bipartite graph $K_{6,m}$ is a minimally 6-connected graph which has a vertex v for which $N(v) \cap V_6 = \varnothing$. Hence, neither '6-contraction-critical' nor 'minimally 6-connected' guarantees that every vertex of a 6-connected graph has a neighbour of degree 6. Recently, Ando, Fujita and Kawarabayashi [11] showed that 'minimally 6-contraction-critical' is sufficient.

Theorem 5.2 *Every vertex of a minimally* 6*-contraction-critical graph has a neighbour of degree* 6.

When $k > 4$, there is a k-contraction-critical graph that is not k-regular. However, from Theorem 4.4 (or Theorem 4.5), we know that the minimum degree of a k-contraction-critical graph is k, for $k = 5$, 6 or 7. So the following problem has been posed.

Problem *Let $k = 5$, 6 or 7. Determine the largest value of c_k for which $n_k \geq c_k n$ holds for every k-contraction-critical graph G of order n.*

For $k = 5$, this problem was solved independently by Ando and Iwase [12], and Li and Su [37] (also see [45]).

Theorem 5.3 *Every 5-connected graph with no 5-contractible edge of order n has at least $\frac{1}{2}n$ vertices of degree 5.*

Since [12] has a sequence of 5-contraction-critical graphs $\{G_i\}$ for which $\lim_{i\to\infty} |V_5(G_i)|/|V(G_i)| = \frac{1}{2}$, the constant c_5 is $\frac{1}{2}$. For $k = 6$ and 7, we know that $\frac{1}{5} \le c_6 \le \frac{1}{2}$ and $\frac{1}{22} \le c_7 \le \frac{6}{13}$ (see [13], [50] and [39]).

If we restrict ourselves to minimally 5-contraction-critical graphs G, then we can expect G to have more vertices of degree 5 (see [17]).

Theorem 5.4 *Every minimally 5-contraction-critical graph of order n has at least $\frac{2}{3}n$ vertices of degree 5.*

There is a minimally 5-contraction-critical graph of order n which has exactly $\frac{4}{5}n$ vertices of degree 5. Hence we cannot expect the bound in Theorem 5.4 to exceed $\frac{4}{5}n$.

Very recently, we have obtained the following results on the average degree of minimally k-contraction-critical graphs, for $k = 5$ and 6 (see [10], [5]).

Theorem 5.5 *The average degree of every minimally contraction-critical 5-connected graph is at most $\frac{15}{2}$. For any positive value ε, there is a minimally contraction-critical 5-connected graph whose average degree is greater than $\frac{15}{2} - \varepsilon$.*

Theorem 5.6 *The average degree of every minimally contraction-critical 6-connected graph is at most $\frac{32}{3}$.*

So far we have been unable to construct a minimally contraction-critical 6-connected graph whose average degree exceeds 10. We think that the sharp bound in Theorem 5.6 should not be $\frac{32}{3}$, but 10.

An edge $e = vw$ of a k-connected graph is said to be *trivially non-contractible* if v and w have a common neighbour of degree k. Theorem 3.1 says that each edge in a contraction-critical 4-connected graph is trivially non-contractible. Every 5-contraction-critical graph has many vertices of degree 5; furthermore, Theorem 4.3 guarantees that every 5-contraction-critical graph has many triangles. It thus seems natural to expect every 5-contraction-critical graph to have many trivially non-contractible edges. Actually, Li and Su [38] have proved the following (also see [3]).

Theorem 5.7 *Every 5-connected graph with no 5-contractible edge of order n has at least $\frac{3}{2}n$ trivially non-contractible edges.*

There is a 5-contraction-critical graph on n vertices with exactly $2n$ trivially non-contractible edges. We cannot expect the bound in Theorem 5.7 to exceed $2n$.

To conclude this section, we present a theorem for 5-connected graphs concerning the local structure around a vertex v near which there are no 5-contractible edges.

Let $Edge^{(i)}(v)$ be the set of edges whose distance from v is i or less. We also need to introduce three specified configurations in 5-connected graphs. A configuration that consists of two triangles with no vertices in common but v is called a *v-bowtie*; so a v-bowtie is isomorphic to $2K_2 + K_1$, and $\deg v = 4$. Next, K_4^- is called a *reduced v-bowtie* if $\deg v = 3$. If, in each triangle of a v-bowtie, there is a vertex of degree 5 other than v, then the v-bowtie is said to be a *v^*-bowtie*. If a reduced v-bowtie has at least two vertices of degree 5 other than v, then it is called a *reduced v^*-bowtie*. Hence, in Fig. 1, (a) is a v^*-bowtie if neither $\{w_1, w_2\} \cap V_5$ nor $\{z_1, z_2\} \cap V_5$ is empty, and (b) is a reduced v^*-bowtie if $|\{w_1, w_2, w_3\} \cap V_5| \geq 2$.

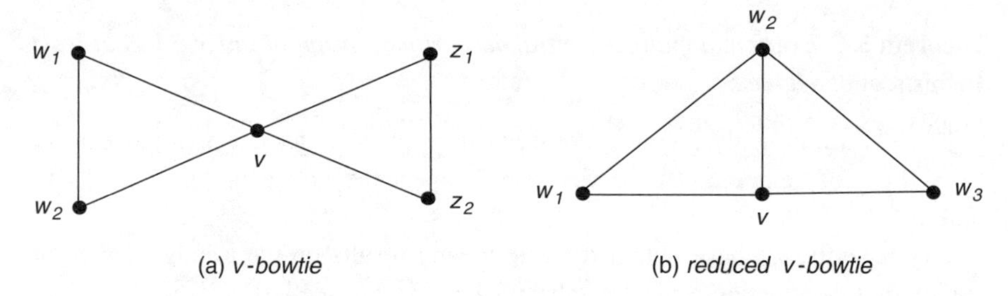

Fig. 1.

Let $S = \{a_1, a_2, v, b_1, b_2\}$ be a 5-cutset of a 5-connected graph G, and let A be a component of $G - S$ for which $V(A) \subseteq V_5(G)$, $|V(A)| = 4$, and $\langle A \rangle \cong K_4^-$ – say, $A = \{x_1, x_2, y_1, y_2\}$, with edges within A and between A and S exactly as in Fig. 2; there may be edges between the vertices of S. We call this configuration $\langle V(A) \cup S \rangle$, a *$K_4^-$-configuration with centre v*. Note that $\{x_1, x_2, y_1, y_2\} \subseteq V_5$, and that the edges in Fig. 2 other than vx_1 and vy_1 are all trivially non-contractible. Moreover, we can find two non-trivial 5-cutsets, $\{x_1, x_2, v, b_1, b_2\}$ and $\{y_1, y_2, v, a_1, a_2\}$, that contain $V(vx_1)$ and $V(vy_1)$, respectively. Hence all edges in Fig. 2 are non-contractible. Note finally that if there is an edge between vertices of S, then it is non-contractible, since S is a 5-cutset of G.

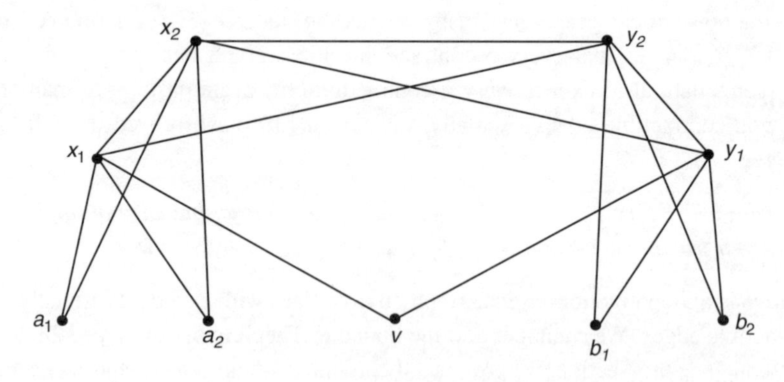

Fig. 2. A K_4^--configuration with centre v

The following local structure theorem was proved in [4]. It guarantees that, for every vertex v of a 5-contraction-critical graph, there exists one of three configurations: a v^*-bowtie, a reduced v^*-bowtie, or a K_4^--configuration with centre v.

Theorem 5.8 *Let v be a vertex of a 5-connected graph G for which $Edge^{(2)}(v) \cap E_c^5 = \varnothing$. If G has neither a v^*-bowtie nor a reduced v^*-bowtie, then G has a K_4^--configuration with centre v.*

6. Local structure and contractible edges

In this section we take a brief look at contractible edges in certain local structures of graphs and give some applications of contractible edges.

Concerning the distribution of contractible edges in longest cycles of 3-connected graphs, Dean, Hemminger and Ota [19] have shown the following.

Theorem 6.1 *Let G be a 3-connected graph. If $G \not\cong K_4$ and $G \not\cong \bar{C}_6$, then a longest cycle of G contains at least 3 contractible edges.*

Aldred, Hemminger and Ota [2] characterized these graphs with precisely three contractible edges on some longest cycle; they are all Hamiltonian. Thus, Theorem 6.1 assures us that every longest cycle of a non-Hamiltonian 3-connected graph has at least four contractible edges. Ellingham, Hemminger and Johnson [27] improved this, by proving the following.

Theorem 6.2 *Every longest cycle of a non-Hamiltonian 3-connected graph contains at least 6 contractible edges.*

Aldred and Hemminger [1] characterized these graphs with precisely four or five contractible edges on some longest cycle.

There is an interesting conjecture, due to Ota [44].

Conjecture C *There is a constant $\alpha > 0$ for which every 3-connected graph G of order at least 5 contains a longest cycle C satisfying $|E(C) \cap E_c^3| \geq \alpha |E(C)|$.*

It was partially solved by Fujita [29], in the case when G is Hamiltonian.

Theorem 6.3 *Every 3-connected Hamiltonian graph G of order at least 5 contains a Hamiltonian cycle C satisfying $|E(C) \cap E_c^3| \geq \frac{1}{8}|E(C)| + \frac{9}{8}$.*

We next take a look at some applications of contractible edges. Using Theorem 4.1, Thomassen [48] proved the following.

Theorem 6.4 *Every k-connected graph has an induced cycle C for which $G - V(C)$ is $(k - 3)$-connected.*

In his proof, Thomassen used induction on the order of G. If the graph G has a triangle, then we can delete the triangle and the resulting graph is $(k-3)$-connected. Hence, we may assume that G is triangle-free, and Theorem 4.1 tells us that G has a k-contractible edge e; we can then apply the induction hypothesis to the graph G/e, obtained from G by contracting the edge e.

If G is a k-connected graph with girth at least 4, Egawa [20] proved that G has an induced cycle C for which $G - V(C)$ is $(k-2)$-connected. Kawarabayashi [32] then proved the following. Since K_4^- has a triangle, this implies Egawa's result.

Theorem 6.5 *Every K_4^--free k-connected graph has an induced cycle C for which $G - V(C)$ is $(k - 2)$-connected.*

When the girth of a k-connected graph G is at least 5, Thomassen [48] conjectured that there is an induced cycle C for which $G - V(C)$ is $(k - 1)$-connected. This challenging conjecture has been attacked by several researchers. After conquering all difficulties, Egawa [22] obtained the following marvellous result.

Theorem 6.6 *Every k-connected graph with girth at least* 5 *has an induced cycle C for which $G - V(C)$ is $(k - 1)$-connected.*

In the proofs of Theorems 6.5 and 6.6, it was important to investigate the distribution of k-contractible edges.

Theorem 4.2 guarantees that a k-connected graph with girth at least 4 has many k-contractible edges. So it is natural to expect that every k-connected graph with girth at least 4 has a non-separating cycle C for which $E(C) \subseteq E_c^k$. Motivated by this, Dean [18] proved that every k-connected graph with girth at least 4 has an induced cycle C for which $E(C) \subseteq E_c^k$ and $G - V(C)$ is connected. Egawa and Saito [26] extended Dean's result and proved that every k-connected graph with girth at least 4 has an induced cycle C for which $E(C) \subseteq E_c^k$ and $G - V(C)$ is $(k - 3)$-connected.

Recently, Egawa, Inoue and Kawarabayashi [24] made a big step forward in this direction.

Theorem 6.7 *Every k-connected graph with girth at least* 4 *has an induced cycle C for which $E(C) \subseteq E_c^k$ and $G - V(C)$ is $(k - 2)$-connected.*

In the same paper, they gave the following very challenging conjecture.

Conjecture D *Every K_4^--free k-connected graph has an induced cycle C for which either $G - V(C)$ is $(k - 1)$-connected, or $G - V(C)$ is $(k - 2)$-connected and $E(C) \subseteq E_c^k$.*

7. Concluding remarks

We have presented some historical results and many recent results on contractible edges in k-connected graphs. We have not touched on either the contractions of much larger subgraphs or graph minors.

In the last decade, much knowledge has been gained of the distribution of 4-contractible edges and the structural properties of 5-contraction-critical graphs. It is possible that our methods can be applied to investigate problems on 4-connected and 5-connected graphs. There are many attractive and challenging problems in this area. Among these, it is a major problem to investigate the structure of graphs of connectivity 6 or more.

References

1. R. E. L. Aldred and R. L. Hemminger, On hamiltonian cycles and contractible edges in 3-connected graphs, *Australas. J. Combin.* **11** (1995), 3–24.
2. R. E. L. Aldred, R. L. Hemminger and K. Ota, The 3-connected graphs having a longest cycle containing only three contractible edges, *J. Graph Theory* **17** (1993), 361–371.
3. K. Ando, Trivially noncontractible edges in a contraction critically 5-connected graph, *Discrete Math.* **293** (2005), 61–72.
4. K. Ando, A local structure theorem on 5-connected graphs, *J. Graph Theory* **60** (2009), 99–129.
5. K. Ando, Average degree of minimally contraction-critically 6-connected graphs, *Discrete Math.* (to appear).
6. K. Ando and Y. Egawa, Contractible edges in a 4-connected graph with vertices of degree greater than four, *Graphs Combin.* **23** (2007), 99–115.
7. K. Ando and Y. Egawa, Edges not contained in triangles and the distribution of contractible edges in a 4-connected graph, *Discrete Math.* **308** (2008), 3449–3460.
8. K. Ando and Y. Egawa, Edges not contained in triangles and the number of contractible edges in a 4-connected graph, *Discrete Math.* **308** (2008), 3463–3472.
9. K. Ando, Y. Egawa, K. Kawarabayashi and M. Kriesell, On the number of 4-contractible edges in 4-connected graphs, *J. Combin. Theory (B)* **99** (2009), 97–109.
10. K. Ando, Y. Egawa and M. Kriesell, The average degree of minimally contraction-critically 5-connected graphs, preprint.
11. K. Ando, S. Fujita and K. Kawarabayashi, Minimally contraction-critically 6-connected graphs, *Discrete Math.* **312** (2012), 671–679.
12. K. Ando and T. Iwase, The number of vertices of degree 5 in a contraction-critically 5-connected graph, *Discrete Math.* **311** (2011), 1925–1939.
13. K. Ando, A. Kaneko and K. Kawarabayashi, Vertices of degree 6 in a contraction critically 6-connected graph, *Discrete Math.* **273** (2003), 55–69.
14. K. Ando, A. Kaneko and K. Kawarabayashi, Vertices of degree 5 in a contraction critically 5-connected graph, *Graphs Combin.* **21** (2005), 27–37.
15. K. Ando, A. Kaneko, K. Kawarabayashi and K. Yoshimoto, Contractible edges and bowties in a k-connected graph, *Ars Combin.* **64** (2002), 239–247.
16. K. Ando and K. Kawarabayashi, Some forbidden subgraph conditions for a graph to have a k-contractible edge, *Discrete Math.* **267** (2003), 3–11.
17. K. Ando and C. Qin, Some structural properties of minimally contraction-critically 5-connected graphs, *Discrete Math.* **311** (2011), 1084–1097.

18. N. Dean, Distribution of contractible edges in k-connected graphs, *J. Combin. Theory (B)* **48** (1990), 1–5.
19. N. Dean, R. L. Hemminger and K. Ota, Longest cycles in 3-connected graphs contain 3 contractible edges, *J. Graph Theory* **12** (1989), 17–21.
20. Y. Egawa, Cycles in k-connected graphs whose deletion results in a $(k-2)$-connected graph, *J. Combin. Theory (B)* **42** (1987), 371–377.
21. Y. Egawa, Contractible edges in n-connected graphs with minimum degree greater than or equal to $[\frac{5n}{4}]$, *Graphs Combin.* **7** (1991), 15–21.
22. Y. Egawa, Contractible cycles in graphs with girth at least 5, *J. Combin. Theory (B)* **74** (1998), 213–264.
23. Y. Egawa, H. Enomoto and A. Saito, Contractible edges in triangle-free graphs, *Combinatorica* **6** (1986), 269–274.
24. Y. Egawa, K. Inoue and K. Kawarabayashi, Nonseparating induced cycles consisting of contractible edges in k-connected graphs, *SIAM J. Discrete Math.* **21** (2008), 1061–1070.
25. Y. Egawa, K. Ota, A. Saito and X. Yu, Noncontractible edges in a 3-connected graph, *Combinatorica* **15** (1995), 357–364.
26. Y. Egawa and A. Saito, Contractible edges in nonseparating cycles, *Combinatorica* **11** (1991), 389–392.
27. M. N. Ellingham, R. L. Hemminger and K. E. Johnson, Contractible edges in longest cycles in non-hamiltonian graphs, *Discrete Math.* **133** (1994), 89–98.
28. M. Fontet, Graphes 4-essentiels, *C. R. Acad. Sci. Paris* **287** (1978), 289–290.
29. K. Fujita, Maximum number of contractible edges on hamiltonian cycles of a 3-connected graph, *Graphs Combin.* **18** (2002), 447–478.
30. R. Halin, Untersuchungen über minimale n-fach zusammenhängende Graphen, *Math. Ann.* **182** (1969), 175–188.
31. K. Kawarabayashi, Note on k-contractible edges in k-connected graphs, *Australas. J. Combin.* **24** (2001), 165–168.
32. K. Kawarabayashi, Contractible edges and triangles in k-connected graphs, *J. Combin. Theory (B)* **85** (2002), 207–221.
33. M. Kriesell, A degree sum condition for the existence of a contractible edge in a κ-connected graph, *J. Combin. Theory (B)* **82** (2001), 81–101.
34. M. Kriesell, A survey on contractible edges in graphs of a prescribed vertex connectivity, *Graphs Combin.* **18** (2002), 1–30.
35. M. Kriesell, Triangle density and contractibility, *Combin. Probab. Comput.* **14** (2005), 133–146.
36. M. Kriesell, Average degree and contractibility, *J. Graph Theory* **51** (2006), 205–224.
37. T. Li and J. Su, A new lower bound on the number of trivially noncontractible edges in contraction critical 5-connected graphs, *Discrete Math.* **309** (2009), 2870–2876.
38. T. Li and J. Su, A new lower bound of the number of vertices of degree 5 in contraction critical 5-connected graphs, *Graphs Combin.* **26** (2010), 395–406.
39. M. Li, X. Yuan and J. Su, The number of vertices of degree 7 in a contraction-critical 7-connected graph, *Discrete Math.* **308** (2008), 6262–6268.
40. W. Mader, Disjunkte Fragmente in kritisch n-fach zusammenhängende Graphen, *Europ. J. Combin.* **6** (1985), 353–359.
41. W. Mader, Generalizations of critical connectivity of graphs, *Discrete Math.* **72** (1988), 267–283.
42. N. Martinov, Uncontractible 4-connected graphs, *J. Graph Theory* **6** (1982), 343–344.
43. K. Ota, The number of contractible edges in 3-connected graphs. *Graphs Combin.* **4** (1988), 333–354.

44. K. Ota, *Non-critical subgraphs in k-connected graphs*, Ph.D. Dissertation, Univ. of Tokyo, 1989.
45. C. Qin, X. Yuan and J. Su, Some properties of contraction-critical 5-connected graphs, *Discrete Math.* **308** (2008), 5742–5756.
46. J. Su, Vertices of degree 5 in contraction-critical 5-connected graphs, *J. Guangxi Normal University* **3** (1997), 12–16 (in Chinese).
47. J. Su and X. Yuan, A new degree sum condition for the existence of a contractible edge in a κ-connected graph, *J. Combin. Theory (B)* **96** (2006), 276–295.
48. C. Thomassen, Non-separating cycles in k-connected graphs, *J. Graph Theory* **5** (1981), 351–354.
49. W. T. Tutte, A theory of 3-connected graphs, *Indag. Math.* **23** (1961), 441–455.
50. Q. Zhao, C. Qin, X. Yuan and M. Li, Vertices of degree 6 in a contraction-critical 6-connected graph, *J. Guangxi Normal Univ.* **25** (2005), 38–43.

5

Connectivity and cycles

R. J. FAUDREE

Connectivity plays a critical role in the existence of paths and cycles in graphs. We present recent extensions of classical results on the relationship between connectivity and properties of paths, cycles and 2-factors in graphs. We also explore these connectivity-cycle relationships for special classes of graphs, such as regular, planar and claw-free graphs.

1. Introduction

Connectivity is a key graphical parameter in conditions that imply the existence of paths, cycles and 2-factors in graphs. Conditions in which connectivity plays a major role in the existence of such subgraphs are discussed. In some cases, and for some classes of graphs, connectivity conditions alone imply the existence of paths, cycles and 2-factors. For example, it is well known that every 2-connected graph with at least three vertices has a cycle. In fact, by the classical result of Menger [86] (see Chapter 1), every pair of vertices of G lies on a common cycle. Using Menger's theorem and induction, Dirac [32] generalized this result.

Theorem 1.1 *If G is a k-connected graph with $k \geq 2$ and order $n \geq 3$, then every set of k vertices lies on a cycle.*

This result is sharp, as the k connected graph $K_k + (\overline{K}_{k+1} \cup K_{n-2k-1})$ shows: there is no cycle that contains the $k+1$ independent vertices in the $\overline{K}_{k+1}$ part of the graph. Watkins and Mesner [107] characterized those k-connected graphs in which there are $k+1$ vertices that do not lie on a cycle – they are those graphs with a cutset of k vertices whose removal results in a graph with more than k components. The graph just described is one example of this class of exceptions.

It is clear from the previous observations that larger connectivity implies longer cycles, and the nature of this relationship has been investigated. Dirac [31] gave a lower bound on the length of the longest cycle in terms of the minimum degree.

Theorem 1.2 *Every 2-connected graph of order n and minimum degree δ contains a cycle of length at least* $\min\{2\delta, n\}$.

An immediate consequence of this is that if $k \leq \frac{1}{2}n$, then any k-connected graph of order n has a cycle of length at least $2k$, and is Hamiltonian if $k \geq \frac{1}{2}n$. Also, this result is sharp, since for $n \geq 2k$, the k-connected graph $K_k + \overline{K}_{n-k}$ contains no cycle of length longer than $2k$.

Egawa, Glass and Locke [34] gave a common generalization of Theorems 1.1 and 1.2.

Theorem 1.3 *If G is a k-connected graph with $k \geq 2$, minimum degree δ, and order n, then every set of k vertices is on a cycle of length at least* $\min\{2\delta, n\}$.

The existence of disjoint cycles was considered by Corradi and Hajnal [29], who proved the following result.

Theorem 1.4 *For $k \geq 2$, if G is a graph with minimum degree at least $2k$ and order at least $3k$, then G contains k vertex-disjoint cycles.*

This result is sharp in that the graph $K_{2k-1} + \overline{K}_{n-2k+1}$ has minimum degree $2k-1$ (in fact, its connectivity is $2k-1$), yet it does not contain k disjoint cycles.

There are many generalizations of these classical results on connectivity and cycles in graphs. In the next section, these are explored and expanded to include paths, cycles with special properties and 2-factors. Surveys on cycles and paths in graphs include Bondy [16], Gould [46] and Faudree [40].

2. Generalizations of classical results

A generalization of Dirac was considered by Kaneko and Saito [67]. For $r \geq s$, a graph satisfies property $P(r, s)$ if for any set R of r vertices, there is a cycle C such that $|C \cap R| = s$. Thus, Dirac's theorem implies that every k-connected graph satisfies $P(k, k)$. Kaneko and Saito proved that every k-connected graph satisfies $P(k+t, k)$

if $t \leq \frac{1}{4}(-1+\sqrt{8k+9})$. This was improved recently by Kawarabayashi [71] in the following result.

Theorem 2.1 *Let G be a k-connected graph with $k \geq 3$. Then for any set S of s vertices with $k \leq s \leq \frac{3}{2}k$, G contains a cycle with precisely k vertices of S.*

A natural question arises: *Can vertices be replaced by edges in this result?* The k edges would then have to form a union of disjoint paths, which we call a *path system*. It is straightforward to verify that each pair of edges in a 2-connected graph lies on some cycle. However, this cannot be extended to the 3-connected case, since it is possible for three edges to form an edge-cut in a graph, and thus there cannot be a cycle containing all three.

Lovász [79] conjectured that this structure – an odd number of edges forming a path system that is also an edge-cut – is the only exception to the existence of a cycle. This was shown to be the case for graphs of low connectivity.

Theorem 2.2 *For $2 \leq k \leq 7$ and any k-connected graph G, if S is a set of k edges that form a path system, then there is a cycle that contains the edges of S, unless k is odd and the set S is an edge-cut of G.*

The case $k=3$ was proved by Lovász [80], the case $k=4$ by Erdős and Győri [38], and the case $k=5$ by Sanders [97]. The cases $k=6$ and 7 were established by Kawarabayashi [70], who also outlined a proof of Lovász's conjecture and said that details would be provided in a series of three papers. Additionally, he showed that if the set of edges is not on a single cycle, then two cycles will do.

Theorem 2.3 *If G is a k-connected graph with $k \geq 2$, and if S is a set of k edges that form a path system, then S is contained either in a single cycle or in the union of two cycles, unless k is odd and the set S is an edge-cut of G.*

The following result of Häggkvist and Thomassen [49] lends additional support to Lovász's conjecture and proves a conjecture of Woodall [110].

Theorem 2.4 *If G is a k-connected graph with $k \geq 2$ and if S is a set of $k-1$ edges that form a path system, then there is a cycle in G that contains the edges of S.*

Very high connectivity is needed to guarantee the existence of a 2-factor in a graph, since the nearly regular complete bipartite graph $K_{\lfloor\frac{1}{2}(n-1)\rfloor,\lceil\frac{1}{2}(n-1)\rceil}$ has no 2-factor and is $\lfloor\frac{1}{2}(n-1)\rfloor$-connected. However, $2k$-connectedness in a sufficiently large graph does imply the existence of k disjoint cycles. This was established by Corradi and Hajnal [29], using only the additional assumption of minimum degree $2k$ (Theorem 1.4), a result that was extended by Egawa [33].

Theorem 2.5 *For each $k \geq 3$, every graph with minimum degree at least $2k$ and sufficiently many vertices contains k disjoint cycles all of the same length.*

A natural question that arises from Theorem 2.5 is: *What connectivity is needed to guarantee the existence of k disjoint cycles that 'separate' any specified set of k vertices?* More precisely, given any set X of k vertices in a graph G, *what is the minimum connectivity that guarantees that there is a set of k disjoint cycles each with precisely one vertex of X*?

At the other extreme is the intersection of longest cycles in a graph. It is easy to see that any two longest cycles in a 2-connected graph share at least two vertices. For example, observe that if C and C' are vertex-disjoint cycles, then 2-connectivity implies the existence of vertex-disjoint paths P and P' between C and C'. The union of these paths and cycles contains a longer cycle than either C or C'. The same argument also applies if the cycles share precisely one vertex. This is the 2-connected case of a conjecture attributed by Grötschel [47] to Scott Smith.

Conjecture A *For $k \geq 2$, any two longest cycles in a k-connected graph intersect in at least k vertices.*

This conjecture cannot be sharpened, since the longest cycles in the graph $K_k + \overline{K}_{n-k}$ are of length $2k$, and for $n \geq 3k$ some pairs of longest cycles intersect in just k vertices. The results in [47] imply that this conjecture holds for $k \leq 5$, and a comment is made as to the conjecture having been verified for $k \leq 10$. S. Burr and T. Zamfirescu (unpublished) showed that every pair of longest cycles in a k-connected graph meet in at least $\sqrt{k} - 1$ vertices. This bound was improved by Chen, Faudree and Gould [23].

Theorem 2.6 *For $k \geq 2$, every pair of longest cycles in a k-connected graph intersect in at least $(k/(4\sqrt[3]{4} + 3))^{3/5}$ vertices.*

If all of the longest cycles are considered, then it is possible for their intersections to be empty. This question was explored by Jendrol and Skupień [61], who proved the following result.

Theorem 2.7 *For $m \geq 7$, there is a 2-connected graph with m longest cycles whose intersection is empty, but in which every set of $m - 1$ longest cycles has non-empty intersection.*

The relationship between connectivity and cycles and paths in graphs is much richer than the classical results and their extensions discussed here. Some of the richness of this relationship is explored in the following sections, on special classes of graphs. The next section is devoted to the relationships between cycle lengths and path lengths.

3. Relative lengths of paths and cycles

It is reasonable to expect that if a 2-connected graph has long paths then it must have a long cycle. To explore this relationship, we let $l_k(p)$ be the smallest integer for which every k-connected graph with a path of length p contains a cycle of length at least $l_k(p)$. In his classic paper [31], Dirac proved the following result, and provided examples verifying the sharpness (see Fig. 1).

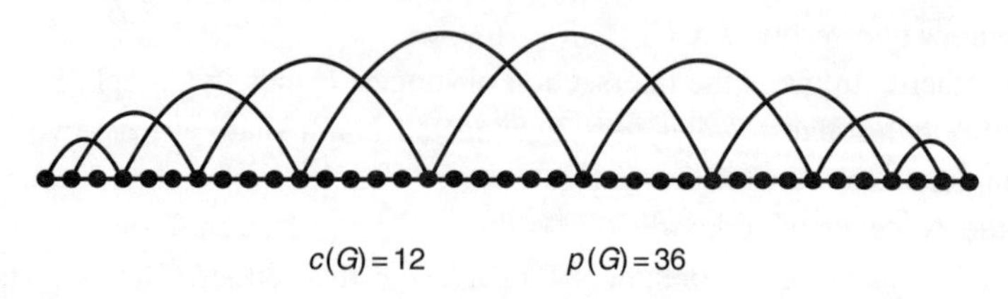

Fig. 1.

Theorem 3.1 $l_2(p) \geq 2\sqrt{p}$, *and this bound is sharp.*

Bondy and Locke [18] proved corresponding results for 3-connected graphs, and they also considered the special class of cubic graphs. We let $l_3^*(p)$ be the smallest integer p for which every 3-connected cubic graph with a path of length p contains a cycle of length at least $l_3^*(p)$.

Theorem 3.2 *With the above notation,*

$\frac{2}{5}p + 2 \leq l_3(p) \leq \frac{1}{2}p + O(p^\alpha)$, *for some* $\alpha < 1$, *and* $\frac{2}{3}p + 2 \leq l_3^*(p) \leq \frac{7}{8}p + 3$.

Graphs for which the lower bounds in both results are sharp involve the concept of a 'vine'. This is a series of overlapping paths attached to a fixed path, and an example is shown in Fig. 1. The upper bound for $l_3^*(p)$ is achieved by the graph F_5 in Fig. 2, which is derived from the Petersen graph. The general graph F_m (the case $m = 5$ is Fig. 2) has a Hamiltonian path of length $8m + 1$, but has no cycle longer than $7m + 2$.

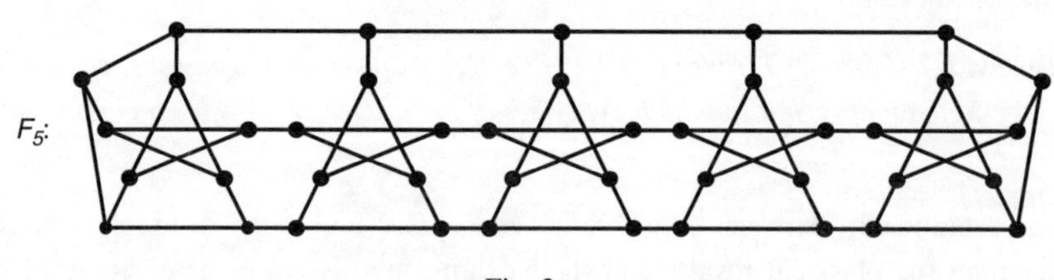

Fig. 2.

Bondy and Locke [18] gave a general upper bound for $l_k(p)$ and conjectured the nature of $l_k(p)$ as $k \to \infty$. The upper bound was later generalized by Locke [78], as follows.

Theorem 3.3 $l_k(p) \leq p(2k-4)/(3k-4)$.

Conjecture B *There exists a sequence of constants* $c_3, c_4, \ldots$ *converging to* 1, *for which* $l_k(p) \geq c_k p$ *for all* k *and* p.

In graphs with many edges, the lengths of the longest cycle and the longest path are essentially the same. We define a graph to be *cycle-tight* if the order of a longest path is at most 1 more than the order of a longest cycle. The following result of Liu, Lu and Tian [77] presents some of these graphs – here, $\sigma_4(G)$ is the minimum sum of the degrees of any four independent vertices.

Theorem 3.4 *Every 3-connected graph of order n with $\sigma_4(G) \geq \frac{1}{3}(4n+5)$ is cycle-tight, and this bound is sharp.*

Corollary 3.5 *Every graph G of order n with $\kappa(G) > \lceil \frac{1}{3}n \rceil$ is cycle-tight.*

4. Regular graphs

Regularity plays a strong role in forcing the circumference of a graph to be large. Jackson [58] gave a minimum-degree condition in a regular graph G of order n that guarantees Hamiltonicity (see part (a) of the following theorem) that is significantly less than the degree condition $\delta(G) \geq \frac{1}{2}n$ of Dirac [31]. The extensions (b) and (c) of this result are due to Zhu, Liu and Yu [114].

Theorem 4.1 *An r-regular 2-connected graph G of order n is Hamiltonian if any of the following conditions holds:*

(a) $n \leq 3r$;
(b) $n = 3r + 1$ *and G is not the Petersen graph;*
(c) $n = 3r + 2$ *or* $3r + 3$, *and* $r \geq 6$.

We turn now to results on the circumference $c(G)$ of a graph G. There are also stronger versions for regular graphs of Dirac's circumference result [31]. In the following theorem, the first part is due to Fan [39] and the second part is due to Aung [4].

Theorem 4.2 *Let G be an r-regular graph of order n.*

(a) *If G is 3-connected, then* $c(G) \geq \min\{3r, n\}$.
(b) *If G is 4-connected, then* $c(G) \geq \min\{4(r-1), \frac{1}{2}(n+3r-2)\}$.

For claw-free regular graphs, there is a still stronger result, as proved by Li [76].

Theorem 4.3 *If G is a 2-connected claw-free r-regular graph of order n, then* $c(G) \geq \min\{4r-2, n\}$.

For positive integers $r \geq 3$ and $s \leq r-3$, the graph $K_2+(s-1)K_{r+1}$ has a spanning subgraph that is 2-connected and r-regular, and has order $n = (s-1)(r+1)+2$ and circumference $2r+4$ (which equals $2(n+s-3)/(s-1)$). This example was the basis for a conjecture of Bondy [15] and led to the following result of Wei [108].

Theorem 4.4 *If G is a 2-connected r-regular graph of sufficiently large order $n \leq sr$ with $s \geq 3$, then $c(G) \geq 2(n+s-3)/(s-1)$.*

In general, for $r \geq 3$, the combination of being r-connected and r-regular is not enough to guarantee that a graph be Hamiltonian; Meredith [87] was the first to construct a family of such graphs. This family led Jackson and Parsons [60] to the following upper bound on the circumference.

Theorem 4.5 *For $r \geq 3$, there exists a number $\varepsilon(r)$ between 0 and 1 for which, if n is sufficiently large, there is an r-connected r-regular graph of order n and circumference less than $n^{\varepsilon(r)}$.*

In fact, there is a wide gap between the best-known upper bounds $n^{\varepsilon(r)}$ and the lower bounds (linear in r) for the circumference of such graphs. However, every r-connected r-regular graph has a 2–factor, as the following results show. In many cases these graphs are *2-factorable* – that is, their edge-sets can be partitioned into 2-factors. Petersen [93] investigated this class of graphs, and showed in particular that while not all 3-connected cubic graphs are Hamiltonian, all have a 2-factor.

Theorem 4.6 *Every cubic graph without a cut-edge is the edge-disjoint union of a 1-factor and a 2-factor.*

Rosenfeld [94] showed that the number of components in 2-factors of 3-connected cubic graphs is not bounded.

Theorem 4.7 *There are arbitrarily large 3-connected cubic graphs G in which every 2-factor has at least $\frac{1}{10}n$ components, where n is the order of G. For 3-connected cubic planar graphs, the number of components is at least $\frac{1}{28}n$.*

The results of Petersen [93] apply in general to k-connected k-regular graphs.

Theorem 4.8 *Let G be a k-connected k-regular graph with $k \geq 2$.*

(a) *If k is even, then G is 2-factorable.*
(b) *If k is odd, then G is the union of a 1-factor and $\frac{1}{2}(k-1)$ 2-factors.*

In a k-connected graph, any k vertices lie on a common cycle (by Theorem 1.1). It is natural to investigate whether adding a regularity condition increases the number of vertices that are always on a common cycle. With that in mind, we make the following definition. For integers k and r with $2 \leq k \leq r$, let $g(k,r)$ be the largest integer l for which every collection of l vertices in an r-regular k-connected graph lies on some cycle.

The case $r = k = 2$ is trivial, since every such graph is a cycle. Holton and Plummer [57] gave examples showing that $g(2,r) = 2$ for all r. The cases $r = 3$ and 4 are shown in Fig. 3, and these give a clear pattern for the general case. Note that the Petersen graph shows that $g(3,3) \leq 9$, and infinitely many examples can be obtained

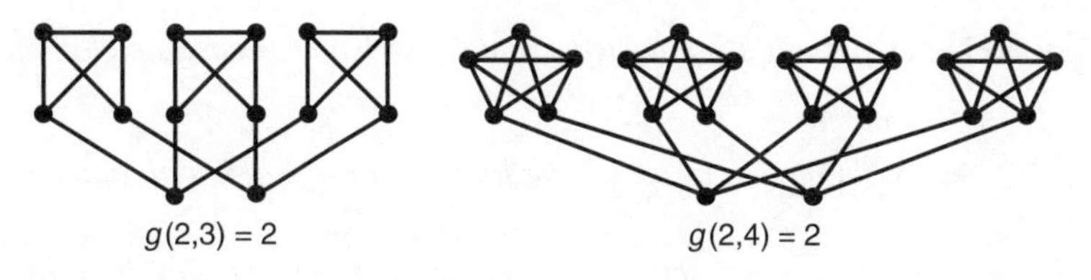

Fig. 3.

by inflating its vertices. Holton *et al.* [56] showed that equality holds here., Much is known about the cases when $r > k$, as a result of examples given by Holton [52].

Theorem 4.9

- $g(2, r) = 2$ *for all* $r \geq 2$.
- $g(3, 3) = 9$.
- $g(k, k+1) = k$ *if* k *is even.*
- $g(k, r) = k$ *if* $r \geq k+2$.

Ellingham, Holton and Little [35] proved an extension of the second part of Theorem 4.9, showing that each set of ten vertices in a 3-connected cubic graph lies on a cycle unless the graph can be contracted to the Petersen graph. They also showed that, for any set of five vertices and one edge in a 3-connected cubic graph, there is a cycle that contains the vertices and avoids the edge. Aldred [1] proved that in every 3-connected cubic graph each set of up to 13 vertices is contained in some path.

For $k \geq 4$, Theorem 4.9 leaves unsettled only the cases of $g(k, k)$ for all k, and $g(k, k+1)$ for k odd. In the latter case, Holton [52] proved that $g(k, k+1) \geq k+2$, but an upper bound was not given. Much attention has been given to the difficult case of $g(k, k)$. Holton [52] and Kelmans and Lomonosov [73] proved that $g(k, k) \geq k+4$ for $k \geq 3$, but this bound is not exact for $k = 3$, since $g(3, 3) = 9$. Meredith [87] proved that $g(k, k) \leq 10k - 11$. However, for k even, McCuaig and Rosenfeld [85] gave examples showing that $g(k, k) \leq 6k - 4$ for $k \equiv 0 \pmod 4$ and that $g(k, k) \leq 8k - 5$ for $k \equiv 2 \pmod 4$. The results for $r = k$ and $r = k+1$ when k is odd are summarized in the following theorem. There is a general cyclability survey by Bau and Holton [7].

Theorem 4.10 *For* $r \geq k \geq 3$,

- $k+4 \leq g(k, k) \leq 6k-4$ *if* $k \equiv 0 \pmod 4$;
- $k+4 \leq g(k, k) \leq 8k-5$ *if* $k \equiv 2 \pmod 4$;
- $k+4 \leq g(k, k) \leq 10k-11$ *if* $k \geq 5$ *is odd*;
- $k+2 \leq g(k, k+1)$ *if* k *is odd.*

Saito [96] addressed the issue of the lengths of cycles containing specified sets of vertices, which was not considered in determining $g(k, r)$.

Theorem 4.11 *Let G be a k-connected graph of order at least $2k$ and circumference c.*

- *For $m < k$, every set of m vertices is contained in a cycle of length at least $2m + c(k - m)/k$.*
- *If G is k-regular, then every set of k vertices is contained in a cycle of length at least $c/3k + \frac{2}{3}(k + 2)$.*
- *If G is planar and 3-connected, then every set of three vertices is contained in a cycle of length at least $\frac{1}{4}c + 3$.*

5. Bipartite graphs

Frequently the same minimum degree and connectivity guarantee longer cycles in bipartite graphs than in graphs in general, and in some cases the guaranteed length is doubled. The following result of Bauer *et al.* [10] illustrates this. A cycle C in a graph is *dominating* if each edge of the graph has at least one of its endpoints on C.

Theorem 5.1 *If G is a 2-connected triangle-free graph of order n, circumference c and minimum degree δ, then either $c \geq \min\{n, 4\delta\}$ or every longest cycle is dominating.*

The following result on bipartite graphs was proved by Jackson [59].

Theorem 5.2 *Let G be a 2-connected $r \times s$ bipartite graph with $r \leq s$, let k and l be the minimum degrees in the partite sets of orders r and s, respectively, and let c be the circumference of G. Then*

- $c \geq 2\min\{s, k + l - 1, 2k - 2\}$;
- $c \geq 2\min\{r, 2k - 2\}$ *if* $l = k$;
- $c \geq 2\min\{r, 2k - 1\}$ *if* $r = s$ *and* $l = k$.

Note that the last part of Theorem 5.2 implies that a 2-connected balanced bipartite graph with minimum degree k in each part is either Hamiltonian or has a cycle of length at least $4k - 2$. This is an example of the cycle length in a bipartite graph being approximately twice the length in the class of all graphs.

More generally, Wang [105] considered an Ore-type condition (on the minimum sum of degrees of non-adjacent vertices) in bipartite graphs. This resulted in the following corollary.

Theorem 5.3 *If G is a 2-connected $r \times s$ bipartite graph with minimum degree δ, then its circumference is at least* $2\min\{r, s, 2\delta - 1\}$.

Further, Wang [106] considered disjoint cycles in balanced bipartite graphs and established a minimum degree condition for a balanced bipartite graph to have a 2-factor. He also showed that the minimum-degree condition is sharp.

Theorem 5.4 *Every 2-connected $r \times r$ bipartite graph with $r \geq 2k+1$ and minimum degree $\delta > s \geq k \geq 2$ has k disjoint cycles of total length at least* $\min\{2r, 4s\}$.

6. Claw-free graphs

In this section we consider the impact of connectivity on the circumference of claw-free graphs – that is, graphs without $K_{1,3}$ as an induced subgraph. Additional information on these graphs appears in a survey by Faudree, Flandrin and Ryjáček [41]. Much of the research on connectivity and claw-free graphs was motivated by the following conjecture by Matthews and Sumner [84].

Conjecture C *Every 4-connected claw-free graph is Hamiltonian.*

Related to this conjecture is a classical conjecture of Chvátal [26] on toughness, where the *toughness* $\tau(G)$ is the minimum ratio of the order of a cutset and the number of components left after the deletion of the cutset.

Conjecture D *There is a t_0 such that every t_0-tough graph is Hamiltonian.*

In general, $\tau(G) \leq \frac{1}{2}k(G)$, since the deletion of a set of cutvertices leaves at least two components. However, in claw-free graphs the deletion of any minimal cut results in precisely two components, so $\tau(G) = \frac{1}{2}k(G)$ for claw-free graphs. Thus, for claw-free graphs, Conjecture **D** when $t_0 = 2$ is equivalent to Conjecture **C**. Results of Enomoto *et al.* [36] showed that $t_0 \geq 2$ would be needed to imply Hamiltonicity, and more recently Bauer, Broersma and Veldman [9] exhibited an example of non-Hamiltonian graphs with $\tau(G) = (9/4 - \varepsilon)$ for any $\varepsilon > 0$. Bauer, Broersma and Schmeichel [8] have a survey on toughness.

A special subclass of claw-free graphs is the class of line graphs, and Thomassen [102] conjectured the following:

Conjecture E *Every 4-connected line graph is Hamiltonian.*

Since line graphs are claw-free, Conjecture C implies Conjecture E. However, we now know that the two are equivalent. This is a consequence of a closure concept for claw-free graphs introduced by Ryjáček [95], an operation that is similar in form and application to the closure operation introduced by Bondy and Chvátal [17].

In a claw-free graph, the neighbourhood $N(v)$ of a vertex has independence number at most 2. Thus, the induced graph $\langle N(v) \rangle$ is either connected or the union of two complete components. If it is connected, then the 'local closure' at v is the graph obtained by replacing $N(v)$ by a complete graph with the same vertex-set. Doing this recursively until every vertex has a neighbourhood that is either complete or the union of two complete graphs yields a graph $\mathrm{cl}(G)$ called the *Ryjáček closure*. Fig. 4 shows a graph G for which $\mathrm{cl}(G)$ is the complete graph; it is the result of taking two local closures. Clearly, the graph $\mathrm{cl}(G)$ is always claw-free and, as Ryjáček [95] showed, the operation is well defined. As before,

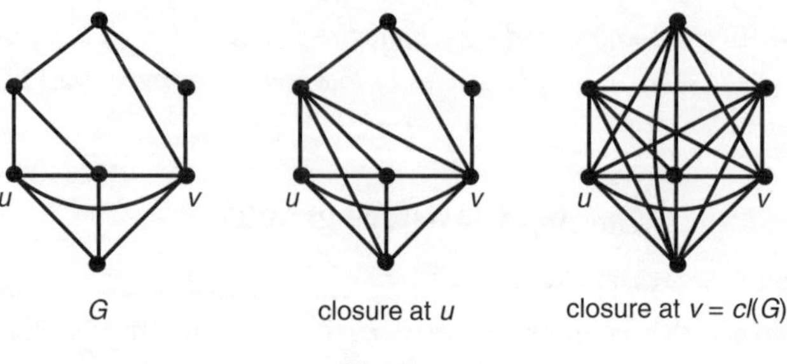

Fig. 4.

$c(G)$ denotes the circumference of the graph G and $p(G)$ is the order of a longest path (see [19]).

Theorem 6.1 *Let G be a claw-free graph. Then*

- *there is a triangle-free graph whose line graph is* $\mathrm{cl}(G)$;
- $c(G) = c(\mathrm{cl}(G))$ *and* $p(G) = p(\mathrm{cl}(G))$.

The Ryjáček closure is a useful tool in the study of cycles in claw-free graphs. For example, the determination of the circumference of a claw-free graph can be reduced to considering an appropriate line graph. Specifically, this implies that Conjectures C and E are equivalent.

Cycles in line graphs have been studied extensively, with one of the useful tools being a result of Harary and Nash-Williams [50] which gives a necessary and sufficient condition for the Hamiltonicity of the line graph $L(G)$ in terms of a dominating Eulerian subgraph of the graph G. An Eulerian subgraph H in a graph G is *dominating* if every edge of the graph has at least one of its endpoints in H.

Theorem 6.2 *The line graph $L(G)$ of a graph G with at least three edges is Hamiltonian if and only if G has a dominating Eulerian subgraph.*

Using this theorem, Zhan [112] proved the following result.

Theorem 6.3 *Every 7-connected line graph is Hamiltonian.*

An immediate consequence of Theorem 6.3 and the Ryjáček closure is that every 7-connected claw-free graph is Hamiltonian. Thus, the gap is between 4 and 7 for the connectivity sufficient to guarantee Hamiltonicity.

The latest result on connectivity and claw-free graphs is by Kaiser and Vrána [66], who show that 5-connectedness with a minimum-degree condition is sufficient for being Hamiltonian.

Theorem 6.4 *Every 5-connected claw-free graph with minimum degree at least 6 is Hamiltonian (and in fact Hamiltonian-connected).*

Generalizing Conjecture C, Jackson and Wormald [62] asked whether, for $r \geq 4$, every r-connected $K_{1,r}$-free graph is Hamiltonian. It is not even known whether some assumption of sufficiently large connectivity is enough.

If the Hamiltonian condition is relaxed to the existence of a 2-factor, then an assumption of being 4-connected and claw-free was shown by Choudum and Paulraj [25] to be sufficient.

Theorem 6.5 *Every 4-connected claw-free graph has a 2-factor.*

Yoshimoto [111] proved the following result on the number of components in a 2-factor in a claw-free graph.

Theorem 6.6 *Let G be a claw-free graph of order n and minimum degree 4. If each edge of G lies on a triangle, then G has a 2-factor with no more than $\frac{1}{4}(n-1)$ components, and this bound is sharp.*

We note that if Conjecture C is true then, for 4-connected graphs, there is a 2-factor with just one component. The question still remains as to the connectivity needed to ensure that any k vertices of a claw-free graph can be separated by k disjoint cycles. Connectivity of at least $2k$ is needed, since the k vertices could all be in the closed neighbourhood of one of the vertices.

There are infinite families of non-Hamiltonian 3-connected claw-free graphs. One example, due to Matthews and Sumner [84], is the graph in Fig. 5; it is the result of subdividing the edges of a perfect matching in the Petersen graph and then taking the line graph.

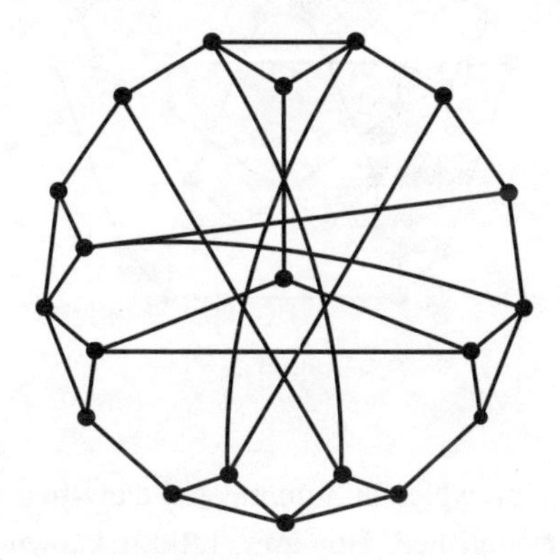

Fig. 5.

Jackson and Wormald [62] found a lower bound on the circumference of 3-connected $K_{1,r}$-free graphs in general, and claw-free graphs in particular.

Theorem 6.7 *Every 3-connected $K_{1,r}$-free graph of order n has a cycle of length at least n^{ε} where $\varepsilon = (\log_2 6 + 2\log_2(2r-1))^{-1}$.*

By taking the inflations of appropriate 3-regular graphs (replacing vertices by triangles), we deduce that this bound is of the correct order of magnitude. Whether there is a better value of the constant ε is not known.

Although not every 3-connected claw-free graph is Hamiltonian, Jackson and Yoshimoto [64] showed when the minimum degree is greater than 3, such a graph does have a 2-factor.

Theorem 6.8 *Every 3-connected claw-free graph G with $\delta(G) \geq 4$ has a 2-factor with at most $\frac{2}{15}n$ components.*

Inflations of certain cubic graphs, such as the Petersen graph, are examples of graphs with restrictions on the number of vertices that can be together on a cycle. This topic was investigated by Győri and Plummer [48], who proved the following theorem.

Theorem 6.9 *If G is a 3-connected claw-free graph, then every set of up to nine vertices lies on a cycle in G.*

This result is sharp, as the graph in Fig. 6 shows. This is an inflation of the Petersen graph, and since the Petersen graph is not Hamiltonian, there is no cycle that contains any set of ten vertices with one from each triangle.

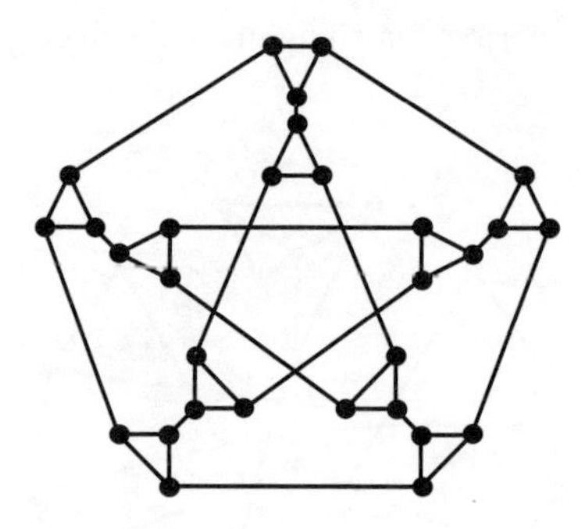

Fig. 6.

The existence of disjoint cycles in 3-connected claw-free graphs, and in particular 2-factors, has been established. However, little is known about the existence of disjoint cycles that separate specified vertices. For example, in the above graph there are pairs of vertices that are not separated by two cycles. Thus, no set of two or more vertices can always be separated by disjoint cycles in 3-connected claw-free graphs, unless some conditions are placed on the vertices.

The circumference of 2-connected claw-free graphs was investigated by Broersma *et al.* [21]. In proving their results, they used the relationship between toughness and connectivity.

Theorem 6.10 *If G is a 2-connected claw-free graph of order n and circumference c, then there is a constant C for which*

$$c(G) \geq \frac{4 \log n}{\log 2} - C.$$

Furthermore, there exists a 2-connected claw-free graph of order n with circumference less than $8 \log(n+6) - 8 \log 3 - 2$.

This theorem is a special case of their more general result on $K_{1,r}$-free graphs.

Theorem 6.11 *If G is a 2-connected $K_{1,r}$-free graph of order n with circumference c, then there is a constant C_r such that*

$$c(G) \geq \frac{4 \log n}{\log(r-1)} - C_r.$$

They also gave a construction that shows that, for given $r \geq 4$ and sufficiently large n, there exists a 2-connected $K_{1,r}$-free graph of order n whose circumference is less than $4 \log n/(\log(r-2)+4)$. Hence, the order of magnitude of the lower bound in Theorem 6.11 is correct.

Jackson and Yoshimoto [63] proved that every 2-connected claw-free graph with a sufficiently high minimum degree has a 2-factor, even though not all such graphs are Hamiltonian.

Theorem 6.12 *Every 2-connected claw-free graph G with $\delta(G) \geq 4$ has a 2-factor with at most $\frac{1}{4}(n+1)$ components.*

We conclude this section with some results relating cycles and connectivity in certain families of graphs.

A graph is *locally connected* if the neighbourhood of each vertex is connected. Local connectivity in claw-free graphs implies the existence of many cycles, a fact first observed by Oberly and Sumner [92].

Theorem 6.13 *Every connected locally connected claw-free graph with at least three vertices is Hamiltonian.*

A graph is *cycle-extendible* if each cycle can be extended to a cycle with one more vertex, and is *fully cycle-extendible* if each vertex is also on a triangle. The Oberly–Sumner result was extended by Hendry [51].

Theorem 6.14 *Every connected locally connected claw-free graph with at least three vertices is fully cycle-extendible.*

A graph G of order n is *panconnected* if, between each pair of vertices v and w of G and for each l satisfying $d(v,w) \leq l \leq n-1$, there is a v–w path of length l. Stronger local connectivity conditions were shown to imply panconnectedness by Kanetkar and Rao [69].

Theorem 6.15 *Every connected locally 2-connected claw-free graph is panconnected.*

Chartrand, Gould and Polimeni [22] gave a local connectivity condition that implies the existence of many long cycles.

Theorem 6.16 *If G is a connected locally k-connected claw-free graph, then the removal of any set of fewer than k vertices leaves a Hamiltonian graph.*

An immediate consequence of this theorem is that every connected locally $(3r-1)$-connected claw-free graph has a 2-factor with precisely r components. However, there are many questions left unanswered on what connectivity and local connectivity is required to guarantee the existence of cycles of prescribed lengths, 2-factors and disjoint cycles that separate specified vertices.

7. Planar graphs

Connectivity has an especially significant impact on the orders of cycles in planar graphs. One of the earliest results illustrating this was a theorem of Tutte [104].

Theorem 7.1 *Every 4-connected planar graph is Hamiltonian.*

This has led to considerable investigations into more general results. In our discussion, it is convenient to have some additional definitions. A graph G of order n is *pancyclic* if it has cycles of all lengths from 3 to n, and is 4-*almost pancyclic* if it has cycles of all these lengths except 4. For $r \leq n$, G is r-*ordered Hamiltonian* if, for any r vertices, there is a Hamiltonian cycle with those vertices in the given order.

Before moving on to the broader 3-connected case, we consider 4-connected graphs. Bondy [14] conjectured that every such planar graph is either pancyclic or nearly so.

Conjecture F *Every 4-connected planar graph contains cycles of every length, except possibly for one even length.*

A related conjecture was made by Malkevitch [83], who exhibited a 4-connected graph with cycles of all lengths, except 4 (see Fig. 7).

Conjecture G *Every 4-connected planar graph with a 4-cycle is pancyclic.*

In support of these conjectures, it has been shown that every 4-connected planar graph G of large order n has a cycle of each length from $n-7$ to $n-1$. For length $n-1$, this follows from a result of Tutte. A series of authors extended the result, and a summary can be found in [30] by Cui, Hu and Wang.

In connection with Malkevitch's conjecture, Trenkler [103] found those values of n for which there is a planar graph of order n that is 4-connected and

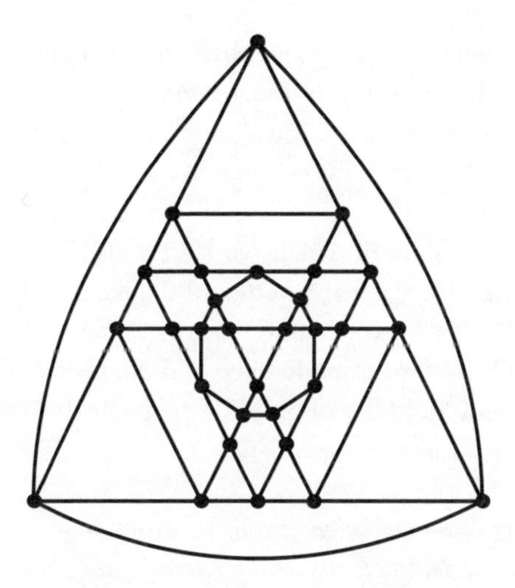

Fig. 7.

4-almost pancyclic. It follows from his result that there is such a graph of each order $n \geq 48$.

Theorem 7.2 *There exists a* 4*-connected* 4*-almost pancyclic planar graph of order* n *if and only if one of the following holds*:

- *for* $n \equiv 0$ *(mod 3)*, $n = 30$ *or* $n \geq 36$;
- *for* $n \equiv 1$ *(mod 3)*, $n \geq 46$;
- *for* $n \equiv 2$ *(mod 3)*, $n = 44$ *or* $n \geq 50$.

Sanders [98] showed that each pair of edges is on a Hamiltonian cycle, a consequence of the following theorem.

Theorem 7.3 *For any pair of vertices* v *and* w *and any edge* $e \neq vw$ *in a* 4*-connected planar graph, there is a Hamiltonian* v*–*w *path containing* e.

This is a generalization of the result of Thomassen [101] that in a 4-connected planar graph, each pair of vertices is joined by a Hamiltonian path. Goddard [44] proved that more can be said when the graph triangulates the plane.

Theorem 7.4 *Every* 4*-connected maximal planar graph is* 4*-ordered Hamiltonian.*

The 'double pyramid' graph $H_n = \overline{K}_2 + C_{n-2}$ is 4-connected and planar but not 5-ordered, showing that Theorem 7.4 cannot be extended. Likewise, since H_n has some path systems of three edges that do not lie on a Hamiltonian cycle, Sanders' result cannot be extended. We also note that every 4-connected planar graph of order at least 6 has a pair of disjoint cycles, since it has a cycle disjoint from any given triangle. The graph H_n also shows that there are 4-connected planar graphs of arbitrarily large order that do not have three disjoint cycles.

We now turn to 3-connected planar graphs. Although not all of them are Hamiltonian, Holton and McKay [55] showed that every 3-connected cubic graph of order 36 or less is Hamiltonian, and found those of order 38 that are not. In fact, there is no positive constant C for which every 3-connected planar graph G of order n has circumference $c(G) \geq Cn$.

Moon and Moser [89] showed that if we start with a 3-connected planar graph (such as K_4) and successively insert a vertex of degree 3 inside each face, we get a 3-connected planar graph whose circumference is at most $n^{\log_3 2}$. They conjectured this order of magnitude for the circumference of 3-connected planar graphs, and this was proved by Chen and Yu [24], not only for the plane, but also for the other three surfaces of non-negative Euler characteristic.

Theorem 7.5 *If G is a 3-connected graph of order n that is embeddable in the sphere, the projective plane, the torus or the Klein bottle, then $c(G) = \Omega(n^{\log_3 2})$.*

More is known about cycles of 3-connected cubic planar graphs. Aldred *et al.* [2] verified the existence of cycles containing arbitrary sets with up to 23 vertices.

Theorem 7.6 *In every 3-connected cubic planar graph, each set of up to 23 vertices is contained in some cycle.*

The bound in Theorem 7.6 is sharp: Holton [53] exhibited a 3-connected cubic planar graph with a set of 24 vertices that is not contained in any cycle. With a bipartite restriction added to the class of 3-connected cubic planar graphs, Barnette [6] made the following conjecture, which is still open.

Conjecture H *Every 3-connected cubic bipartite planar graph is Hamiltonian.*

This conjecture was shown to be true by Holton, Manvel and McKay [54] for graphs with up to 64 vertices, and has also been verified for some infinite classes of graphs. For example, Goodey ([45]) showed that every 3-connected bipartite graph which can be embedded in the plane with every face either a quadrilateral or a hexagon is Hamiltonian.

The complete bipartite graph $K_{2,n}$ is planar and 2-connected, and all cycles have length 4, so 2-connectedness does not imply the existence of long cycles in planar graphs. If other parameters are considered, such as the toughness, then more can be said about the circumference. Note that the toughness of $K_{2,n}$ is $2/n$ and this approaches 0 as $n \to \infty$. If a bound is placed on the toughness, then more can be said about the length of longest cycles. The next result of Böhme, Broersma and Veldman [11] gives a lower bound for the circumference of 2-connected planar graphs in terms of the toughness.

Theorem 7.7 *If G is a 2-connected planar graph of toughness m, then there is a constant d (depending on m) for which $c(G) \geq d \log n$.*

8. The Chvátal–Erdős condition

A classical paper of Chvátal and Erdős [27] on Hamiltonian properties explored implications of the relationship between the connectivity $\kappa(G)$ and the independence number $\alpha(G)$ of a graph G. Because of their results, we say that G satisfies the *Chvátal–Erdős condition* if $\kappa(G) \geq \alpha(G)$.

Theorem 8.1 *Let G be a graph of order $n \geq 3$.*

- *If $\kappa(G) \geq \alpha(G) - 1$, then G has a Hamiltonian path.*
- *If $\kappa(G) \geq \alpha(G)$, then G has a Hamiltonian cycle.*
- *If $\kappa(G) \geq \alpha(G) + 1$, then G is Hamiltonian-connected.*

This theorem has given rise to many generalizations. For example, using classical results from Ramsey theory, Flandrin *et al.* [43] proved the following variation of the second part.

Theorem 8.2 *Every graph of sufficiently large order (depending on the independence number) that satisfies the Chvátal–Erdős condition is pancyclic.*

Häggkvist and Thomassen [49] showed that an appropriate Chvátal–Erdős condition implies the existence of a Hamiltonian cycle containing predetermined disjoint paths.

Theorem 8.3 *If G satisfies the Chvátal–Erdős condition, then each set of disjoint paths of length at most $\kappa(G) - \alpha(G)$ is contained in a Hamiltonian cycle.*

Wei and Zhu [109] showed that triangle-free graphs satisfying the Chvatál–Erdős condition also satisfy a strong version of panconnectedness.

Theorem 8.4 *Let G be a triangle-free graph of order n other than C_5 or $K_{n/2,n/2}$ that satisfies the Chvátal–Erdős condition. Then,*

- *if $4 \leq i \leq n$, each edge is in a cycle of length i;*
- *if $4 \leq i \leq n - 1$, each pair of vertices is connected by a path of length i.*

Many of the degree conditions that guarantee that a graph is Hamiltonian also guarantee the existence of a 2-factor with a specified number of cycles. It seems natural to conjecture that the Chvátal–Erdős condition also guarantees this. The case of two cycles was proved by Kaneko and Yoshimoto [68].

Theorem 8.5 *Every 4-connected graph G other than K_5 that satisfies the Chvátal–Erdős condition has a 2-factor with two cycles.*

9. Ordered graphs

In this section, we consider two other concepts and examine how they are related to each other and to connectivity.

A graph G is *k-ordered* if every ordered set of k vertices lies on a cycle in the designated order. This concept, introduced by Ng and Schultz [91], is considerably stronger than cyclability. (In Section 7 we considered the particular case in which the cycle was specified to be Hamiltonian.) Recall also that a graph G is *k-linked* if, given any collection of k pairs of vertices $(x_1, y_1), (x_2, y_2), \ldots, (x_k, y_k)$, there are k internally disjoint paths P_i for which P_i is an x_i–y_i path. The following theorem gives some elementary relationships between these two concepts.

Theorem 9.1 *Let $k \geq 1$. Then,*

(a) *every k-linked graph is k-ordered;*
(b) *every $2k$-ordered graph is k-linked.*

Proof (a) Let G be a k-linked graph, and let $X = (x_1, x_2, \ldots, x_k)$ be an ordered set of k of its vertices. Let $Y = (y_1, y_2, \ldots, y_k)$ with $y_i = x_{i+1}$ for $i \leq k-1$ and $y_k = x_1$. Then from the definition of k-linked, there exist internally disjoint x_i–y_i paths, and their union is a cycle containing the vertices of X in the given order.

(b) This follows from the fact that a cycle containing the vertices of the ordered set $(x_1, y_1, x_2, y_2, \ldots, x_k, y_k)$ of $2k$ vertices contains the required k paths for a k-linkage. ■

It has been known for some time that sufficiently high connectivity implies k-linkage, due to results of Jung [65] and Larman and Mani [75]. The much sharper bound that every $22k$-connected graph is k-linked was proved by Bollobás and Thomason [13], and this was improved substantially by Thomas and Wollan [100].

Theorem 9.2 *Every $10k$-connected graph is k-linked.*

It is likely that the connectivity needed to imply k-linkage is significantly less than $10k$. On the other hand, it needs to be at least $3k-1$, as is shown by the graph obtained from K_{3k-1} by deleting k independent edges, which has connectivity $3k-2$ but is not k-linked. In [100] the stronger result was proved that every $2k$-connected graph of order n and at least $10kn$ edges is k-linked. The following was also conjectured, and it was noted that the conjecture is true for $k \leq 3$.

Conjecture I *Every $2k$-connected graph of order n which has at least $(2k-1)n - \frac{1}{2}(3k+1)k + 1$ edges is k-linked.*

Corresponding questions can also be posed for k-ordered graphs. The graph $H_2 = K_{2k-1} - C_k$ has connectivity $2k-4$, but is not k-ordered, because if the k vertices on the 'missing' cycle are chosen in the natural order, then there is no cycle meeting our requirement. Thus, for $k \geq 5$, at least $(2k-3)$-connectedness is required to imply k-ordered.

Jung [65] found a bound on the connectivity that guarantees that a graph is 2-linked.

Theorem 9.3 *Every* 6*-connected graph is* 2*-linked.*

This result is sharp. The graph H_3 in Fig. 8 is a 5-regular 5-connected planar graph that is not 2-linked, since there is no linkage for the pairs (x_1, y_1) and (x_2, y_2). Also, it is not 4-ordered, since there is no cycle containing the ordered set (x_1, y_1, x_3, y_2).

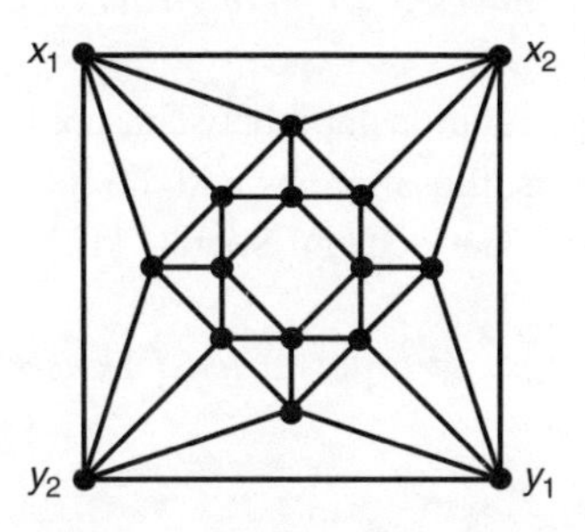

Fig. 8.

The least connectivity that implies 4-orderedness is not known, but the following is an interesting question: *Is every* 6*-connected graph* 4*-ordered?*

No graph G with connectivity $2k - 2$ can be k-linked, since if the collection of k pairs contains a minimum cutset S and the two vertices of some pair are in different components of $G - S$, then there cannot exist the path system required for k-linkage. Thus, every k-linked graph is $(2k - 1)$-connected, but may not be $2k$-connected, as K_{2k} shows. Likewise, every k-ordered graph is $(k - 1)$-connected, and Mészáros [88] exhibited, for k odd, an infinite family of $(k - 1)$-regular graphs that are k-ordered Hamiltonian. Recall that the graph H_2 is $(k - 1)$-linked, but not k-ordered. Also, for $n \geq 2k$ and k even, the complete bipartite graph $K_{k,n-k}$ is k-ordered but is not $(\frac{1}{2}k + 1)$-linked. We summarize the known results about the relationship between linkage, ordered and connectivity in the following theorem.

Theorem 9.4 *Let* $k \geq 3$*. Then,*

- *being k-linked implies being $(2k - 1)$-connected, but does not imply being $2k$-connected;*
- *being k-ordered implies being $(k - 1)$-connected, but does not imply being k-connected;*
- *being k-linked implies being k-ordered, but does not imply being $(k + 1)$-ordered;*
- *being $2k$-ordered implies being k-linked, but does not imply being $(k + 1)$-linked;*
- *being $10k$-connected implies being k-linked, but being $(3k - 3)$-connected does not;*
- *being $10k$-connected implies being k-ordered, but being $(2k - 4)$-connected does not.*

It would be of interest to know the sharpest relationships between the pairs of these parameters. With girth conditions placed on the graph, some sharp results between connectivity and linkage have been established. Improving a result of Mader [82], Kawarabayshi [72] showed the following:

Theorem 9.5 *Let $k \geq 1$. Then,*

- *for $k = 4$ or 5, every $2k$-connected graph of girth at least 19 is k-linked;*
- *for $k \neq 4$ or 5, every $2k$-connected graph of girth at least 11 is k-linked.*

More is known about the relationships between connectivity and linkage in the case of chordal graphs (that is, those graphs with no induced cycle of length greater than 3). In particular, Böhme, Gerlach and Stiebitz [12] proved the following result.

Theorem 9.6 *A chordal graph of order at least $2k$ is k-linked if and only if it is $(2k-1)$-connected.*

Connectivity conditions for $K_{1,r}$-free graphs – in particular, claw-free graphs – were established by Faudree *et al.* [42].

Theorem 9.7 *For $r \geq 3$, every $((2r-2)(k-1)+1)$-connected $K_{1,r}$-free graph is k-linked.*

Corollary 9.8 *Every $(4k-3)$-connected claw-free graph is k-linked.*

10. Numbers of cycles

To conclude this chapter we look at how the number of different cycle lengths and the total number of cycles vary with the connectivity.

Theorem 10.1 *For $k \geq 1$, the minimum number of different lengths of cycles in any k-connected graph is $k-1$.*

Proof We first observe that the cycles in every k-connected graph have at least $k-1$ different lengths. This follows from the fact that if P is a longest path in a k-connected graph G, then the chords of P from the first vertex form cycles of $k-1$ different lengths. That this bound is sharp follows from the fact that the complete bipartite graph $K_{k,k}$ is k-connected and has only cycles of the $k-1$ lengths $4, 6, \ldots, 2k$. ■

If additional requirements are made, then there may be cycles of other lengths. An example of this is the following result of Erdős *et al.* [37].

Theorem 10.2 *Let $\lambda_g(k)$ be the minimum number of different lengths of cycles in any graph G of girth g and minimum degree $\delta(G) \geq k$. Then*

- $\lambda_5(k) \geq \frac{1}{4}(k^2 - k - 2)$;
- *there exists a positive constant C_7 for which $\lambda_7(k) \geq C_7 k^{5/2}$;*
- *there exists a positive constant C_9 for which $\lambda_9(k) \geq C_9 k^3$;*
- *for $t \geq 3$, there exists a positive constant C_{4t-1} for which $\lambda_{4t-1}(k) \geq C_{4t-1} k^{t/2}$.*

In fact, using only the average degree, Sudakov and Verstraëte [99] strengthened this result with information about the distribution of the cycle lengths.

Theorem 10.3 *There is a constant C for which every k-connected graph of girth g has a set of at least $Ck^{(g-1)/2}$ consecutive even cycle lengths.*

We now turn to the total number of cycles, rather than the number of lengths of cycles. Let $\psi_n(k)$ be the minimum number of different cycles in any k-connected graph of order n. By considering trees, cycles, wheels and regular complete bipartite graphs, we see that $\psi_n(1) = 0$, $\psi_n(2) = 1$, and $\psi_n(3) \leq c_3 n^2$. Given a function $f(n)$ and a family of graphs $\mathcal{F}$, we say that $\psi_n(k)$ is *$f(n)$-bounded* on $\mathcal{F}$ if there exist constants A and B for which $\psi_n(k)$ lies between $Af(n)$ and $Bf(n)$, for all graphs of order n in $\mathcal{F}$. Knor [74] proved the following results.

Theorem 10.4 *Let $\psi_n(k)$ be the minimum number of cycles in any k-connected graph of order n. Then*

- *$\psi_n(2) = 1$, and is n^2-bounded on the family of graphs of minimum degree at least 3;*
- *$\psi_n(3)$ is n^2-bounded, and is n^3-bounded on the family of graphs of minimum degree at least 5.*

Knor also conjectured that $\psi_n(3)$ is n^k-bounded on the family of graphs with connectivity k, minimum degree δ and maximum degree Δ for which $k < \delta < \Delta$.

Clark and Entringer [28] determined the minimum number of cycles in graphs with small connectivity in terms of the cycle rank of the graph.

Theorem 10.5 *Let $\rho_k(r)$ be the minimum number of cycles in any k-connected graph with cycle rank r. Then*

- *for $r \geq 0$, $\rho_1(r) = r$;*
- *for $r \geq 1$, $\rho_2(r) = \frac{1}{2}r(r+1)$;*
- *for $r \geq 3$, $\rho_3(r) = r^2 - r + 1$.*

For cubic graphs, more is known. For example, Barefoot, Clark and Entringer [5] discovered the following results for the minimum number of cycles in cubic graphs of low connectivity.

Theorem 10.6 *Let $\nu_k(n)$ be the minimum number of cycles in any k-connected cubic graph of order n. Then*

- *for $n \geq 14$, $\nu_1(n) = 3\lfloor \frac{1}{4}n \rfloor + 8$;*
- *for $n \geq 8$, $\nu_2(n) = \lceil \frac{1}{8}n(n+14) \rceil$.*

They also conjectured that, for greater connectivity, the number $\nu_k(n)$ is superpolynomial in n, which follows from the next result of Aldred and Thomassen [3].

Theorem 10.7 *The minimum number $\nu_3(n)$ of cycles in a 3-connected cubic graph of order n satisfies*

$$2^{n^{0.17}} < \nu_3(n) < 2^{n^{0.95}}.$$

This implies that the minimum number of different cycle lengths in 3-connected cubic graphs is not bounded, as it is for 3-connected graphs in general, but that it is a function of the order of the graph.

References

1. R. E. L. Aldred, Paths through m vertices in 3-connected cubic graphs, *Ars Combin.* **17** (1984), 85–92.
2. R. E. L. Aldred, S. Bau, D. A. Holton and B. McKay, Cycles through 23 vertices in 3-connected cubic planar graphs, *Graphs Combin.* **15** (1999), 373–376.
3. R. E. L. Aldred and C. Thomassen, On the number of cycles in 3-connected cubic graphs, *J. Combin. Theory (B)* **71** (1997), 79–84.
4. M. Aung, Circumference of a regular graph, *J. Graph Theory* **13** (1989), 149–155.
5. C. A. Barefoot, L. Clark and R. Entringer, Cubic graphs with the minimum number of cycles, *Congr. Numer.* **53** (1986), 49–62.
6. D. W. Barnette, Conjecture 5, *Recent Progress in Combinatorics* (ed. W. T. Tutte), Academic Press (1969), 343.
7. S. Bau and D. H. Holton, Cycles in regular graphs, *Ars Combin.* **29** (1990), 175–183.
8. D. Bauer, H. Broersma and E. Schmeichel, Toughness in graphs – a survey, *Graphs Combin.* **22** (2006), 1–35.
9. D. Bauer, H. Broersma and H. J. Veldman, Not every 2-tough graph is hamiltonian, *Discrete Appl. Math.* **99** (2000), 317–321.
10. D. Bauer, N. Kahl, L. McGuire and E. Schmeichel, Long cycles in 2-connected triangle-free graphs, *Ars Combin.* **86** (2008), 295–304.
11. T. Böhme, H. J. Broersma and H. Veldman, Toughness and longest cycles in 2-connected graphs, *J. Graph Theory* **239** (1996), 257–263.
12. T. Böhme, T. Gerlach and M. Stiebitz, Ordered and linked chordal graphs, *Discuss. Math. Graph Theory* **28** (2008), 367–373.
13. B. Bollobás and A. Thomason, Highly linked graphs, *Combinatorica* **16** (1996), 313–320.
14. J. A. Bondy, Pancyclic graphs: recent results, *Infinite and Finite Sets, Colloq. Math. Soc. János Bolyai* **10** (1975), 181–187.
15. J. A. Bondy, Hamiltonian cycles in graphs and digraphs, *Congr. Numer.* **21** (1978), 3–28.
16. J. A. Bondy, Basic graph theory: paths and cycles, *Handbook of Combinatorics* (eds. R. Graham, M. Grötschel and L. Lovász), MIT Press and North Holland (1995), 3–110.

17. J. A. Bondy and V. Chvátal, A method in graph theory, *Discrete Math.* **15** (1976), 111–135.
18. J. A. Bondy and S. C. Locke, Relative lengths of paths and cycles in 3-connected graphs, *Discrete Math.* **33** (1981), 111–122.
19. S. Brandt, O. Favaron and Z. Ryjáček, Closure and stable properties in claw-free graphs, *J. Graph Theory* **34** (2000), 30–41.
20. H. J. Broersma, Z. Ryjáček and I. Schiermeyer, Closure concepts: a survey, *Graphs Combin.* **16** (2000), 17–48.
21. H. J. Broersma, J. Van den Heuvel, H. A. Jung and H. Veldman, Long paths and cycles in tough graphs, *Graphs Combin.* **9** (1993), 3–17.
22. G. Chartrand, R. J. Gould and A. D. Polimeni, A note on locally connected and hamiltonian-connected graphs, *Israel J. Math.* **33** (1979), 5–8.
23. G. Chen, R. Faudree and R. Gould, Intersections of longest cycles in k-connected graphs, *J. Combin. Theory (B)* **72** (1998), 143–149.
24. G. Chen and X. Yu, Long cycles in 3-connected graphs, *J. Combin. Theory (B)* **86** (2002), 80–99.
25. S. A. Choudum and M. S. Paulraj, Regular factors in $K_{1,3}$-free graphs, *J. Graph Theory* **15** (1991), 259–265.
26. V. Chvátal, Tough graphs and hamiltonian circuits, *Discrete Math.* **5** (1973), 215–228.
27. V. Chvátal and P. Erdős, A note on Hamiltonian circuits, *Discrete Math.* **2** (1972), 111–113.
28. L. H. Clark and R. C. Entringer, The minimum number of cycles in graphs with given cycle rank and small connectivity, *J. Combin. Math. Combin. Comput.* **3** (1988), 169–181.
29. K. Corradi and A. Hajnal, On the maximal number of independent circuits of a graph, *Acta Math. Acad. Sci. Hungar.* **14** (1963), 423–443.
30. Q. Cui, Y. Hu and J. Wang, Long cycles in 4-connected planar graphs, *Discrete Math.* **309** (2009), 1051–1059.
31. G. A. Dirac, Some theorems on abstract graphs, *Proc. London Math. Soc.* **27** (1952), 69–81.
32. G. A. Dirac, In abstrakten Graphen vorhandene vollständige 4-Graphen und ihre Unterteilungen, *Math. Nachr.* **22** (1960), 61–85.
33. Y. Egawa, Vertex disjoint cycles of the same length, *J. Combin. Theory (B)* **66** (1996), 168–200.
34. Y. Egawa, R. Glas and S. C. Locke, Cycles and paths through specified vertices in k-connected graphs, *J. Combin. Theory (B)* **52** (1991), 20–91.
35. M. N. Ellingham, D. A. Holton and C. H. C. Little, Cycles through ten vertices in 3-connected cubic graphs, *Combinatorica* **4** (1984), 265–273.
36. H. Enomoto, B. Jackson, P. Katerinis and A. Saito, Toughness and the existence of k-factors, *J. Graph Theory* **9** (1985), 87–95.
37. P. Erdős, R. J. Faudree, C. C. Rousseau and R. H. Schelp, The number of cycle lengths in graphs of given minimum degree and girth, *Discrete Math.* **200** (1999), 55–60.
38. P. Erdős and E. Győri, Any four independent edges of a 4-connected graph are contained in a circuit, *Acta Math. Hungar.* **46** (1985), 311–313.
39. G. Fan, Longest cycles in regular graphs, *J. Combin. Theory (B)* **39** (1985), 325–345.
40. R. J. Faudree, Connectivity and cycles in graphs, *Congr. Numer.* **187** (2007), 97–131.
41. R. J. Faudree, E. Flandrin and Z. Ryjáček, Claw-free graphs – A survey, *Discrete Math.* **164** (1997), 87–147.

42. R. J. Faudree, R. J. Gould, T. Lindquester and R. H. Schelp, On k-linked graphs, *Combinatorics, Graph Theory and Algorithms*, Vol. I, New Issues Press (1999), 387–400.
43. E. Flandrin, H. Li, A. Marczyk, I. Schiermeyer and M. Woźniak, Chvátal–Erdős condition and pancyclism, *Discuss. Math. Graph Theory* **26** (2006), 335–342.
44. W. Goddard, 4-connected maximal planar graphs are 4-ordered, *Discrete Math.* **257** (2002), 405–410.
45. P. R. Goodey, Hamiltonian circuits in polytopes with even sides, *Israel J. Math.* **22** (1975), 52–56.
46. R. Gould, Advances on the Hamiltonian problem – A survey, *Graphs Combin.* **19** (2003), 7–52.
47. M. Grötschel, On intersections of longest cycles, *Graph Theory and Combinatorics* (ed. B. Bollobás), Academic Press (1984), 171–189.
48. E. Győri and M. Plummer, A nine vertex theorem for 3-connected claw-free graphs, *Studia Sci. Math. Hungar.* **38** (2001), 233–244.
49. R. Häggkvist and C. Thomassen, Circuits through specified edges, *Discrete Math.* **41** (1982), 29–34.
50. F. Harary and C. St. J. A. Nash-Williams, On Eulerian and Hamiltonian graphs and line graphs, *Canad. Math. Bull.* **8** (1965), 701–709.
51. G. R. T. Hendry, Extending cycles in graphs. *Discrete Math.* **85** (1990), 59–72.
52. D. A. Holton, Cycles through specified vertices in k-connected regular graphs, *Ars Combin.* **13** (1982), 129–143.
53. D. A. Holton, Cycles in 3-connected cubic planar graphs, *Cycles in Graphs, Math. Stud.* **115** (1985), 219–226.
54. D. A. Holton, B. Manvel and B. D. McKay, Hamiltonian cycles in cubic 3-connected bipartite planar graphs, *J. Combin. Theory (B)* **38** (1985), 279–297.
55. D. A. Holton and B. D. McKay, The smallest non-Hamiltonian 3-connected cubic planar graphs have 38 vertices, *J. Combin. Theory (B)* **45** (1988), 305–319.
56. D. A. Holton, B. D. McKay, M. D. Plummer and C. Thomassen, A nine point theorem for 3-connected graphs, *Combinatorica* **2** (1982), 53–62.
57. D. A. Holton and M. D. Plummer, On the cyclability of k-connected $(k+1)$-regular graphs, *Ars Combin.* **23** (1987), 37–56.
58. B. Jackson, Hamiltonian cycles in regular 2-connected graphs, *J. Combin. Theory (B)* **29** (1980), 27–46.
59. B. Jackson, Long cycles in bipartite graphs, *J. Combin. Theory (B)* **38** (1985), 118–131.
60. B. Jackson and T. Parsons, A shortness exponent for r-regular r-connected graphs, *J. Graph Theory* **6** (1982), 169–176.
61. S. Jendrol and Z. Skupień, Exact numbers of longest cycles with empty intersection, *Europ. J. Combin.* **18** (1997), 575–578.
62. B. Jackson and N. Wormald, Long cycles and 3-connected spanning subgraphs of bounded degree in 3-connected $K_{1,d}$-free graphs, *J. Combin. Theory (B)* **63** (1995), 163–169.
63. B. Jackson and K. Yoshimoto, Even subgraphs of bridgeless graphs and 2-factors of line graphs, *Discrete Math.* **307** (2007), 2775–2785.
64. B. Jackson and K. Yoshimoto, Spanning even subgraphs of 3-edge-connected graphs, *J. Graph Theory* **62** (2009), 37–47.
65. H. A. Jung, Eine Verallgemeinerung des n-fachen Zusammenhangs für Graphen, *Math. Ann.* **187** (1970), 95–103.

66. T. Kaiser and P. Vrána, Hamiltonian cycles in 5-connected line graphs, *Europ. J. Combin.* **33** (5) (2012), 924–947.
67. A. Kaneko and A. Saito, Cycles intersecting a prescribed vertex set, *J. Graph Theory* **15** (1991), 655–664.
68. A. Kaneko and K. Yoshimoto, A 2-factor with two components of a graph satisfying the Chvátal–Erdős condition, *J. Graph Theory* **43** (2003), 269–279.
69. S. V. Kanetkar and P. R. Rao, Connected, locally 2-connected $K_{1,3}$-free graphs are panconnected, *J. Graph Theory* **8** (1984), 347–353.
70. K. Kawarabayashi, One or two disjoint circuits cover independent edges, Lovász-Woodall conjecture, *J. Combin. Theory (B)* **84** (2002), 1–44.
71. K. Kawarabayashi, Cycles through a prescribed vertex set in n-connected graphs, *J. Combin. Theory (B)* **90** (2004), 315–323.
72. K. Kawarabayashi, k-linked graphs with girth condition, *J. Graph Theory* **45** (2004), 48–50.
73. A. K. Kelmans and M. V. Lomonosov, When m vertices in a k-connected graph cannot be walked round along a simple cycle, *Discrete Math.* **38** (1982), 317–322.
74. M. Knor, On the number of cycles in k-connected graphs, *Acta Math. Univ. Comenianae* **63** (1994), 315–321.
75. D. G. Larman and P. Mani, On the existence of certain configurations within graphs and the 1-skeletons of polytopes, *Proc. London Math. Soc.* **20** (1970), 144–160.
76. M. C. Li, Longest cycles in regular 2-connected claw-free graphs, *Discrete Math.* **137** (1995), 277–295.
77. H. Liu, M. Lu and F. Tian, Relative length of longest paths and cycles in graphs, *Graphs Combin.* **23** (2007), 433–443.
78. S. C. Locke, Relative lengths of paths and cycles in k-connected graphs, *J. Combin. Theory (B)* **32** (1982), 206–222.
79. L. Lovász, Problem 5, *Period. Math. Hungar.* **4** (1974), 82.
80. L. Lovász, Problem 6.67, *Combinatorial Problems and Exercises*, North-Holland (1979), 46.
81. W. Mader, Homomorphieeigenschaften und mittlere Kantendichte von Graphen, *Math. Annalen* **174** (1967), 265–268.
82. W. Mader, Topological subgraphs in graphs of large girth, *Combinatorica* **18** (1998), 405–412.
83. J. Malkevitch, On the lengths of cycles in planar graphs, *Proceedings of the Conference on Graph Theory and Combinatorics at St. John's University, Jamaica, NY*, Lecture Notes in Mathematics **186**, Springer (1970), 191–195.
84. M. Matthews and D. Sumner, Hamiltonian results in $K_{1,3}$-free graphs, *J. Graph Theory* **9** (1984), 139–146.
85. W. D. McCuaig and M. Rosenfeld, Cyclability of r-regular r-connected graphs, *Bull. Austral. Math. Soc.* **29** (1984), 1–11.
86. K. Menger, Zur allgemeinen Kurventheorie, *Fund. Math.* **10** (1927), 95–115.
87. G. H. J. Meredith, Regular n-valent n-connected non-Hamiltonian non-n-edge-colorable graphs, *J. Combin. Theory (B)* **14** (1973), 55–60.
88. K. Mészáros, On low degree k-ordered graphs, *Discrete Math.* **308** (2008), 2418–2426.
89. J. W. Moon and L. Moser, Simple paths on polyhedra, *Pacific J. Math.* **13** (1963), 629–631.
90. J. W. Moon and L. Moser, On hamiltonian bipartite graphs, *Israel J. Math.* **1** (1963), 163–165.
91. L. Ng and M. Schultz, k-ordered hamiltonian graphs, *J. Graph Theory* **24** (1997), 45–57.

92. D. J. Oberly and D. P. Sumner, Every connected, locally connected nontrivial graph with no induced claw is hamiltonian, *J. Graph Theory* **3** (1979), 351–356.
93. J. Petersen, Die Theorie der regulären Graphs, *Acta Math.* **15** (1891), 193–220.
94. M. Rosenfeld, The number of cycles in 2-factors of cubic graphs, *Discrete Math.* **84** (1990), 285–294.
95. Z. Ryjáček, On a closure concept in claw-free graphs, *J. Combin. Theory (B)* **70** (1997), 217–224.
96. A. Saito, Long cycles through specified vertices in a graph, *J. Combin. Theory (B)* **47** (1989), 220–230.
97. D. P. Sanders, On circuits through five edges, *Discrete Math.* **159** (1996), 199–215.
98. D. P. Sanders, On paths in planar graphs, *J. Graph Theory* **24** (1997), 341–345.
99. B. Sudakov and J. Verstraëte, Cycle lengths in sparse graphs, *Combinatorica* **28** (2008), 357–372.
100. R. Thomas and P. Wollan, An improved linear edge bound for graph linkages, *Europ. J. Combin.* **26** (2005), 309–324.
101. C. Thomassen, A theorem on paths in planar graphs, *J. Graph Theory* **7** (1983), 169–176.
102. C. Thomassen, Reflections on graph theory, *J. Graph Theory* **10** (1986), 309–324.
103. M. Trenkler, On 4-connected, planar 4-almost pancyclic graphs, *Math. Slovaca* **39** (1989), 13–20.
104. W. T. Tutte, A theorem on planar graphs, *Trans. Amer. Math. Soc.* **82** (1956), 99–116.
105. H. Wang, On long cycles in a bipartite graph, *Graphs Combin.* **12** (1996), 373–384.
106. H. Wang, Maximal total length of k disjoint cycles in bipartite graphs, *Combinatorica* **25** (2005), 367–377.
107. M. E. Watkins and D. M. Mesner, Cycles and connectivity in graphs, *Canad. J. Math.* **19** (1967), 1319–1328.
108. B. Wei, On the circumference of regular 2-connected graphs, *J. Combin. Theory (B)* **75** (1999), 88–99.
109. B. Wei and Y. Zhu, The Chvátal–Erdős condition for panconnectivity of triangle-free graphs, *Discrete Math.* **252** (2002), 203–214.
110. D. R. Woodall, Sufficient conditions for circuits in graphs, *Proc. London Math. Soc.* **24** (1972), 739–755.
111. K. Yoshimoto, On the number of components in 2-factors of claw-free graphs, *Discrete Math.* **307** (2007), 2808–2819.
112. S. Zhan, On hamiltonian line graphs and connectivity, *Discrete Math.* **89** (1991), 89–95.
113. Y. Zhu, Z. Liu and Z. Yu, An improvement of Jackson's result on Hamilton cycles in 2-connected regular graphs, *Cycles in Graphs* (eds. B. R. Alspach and C. D. Godsil), *Ann. Discrete Math.* **27** (1985), 237–247.
114. Y. Zhu, Z. Liu and Z. Yu, 2-connected k-regular graphs on at most $3k+3$ vertices to be hamiltonian, *J. Systems Sci. Math. Sci.* **6** (1986), 36–49.

6

H-linked graphs

MICHAEL FERRARA and RONALD J. GOULD

Given graphs G and H, we say that G is H-linked provided that every injective function $f : V(H) \to V(G)$ can be extended to a subdivision of H. When $H = kK_2$, this is the well-known k-linked problem. In this chapter we trace the development of H-linked graphs and consider a number of natural variations of the problem.

1. Introduction

One of the most natural and central properties in the study of graphs is connectivity. Connectivity is so important that many variants of how a graph is 'put together' have been developed, in an effort to gain further insights into a vast array of problems. Binding number, toughness, integrity and other parameters have been introduced to measure some aspect of connectedness, and all have found uses. However, connectivity remains the most fundamental and useful such property and, as such, it is natural that even stronger variations have been explored.

Connectivity itself has many 'levels'. We say that a graph G of order at least k is *k-connected* if the removal of every set of $k-1$ vertices leaves a non-trivial connected graph. One of the first, and most useful, results about k-connected graphs is due to Dirac [15].

Theorem 1.1 *For $k \geq 2$, if G is a k-connected graph, then each set of k vertices of G lies on some cycle.*

As more uses were found for connectivity, additional formulations were developed. With the aid of Menger's theorem [52], an alternative characterization for k-connectedness, central to our development, exists.

Theorem 1.2 *A graph G is k-connected if and only if, for any two disjoint sets S and T with k vertices, there exist k disjoint $S - T$ paths.*

Here there is no requirement about which vertex of S is connected to which vertex of T, only that the required collection of disjoint paths exists. However, we can certainly ask for an even stronger condition by specifying more about how these path connections are to occur. In one sense, when a graph is k-connected, the collection of paths from the set S to the set T defines a permutation, where the path P_r ($r = 1, 2, \ldots, k$) takes the vertex s_{i_r} to the vertex t_{j_r} (so that the permutation maps i_r to j_r). A stronger condition permits us to prescribe the permutation defined by the path system from S to T. Formally, we have the following definition.

In a graph G, two disjoint ordered sets, $S = \{s_1, s_2, \ldots, s_k\}$ and $T = \{t_1, t_2, \ldots, t_k\}$, of k vertices each, are said to be *k-linked* if there are k disjoint paths $P_1, P_2, \ldots, P_k$ for which each P_i is an s_i–t_i path. The k paths are then said to form an (S, T)-*linkage* in G. Further, we say that G is *k-linked* if every such pair of sets S and T are k-linked.

When G is k-linked, all permutations on k elements can be achieved, since the ends of the paths can be designated in any order we wish. This is a very strong requirement, and clearly every graph with this property is k-connected. The converse does not hold, however, as it is easy to find examples of k-connected graphs that are not k-linked; the cycle C_n ($n \geq 4$) is such an example for $k = 2$.

It is therefore natural to ask what level of connectivity is needed to achieve this stronger condition.

Problem A *Does there exist a connectivity $f(k)$ that ensures that a graph G is k-linked?*

We show later that $f(k)$ is defined for all k. Certainly many other questions could be asked here. For example, given partitions $S = S_1 \cup S_2$ and $T = T_1 \cup T_2$, where $|S_i| = |T_i|$ ($i = 1, 2$), what connectivity allows us to find disjoint paths P_j (for $j = 1, 2, \ldots, |S_1|$) that join S_1 to T_1, and Q_j (for $j = 1, 2, \ldots, |S_2|$) that join S_2 to T_2? Here we are asking for something less than being k-linked, but something more than being k-connected. This opens the door to many other similar questions.

For terms not defined here, see the Preliminaries chapter of this book.

2. k-linked graphs

One of the first results on k-linked graphs was obtained independently by Jung [36] and Larman and Mani [46]. We state a simplified form of their results.

Theorem 2.1 *Every $2k$-connected graph containing a subdivision of K_{3k} is k-linked.*

The technique used to prove this theorem is to use the connectivity and the structure of the subdivided K_{3k} to route the paths between the chosen pairs of vertices. Roughly speaking, a path (possibly trivial) from a vertex s_i to some vertex v_{i_1} in the subdivided K_{3k}, and a path from vertex t_i to some vertex v_{i_2} in the subdivided K_{3k}, are found. Then a carefully chosen path within the subdivided K_{3k} completes the $s_i - t_i$ path. The connectivity allows all paths to be built simultaneously, and the subdivided K_{3k} is dense enough for all the paths to pass through it. This is a useful idea, and one that we shall see again. The subdivided clique uses $3k$ vertices, since $2k$ of them could be the pairs we are linking.

The hunt for subdivisions of cliques (and other graphs) is a major topic in its own right. For example, Mader [48] proved the following result.

Theorem 2.2 *If G is a graph with minimum degree $\delta(G) \geq 2^{\binom{k}{2}}$, then G contains a subdivision of K_k.*

Taken together, Theorems 2.1 and 2.2 imply that the function $f(k)$ in Problem A is well defined, although possibly exponential in k. However, as the following discussion reveals, it is actually linear.

One possible approach to showing that $f(k)$ is linear is to try to improve the minimum degree condition in Theorem 2.2. To this end, define $h(k)$ to be the smallest number for which every graph G with $\delta(G) \geq h(k)$ contains a subdivision of K_k. Note that if $s = \binom{2k}{2}$, then $K_{s,s}$ contains a subdivision of K_{2k}, whereas $K_{s,s-1}$ does not; thus, $h(2k) \geq s$. Szemerédi (see [71]) used this observation in proving the following theorem, which essentially gives the order of growth of $h(k)$.

Theorem 2.3 *There exists a constant $c > 0$ such that $\frac{1}{2}k^2 \leq h(k) \leq ck^2 \log k$, for $k > 1$.*

The exact value of $h(k)$ is known for only a few values. We see that $h(k) = k - 1$ for $k = 2$ and 3, and Dirac [16] showed that this holds for $k = 4$ as well.

Corollary 2.4 *If $v_1, v_2, \ldots, v_k$ and $w_1, w_2, \ldots, w_k$ are vertices (not necessarily distinct) in an $h(3k)$-connected graph G, then G has k disjoint paths $P_1, P_2, \ldots, P_k$ such that P_i is a $v_i - w_i$ path.*

In this corollary, the fact that the chosen vertices are not necessarily distinct is what distinguishes the result from Theorem 2.1. It also implies that in any $h(3k)$-connected graph, any ordered set of k vertices lie on a common cycle in the prescribed order. We call this property *k-ordered*, and more will be said about it later.

Returning to the problem of determining $f(k)$, Robertson and Seymour [61] improved Theorem 2.1 by showing that any $2k$-connected graph G containing a K_{3k}-minor is k-linked. Again, this K_{3k}-minor can be used to control the paths that are needed for the k-linkage. This result, together with bounds on the extremal function for complete minors by Kostochka [41] and Thomason [67], showed that $f(k) = O(k\sqrt{\log k})$.

Bollobás and Thomason [4] made further progress toward proving the linearity of $f(k)$ by showing that it is not necessary for the graph to contain a complete minor. Rather, they showed that a $2k$-connected graph is k-linked provided that it has a sufficiently dense minor. This approach is intuitively reasonable, as such a minor provides enough density to construct the paths in the desired k-linkage.

Theorem 2.5 *Let H be a graph of order p with $\delta(H) > \frac{1}{2}p + 4k - 2$. Then every $2k$-connected graph containing H as a minor is k-linked.*

Bollobás and Thomason used Theorem 2.5 to show that $f(k)$ is linear, and specifically that every $22k$-connected graph is k-linked [4]. This bound has subsequently been improved. The current best-known bound on $f(k)$ is a consequence of the following result by Thomas and Wollan [66].

Theorem 2.6 *Every $2k$-connected graph with average degree at least $10k$ is k-linked.*

Corollary 2.7 *Every $10k$-connected graph is k-linked.*

Let H be a graph with k vertices, $v_1, v_2, \ldots, v_k$, and l edges. A graph G is called *H-linked* if, for any ordered set of k of its vertices it has m internally disjoint paths joining v_i to v_j in G if and only if H has an edge from w_i to w_j; in other words, for every ordered set of k vertices, G contains a subdivision of H with those k designated vertices as the branch vertices. Note that a graph is k-linked if and only if it is kK_2-linked. General H-linkedness is examined in greater detail in Section 8.

The following result of Bollobás and Thomason [4] gives a connectivity bound which ensures that a graph contains a subdivision of a given graph (in fact, many subdivisions).

Theorem 2.8 *If H is a graph with k vertices and m edges, then every $(22m + k)$-connected graph is H-linked.*

As with many difficult problems, it is sometimes possible to determine an exact answer in certain cases. For $f(k)$, small values of k allow this. When $k = 1$, we are merely asking for our graph to be connected, so $f(1) = 1$. Jung [36] showed that $f(2) = 6$. For an example of a 5-connected non-2-linked graph, see Fig. 1. Notice

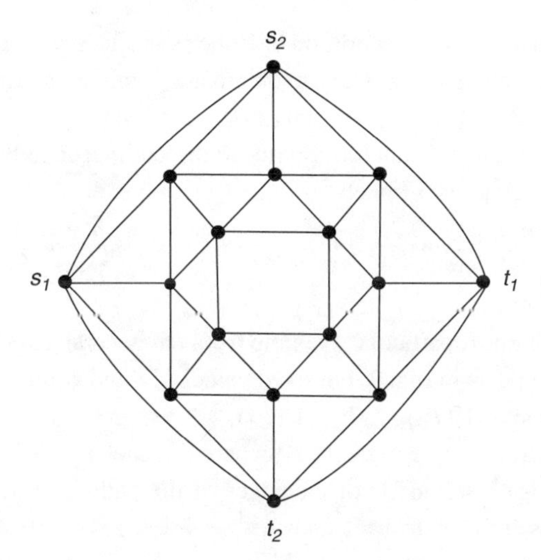

Fig. 1. A 5-connected graph that is not 2-linked

that this graph is planar and 5-regular, but is not maximal planar. Jung [36] actually proved much more.

Theorem 2.9

- *Every maximally planar* 4*-connected graph is* 2*-linked.*
- *Every non-planar* 4*-connected graph is* 2*-linked.*

This result implies that any 4-connected graph of order n with at least $3n-6$ edges is 2-linked. Independently, Seymour [63] and Thomassen [68] found all graphs that are maximally non-2-linked – that is, those maximal graphs with disjoint 2-element sets that are not 2-linked.

We begin with a specific pair of sets $S=\{s_1, s_2\}$ and $T=\{t_1, t_2\}$, and let F be any graph obtained as follows. Start with a plane graph in which the boundary of the unbounded face is a 4-cycle $s_1s_2t_1t_2$, and in which each interior face is a triangle, but with no separating 3-cycle (no vertex appears inside a 3-cycle). Inside each triangle, a complete graph adjacent to all three of its vertices may be added (but of course, the result may no longer be planar). The resulting graph F is called an *(S, T)-web.*

Theorem 2.10 *Let S and T be disjoint sets of pairs of vertices. A graph is maximal with respect to having no (S, T)-linkage if and only if it is an (S, T)-web.*

In [10], the dense minor approach was also taken to study the $k=3$ case. Let K_n^- denote the complete graph on n vertices with one edge missing.

Theorem 2.11 *Every* 6*-connected graph with a K_9^--minor is* 3*-linked.*

This result relaxed a K_9-minor condition of Robertson and Seymour [61]. The condition on the number of edges necessary to obtain a K_9^- minor, together with Theorem 2.11, imply that $f(3) \leq 18$.

Recently, Thomas and Wollan [66] produced the optimal edge bound for ensuring that a 6-connected graph is 3-linked.

Theorem 2.12 *Every 6-connected graph of order n and at least* $5n - 14$ *edges is 3-linked.*

This bound is sharp. A particular example from the general class given by Thomas and Wollan [66] appears in Fig. 2. For fixed values of l and k, the graph is composed of the disjoint union of $V(P_0)$, $V(P_1)$, $V(P_2)$, $V(P_3)$, and a graph $H = K_{k-1}$, along with vertices $\{s_3, s_4, \ldots, s_k, t_3, t_4, \ldots, t_k\}$ where, for each i, P_i is a path on l vertices with $l \geq 2$. The edge-set consists of the edges of the paths and the complete graph H, with the edges from v_j^i to v_j^{i+1} and v_{j+1}^{i+1}, for $1 \leq j \leq l - 1$. Also included are the edges from v_l^i to v_l^{i+1} for $0 \leq i \leq 3$, where the superscript arithmetic is taken modulo 4. Finally, let v_l^i be adjacent to every vertex of H and, for $j \geq 3$, let s_j and t_j be adjacent to all other vertices except each other. These graphs all have $n = 4l + 3k - 5$ vertices and $(2k - 1)n - \frac{1}{2}(3k + 1)k$ edges. In our example, $l = 5$ and $k = 3$, producing a graph with 24 vertices and 105 edges. Further, not all edges are shown as $N(s_3) = N(t_3) = V - \{s_3, t_3\}$.

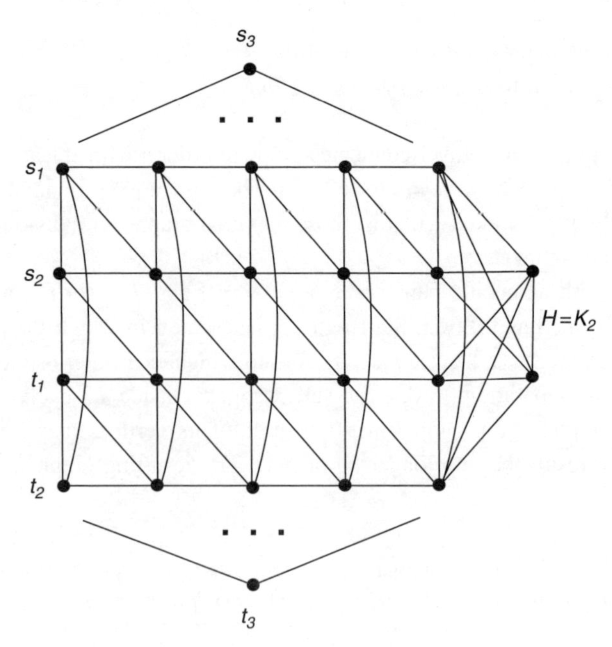

Fig. 2. A maximal non-3-linked 6-connected graph

For $k \geq 3$, the exact value of $f(k)$ is not known. We note, however, a long-standing conjecture.

Conjecture A *Every 8-connected graph is 3-linked.*

In the light of Theorems 2.9 and 2.12 it is reasonable to ask the following question.

Question *For $k \geq 4$, does there exist a function $m(k)$ for which any $2k$-connected graph with at least $m(k)$ edges is k-linked?*

The question of how connected a graph must be to be k-linked does not seem to have a simple answer. The graph H obtained by deleting the edges of a k-matching from the graph K_{3k-1} is not k-linked and has connectivity $3k - 4$. But the problem with H may well be that it is just too small. It has long been speculated that $(2k+2)$-connected graphs of sufficiently large order are k-linked. That $(2k + 1)$-connected graphs fail to be k-linked can be seen by starting with a 5-connected (S, T)-web (not 2-linked and arbitrarily large) and first joining a new vertex s_3 to all vertices of the web and then joining a new vertex t_3 to all vertices of this graph. We now have a 7-connected non-3-linked graph. This process can clearly be continued to produce $(2k + 1)$-connected non-k-linked graphs.

It is natural to expect that, aside from connectivity conditions, enough edges distributed throughout the graph could also imply that a graph is k-linked. Such conditions have been investigated in a variety of ways, and an early result of this type was found by Faudree *et al.* [21]. We say that a graph is *H-free* if it contains no induced subgraph isomorphic to H.

Theorem 2.13 *For $k \geq 2$, every $(4k - 3)$-connected $K_{1,3}$-free graph is k-linked.*

The graph in Fig. 3 is 4-connected and $K_{1,3}$-free, but is easily seen not to be 2-linked.

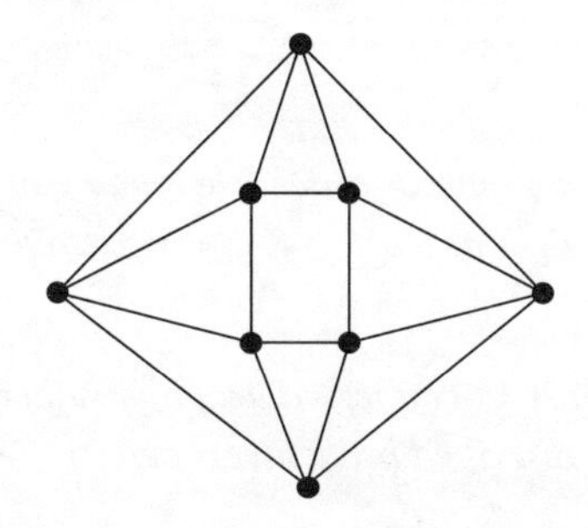

Fig. 3. A 4-connected $K_{1,3}$-free non-2-linked graph

Recall that the *rth power* G^r of a graph G has vertex-set $V(G)$, where vw is an edge if $1 \leq \text{dist}(v, w) \leq r$. Faudree *et al.* [21] proved the following results.

Theorem 2.14

- *If G is a connected graph of order at least $2k$, then G^{2k-1} is k-linked.*
- *If G is an s-connected graph of order at least $4k + 2$, then G^t is k-linked for $t \geq 4k/s$.*

Let $D(n, k)$ be the minimum positive integer d for which every graph of order n and minimum degree at least d is k-linked. Further, let $R(n, k)$ denote the minimum positive integer r for which every graph of order n with $\delta_2(G) \geq r$ is k-linked, where $\delta_2(G)$ is the minimum degree sum of a pair of non-adjacent vertices in G. Kawarabayashi, Kostochka and Yu [39] obtained the following results.

Theorem 2.15 *If $k \geq 2$, then*

$$R(n,k) = \begin{cases} 2n-3, & \text{if } n \leq 3k-1 \\ \lfloor \frac{2}{3}(n+5k) \rfloor - 3, & \text{if } 3k \leq n \leq 4k-2 \\ n+2k-3, & \text{if } n \geq 4k-1 \end{cases}$$

and

$$D(n,k) = \begin{cases} n-1, & \text{if } n \leq 3k-1 \\ \lfloor \frac{1}{3}(n+5k) \rfloor - 1, & \text{if } 3k \leq n \leq 4k-2 \\ \lceil \frac{1}{2}(n-3) \rceil + k, & \text{if } n \geq 4k-1. \end{cases}$$

A combination of connectivity and degree conditions was used by Pfender [60]. In the light of Corollary 2.7, we state an improvement of his result.

Theorem 2.16 *If G is a $2k$-connected graph of order n with $\delta(G) \geq \frac{1}{3}n + 10k$, then G is k-linked.*

Another parameter that has been useful in the study of linkage is girth. The following results are due to Mader [50].

Theorem 2.17 *For every graph H, there is an integer $g(H)$ for which every graph G with $\delta(G) \geq \max\{\Delta(H), 3\}$ and with girth at least $g(H)$ contains a subdivision of H.*

Theorem 2.18 *Let H be a graph of order n without isolated vertices. Then every $2n$-connected graph with girth at least $g(K_{2n,2n})$ contains a subdivision of H with prescribed branch vertices.*

Corollary 2.19 *Every $n(n-1)$-connected graph of sufficiently large girth contains a subdivision of K_n with prescribed branch vertices.*

Later, Kawarabayashi [38] was able to be more specific about how large a girth suffices.

Theorem 2.20

- *For $k = 4$ or 5, every $2k$-connected graph of girth 19 is k-linked.*
- *For $k \geq 6$, every $2k$-connected graph of girth at least 11 is k-linked.*

Note that Theorem 2.20 provides no information for $k = 2$ or 3. However, Theorem 2.9 tells us when a 4-connected graph is 2-linked, and Theorem 2.12 provides a sufficient condition for a 6-connected graph to be 3-linked.

3. Weak linkage

When faced with a difficult problem, it is natural that related questions would also be investigated. Here, the most natural variation to consider is that of edge-connectivity, rather than vertex-connectivity.

We say that a graph G is *weakly k-linked* if, given any two sets of k distinct vertices, $S = \{s_1, s_2, \ldots, s_k\}$ and $T = \{t_1, t_2, \ldots, t_k\}$, we can find pairwise edge-disjoint s_i–t_i paths.

Thomassen [68] conjectured that there exists a function $g(k)$ for which every $g(k)$-edge-connected multigraph is weakly k-linked. An obvious necessary condition for a multigraph to be weakly k-linked is that a graph be k-edge-connected. However, Thomassen showed that this condition is not sufficient if k is even. To see this, consider the multigraph obtained from the cycle $s_1 s_2 \ldots s_k t_1 t_2 \ldots t_k s_1$ by replacing each edge with $\frac{1}{2}k$ edges. This multigraph is k-edge connected, but has no $(s_1, s_2, \ldots, s_k, t_1, t_2, \ldots, t_k)$-linkage. Indeed, once edge-disjoint paths from $s_1, s_2, \ldots, s_{k-1}$ are formed, $\frac{1}{2}k$ of these paths must use one of s_k or t_k; but then no s_k–t_k path is possible without re-using edges.

Thomassen conjectured the value of $g(k)$.

Conjecture B

$$g(k) = \begin{cases} k & \textit{if } k \textit{ is odd,} \\ k+1 & \textit{if } k \textit{ is even.} \end{cases}$$

Early work by a variety of people established $g(k)$ for some small values of k. Okamura [58] showed that every $4k$-edge-connected graph is weakly $3k$-linked.

Hirata, Kubota and Saito [34] proved the following result.

Theorem 3.1 *For $i = 1, 2, 3$, Let (s_i, t_i), be pairs of vertices of G, and let $k \geq 2$. If $\lambda(s_i, t_i) \geq 2k+1$, for (s_1, t_1), (s_2, t_2) and (s_3, t_3), then there exist $2k+1$ edge-disjoint paths such that one path joins s_1 to t_1, another joins s_2 to t_2, and the remaining paths join s_3 to t_3.*

They deduced as a corollary that every $(2k+1)$-edge connected graph is weakly $(k+2)$-linked for $k \geq 2$. It follows that every 5-edge connected graph is weakly 4-linked, meeting the value conjectured by Thomassen, a result also obtained independently by H. Enomoto and A. Saito (unpublished).

To date, the strongest result is due to Huck [35], getting within 1 of Thomassen's conjecture.

Theorem 3.2

$$g(k) \leq \begin{cases} k+1 & \textit{if } k \textit{ is odd,} \\ k+2 & \textit{if } k \textit{ is even.} \end{cases}$$

4. Digraphs

We next turn our attention to similar linkage questions about digraphs. A digraph is *strongly k-linked* if the paths in an (S, T)-linkage can be chosen so that they are directed from the source vertex in S to the sink vertex in T. Bermond and Lovász [3] asked whether every sufficiently strongly connected digraph contains a directed cycle through any two prescribed vertices. Thomassen [68] made a stronger conjecture, that there exists a function $f_d(k)$ for which every strongly $f_d(k)$-connected digraph D is strongly k-linked.

For tournaments we have the following result, also due to Thomassen [69]. As usual, $\kappa(x, y)$ denotes the maximum number of internally disjoint directed $x-y$ paths.

Theorem 4.1 *For each k, there exists a smallest natural number $f_T(k)$ such that, if $s_1, s_2, \ldots, s_k$ and $t_1, t_2, \ldots, t_k$ are distinct vertices in a tournament T with $\kappa(s_i, t_i) \geq f_T(k)$ for each i, then T has k disjoint directed paths $P_1, P_2, \ldots, P_k$ such that P_i is an $s_i - t_i$ path.*

Theorem 4.1 has some interesting corollaries.

Corollary 4.2 *If $s_1t_1, s_2t_2, \ldots, s_kt_k$ are prescribed independent arcs in a tournament T, such that $\kappa(v, w) \geq f_T(k)$ for each $v, w \in \{s_1, s_2, \ldots, s_k, t_1, t_2, \ldots, t_k\}$, then the arcs $s_1t_1, s_2t_2, \ldots, s_kt_k$ are contained in a cycle in the given order.*

Corollary 4.3 *Let D be a digraph with vertices $v_1, v_2, \ldots, v_n$, none of which is isolated, and with at most k arcs. If T is a tournament with $A = \{w_1, w_2, \ldots, w_n\} \subseteq V(T)$ for which $\kappa(w_i, w_j) \geq f_T(k)$ for all i, j, then T contains a subdivision of D in which v_i corresponds to w_i, for each i.*

Thomassen [69] determined a connectivity sufficient to imply a spanning k-linkage in a tournament.

Theorem 4.4 *Let $Z = \{s_1, s_2, \ldots, s_k, t_1, t_2, \ldots, t_k\}$ be a set of distinct vertices in an $h_T(k)$-connected tournament T, where $h_T(k) = f_T(5k) + 12k + 9$. Then T has k disjoint paths $P_1, P_2, \ldots, P_k$ spanning T, for which P_i joins s_i to t_i.*

Natural consequences of this spanning linkage include the following result.

Corollary 4.5 *If $e_1, e_2, \ldots, e_k$ are pairwise independent arcs in an $h_T(k)$-connected tournament T, then T has a Hamiltonian cycle containing $e_1, e_2, \ldots, e_k$ in that cyclic order – that is, the digraph is k-arc-ordered Hamiltonian.*

Theorem 4.6 *Let D be a digraph with vertices $v_1, v_2, \ldots, v_n$, none of which is isolated, and with at most k arcs. If T is an $h_T(k)$-connected tournament containing vertices $\{w_1, w_2, \ldots, w_k\} \subseteq V(T)$, then T has a spanning subdigraph that is a subdivision of D with each w_i corresponding to v_i.*

The k-linked property has also been studied in other special classes of digraphs closely related to tournaments. A digraph is *semi-complete* if, between any pair of vertices v and w, there is at least one of the arcs vw or wv, while it is *locally semi-complete* if both the in-neighbours and the out-neighbours of each vertex induce semi-complete digraphs. A consequence of a slight extension of Theorem 4.1 is the following result due to Bang-Jensen [1].

Corollary 4.7 *Every 5-strong semi-complete digraph is 2-linked.*

Bang-Jensen [1] constructed a 4-strong semi-complete digraph that is not 2-linked, demonstrating that this corollary is sharp. He obtained a result similar to that of Thomassen's for tournaments.

Theorem 4.8 *For every k, there is a natural number $f_{lsc}(k)$ for which every $f_{lsc}(k)$-strong locally semi-complete digraph is k-linked.*

Although tournaments and some other classes of digraphs allow a development similar to that for graphs, the fundamental linkage question for digraphs is quite different. Thomassen [70] has produced a class of examples that show that, for each natural number k, there exists a strongly k-connected digraph that is not 2-linked. Thus, the answer to Problem A in the Introduction is *no* for digraphs. A natural follow-up to these observations is the following result.

Theorem 4.9 *For every k, there exists a strongly k-connected digraph with two vertices that are not on any directed cycle.*

However, Thomassen's example does not end our inquiry into digraphs. It is reasonable to assume that if a digraph has sufficiently many arcs, then it is k-linked, and Manoussakis [51] showed that this is the case.

Theorem 4.10 *Every digraph of order $n \geq 2k$ and with at least $n(n-2)+2k$ arcs is k-linked.*

The proof of this result is based on the following useful lemma from [51].

Theorem 4.11 *If $D - v$ is k-linked, for some vertex v with min $\{d^+(v), d^-(v)\} \geq 2k-1$, then D is k-linked.*

The arc requirement from Theorem 4.10 is a strong one, and so the result is somewhat limited. However, Manoussakis [51] gave an example to show that this bound cannot be improved.

Heydemann and Sotteau [33] considered 2-linkage under somewhat different conditions. Here, $\delta^0(D) = \min\{\delta^+(D), \delta^-(D)\}$.

Theorem 4.12 *Every digraph of order n with $\delta^0 \geq \frac{1}{2}n + 1$ is 2-linked.*

They also produced examples to show the bound on δ^0 cannot be weakened. But it is possible to strengthen their result somewhat (see [2]).

Theorem 4.13 *If D is a digraph of order n with $\delta^0 \geq \frac{1}{2}n + 1$, then, for any four vertices s_1, s_2, t_1, t_2, there exist disjoint s_1–t_1 and s_2–t_2 paths whose union spans D.*

Manoussakis also considered 3-linkages.

Theorem 4.14 *Every digraph of order $n \geq 9$ and $\delta^0 \geq \frac{1}{2}n + 2$ is 3-linked.*

Based upon Theorems 4.12 and 4.14, Manoussakis posed the following problem.

Problem B *Determine the minimum function $f(n, k)$ for which every digraph of order n and $\delta^0 \geq f(n, k)$ is k-linked.*

Hurkens (see [51]) showed that $f(n, 4) = \frac{1}{2}n + 3$ when $n \geq 13$, and Manoussakis asked whether $f(n, k) = \frac{1}{2}n + k - 1$, for $n \geq 4k - 3$. This problem has recently been settled by Kühn and Osthus [45].

Theorem 4.15 *For $k \geq 2$, every digraph of order $n \geq 1600k^3$ with $\delta^0 \geq \frac{1}{2}n + k - 1$ is k-linked.*

Kühn and Osthus went further. It is easy to see that every k-linked digraph is k-ordered. Conversely, every $2k$-ordered digraph is k-linked, since then a cycle exists that encounters the vertices $s_1, t_1, s_2, t_2, \ldots, s_k, t_k$ in that order, and the linkage paths are clear. The next result by Kühn and Osthus [45] extends the previous one.

Theorem 4.16 *For $k \geq 2$, every digraph of order $n \geq 200k^3$ with $\delta^0 \geq \frac{1}{2}(n + k) - 1$ is k-ordered.*

We say that a digraph D is *k-arc ordered* if, for each sequence $e_1, e_2, \ldots, e_k$ of k independent arcs, D contains a cycle containing these arcs in this order. One can easily verify that every k-linked digraph is k-arc ordered, thus implying the following result.

Corollary 4.17 *For $k \geq 2$, every digraph of order $n \geq 1600k^3$ with $\delta^0 \geq \frac{1}{2}n + k - 1$ is k-arc-ordered.*

5. Modulo and parity linkage

In this section we consider strengthening the property of k-linkage to include some control over the lengths of the paths in the linkage. Throughout this section, we let μ be a k-tuple $(m_1, m_2, \ldots, m_k)$ of positive integers. A graph is *modulo-μ-linked* if it is k-linked and, for any k-tuple of integers $(d_1, d_2, \ldots, d_k)$ and for any disjoint sets S and T of k vertices, there exists an (S, T)-linkage with paths $P_1, P_2, \ldots, P_k$ for which the length of P_i is congruent to d_i (modulo m_i), for each i. In [72], Thomassen demonstrated that a graph with sufficiently high connectivity is modulo-$(m_1, m_2, \ldots, m_k)$-linked when each m_i is odd.

Theorem 5.1 *For any positive integers t and p, there exists a function $h(t, p)$ for which every $h(t, p)$-connected graph is modulo-μ-linked, for any t-tuple μ of odd positive integers less than p.*

The value of $h(t, p)$ determined by Thomassen in [72] was extremely large. Chen *et al.* [9] made an improvement by showing that (still with each m_i odd), if S is the sum of the entries in μ and if $t \geq 5$, then $h(t, p) \geq 6S - 4t + 36$.

A graph G is *k-parity-linked* if G is k-linked and the paths $P_1, P_2, \ldots, P_k$ in any k-linkage can be chosen with their lengths having prescribed parities – that is, G is k-parity-linked if and only if G is modulo-(2, 2, ..., 2)-linked. Given that Theorem 5.1 and the subsequent result from [9] involve only odd values of m_i, one might expect that a strict connectivity condition might not suffice to assure that G is k-parity linked. Thomassen [73] demonstrated that this is the case by showing that a graph G with sufficiently high connectivity is k-parity-linked provided that G is, in some sense, 'far enough' from being bipartite. The *bipartite index* of G is the minimum number of vertices whose deletion from G leaves a bipartite graph.

Theorem 5.2 *There exists a function $p(k)$ for which every $p(k)$-connected graph with bipartite index at least $4k - 3$ is k-parity-linked.*

The bound on the bipartite index of G is sharp. To see this, consider the graph obtained from $K_{r,r}$, with r sufficiently large, by adding the edges of a K_{2k-1} to one partite set and the edges of $K_{k(2)}$ to the other partite set. Then it is impossible to find disjoint paths of odd length connecting the pairs of non-adjacent vertices in the copy of $K_{k(2)}$.

Kawarabayashi and Reed [40] improved on the exponential bounds on $p(k)$ from [73] by showing that $p(k) = 50k$ suffices in Theorem 5.2. Also, Chen *et al.* [9] obtained similar results for more general modulo-μ-linked graphs which extended the main result from [12], again under the restriction that the bipartite index of G is sufficiently large.

Theorem 5.3 *Let $\mu = (m_1, m_2, \ldots, m_k)$ be a k-tuple of positive integers with sum S, such that, for some $l \leq k$, each m_i with $l + 1 \leq i \leq k$ is odd, and let $\varepsilon = 0$ if $\min\{m_i\} = 1$ and $\varepsilon = 1$ otherwise. If G is 45S-connected and has bipartite index at least $2k + 2l - 3 + \varepsilon$, then G is modulo-$(2m_1, 2m_2, \ldots, 2m_l, m_{l+1}, m_{l+2}, \ldots, m_k)$-linked.*

We note that if $k = l$ and each $m_i = 1$, then Theorem 5.3 requires G to have bipartite index $4k - 3$ or more, in line with Theorem 5.2. One additional consequence of Theorem 5.2 deals with the existence of a *totally odd* K_4-subdivision, which is a subdivision of K_4 in which every edge corresponds to a path of odd length. Thomassen [74] and Zang [76] independently verified that every 4-chromatic graph contains a totally odd K_4-subdivision, affirming a conjecture of B. Toft from 1974. In [74], Thomassen conjectured that every 4-connected graph with bipartite index at least 2 contains a totally odd K_4-subdivision, and in [74] showed that this holds with the additional hypothesis that G has at least three triangles. As a corollary to Theorem 5.2 he also showed that every 8^{162}-connected graph with bipartite index at most 2 contains a totally odd K_4-subdivision.

6. Disjoint connected subgraphs

For positive integers $s_1, s_2, \ldots, s_k$, a graph G is said to be $(s_1, s_2, \ldots, s_k)$-*linked* if the order of G is at least $\sum_{i=1}^{k} s_i$ and if, for any k vertex-disjoint subsets $S_1, S_2, \ldots, S_k$ with $|S_i| = s_i$, G contains disjoint connected subgraphs $F_1, F_2, \ldots, F_k$ for which $S_i \subseteq V(F_i)$. Note that the case $s_1 = s_2 = \ldots = s_k = 2$ is the original k-linked problem. Little is yet known on this generalization. In [10] the following was shown.

Theorem 6.1 *Every 7-connected graph with a K_9^--minor is (2, 5)-linked.*

Mori [53] obtained the following interesting characterization of (3, 3)-linked planar graphs.

Theorem 6.2 *A planar graph of order at least 6 is (3, 3)-linked if and only if it is 4-connected and maximal planar.*

Clearly, a 4-ordered graph is 2-linked, and hence (2, 2)-linked. Goddard [28] proved that every 4-connected maximal planar graph is 4-ordered. While the converse is not true in general, the following corollary of Theorem 6.2 holds.

Corollary 6.3 *Every (3, 3)-linked planar graph of order at least 6 is 4-ordered.*

We mention that X. Yu (unpublished) showed that every 8-connected graph is (2, 3)-linked.

The following problems are natural, in the light of other results in this chapter.

Problem C (a) *Determine the minimum function $g(i, j)$ for which every $g(i, j)$-connected graph is (i, j)-linked. More generally, determine the minimum function value r for which every r-connected graph is $(s_1, s_2, \ldots, s_k)$-linked.*
(b) *Find connectivity and minor containment conditions that imply that a graph is $(s_1, s_2, \ldots, s_k)$-linked.*

7. The disjoint paths problem

It is natural to ask whether we can actually find an (S, T)-linkage when given two sets of k vertices, a problem known as the *disjoint paths problem*. In this section we take a brief look at this problem.

Disjoint paths problem
Instance: A graph G and pairs $(s_1, t_1), (s_2, t_2), \ldots, (s_k, t_k)$ of vertices in G.
Question: Do there exist k disjoint paths $P_1, P_2, \ldots, P_k$ for which each P_i joins s_i to t_i?

Karp [41] showed that this problem is NP-complete if k is a variable part of the input. However, the question has proved to be more tractable for fixed k. For example, Seymour [63], Shiloach [64] and Thomassen [68] gave polynomial algorithms to solve the disjoint paths problem when $k = 2$. However, Fortune, Hopcroft and Wyllie [27] showed that the corresponding question for digraphs is NP-complete, even for $k = 2$.

Robertson and Seymour [61] provided an $O(n^3)$-time algorithm to solve the disjoint paths problem for fixed k. However, as they pointed out, the algorithm is not practically feasible because it would require the manipulation of huge constants.

Other results of note include the following: Tholey [65] gave the fastest known algorithms for solving the 2-path problem (both vertex and edge-disjoint versions). Broder *et al.* [7] presented a polynomial-time randomized algorithm for finding the paths in random graphs chosen from spaces with edge densities above the connectivity thresholds. Schrijver [62] showed that the k disjoint paths problem for planar digraphs can be solved in polynomial time for fixed k.

8. *H*-linked graphs

In this section, we revisit the idea of an H-linked graph and explore a number of related problems. This concept was introduced by Jung [36] and generalizes not only the class of k-linked graphs but also several other well-studied classes of graphs. The following, more technical, definition of an H-subdivision contained within a graph G is useful.

Let $\mathcal{P}(G)$ denote the set of paths in a graph G, and let H be a multigraph, possibly containing loops. An *H-subdivision* in a graph G is a pair of mappings $f : V(H) \to V(G)$ and $g : E(H) \to \mathcal{P}(G)$ for which

- f is injective;
- for each edge vw in $E(H)$, $g(vw)$ is an $f(v)$–$f(w)$ path in G;
- distinct edges of H map to internally disjoint paths in G.

Recall that the vertices in the range of f are called *branch vertices* of the subdivision. We call the paths in the range of g the *edge-paths* of the subdivision. This definition allows us to give an alternative definition of an H-linked graph: a graph G is *H-linked* if any injective function $f : V(H) \to V(G)$ can be extended to an H-subdivision.

As mentioned above, the notion of an H-linked graph encompasses that of a k-linked graph, since a graph is k-linked if and only if G is kK_2-linked. Additionally, a graph is k-connected if and only if it is $(K_2 \cup \overline{K_{k-1}})$-linked. Also, via the fan lemma of Dirac [15], a graph is k-connected if and only if it is $K_{1,k}$-linked. In a similar manner, a graph is k-ordered if and only if it is C_k-linked.

We first consider some relationships between H-linked graphs. Specifically, we define a partial order $\leq$ on the family of graphs by $H \leq J$ if every J-linked graph is H-linked. Clearly, if $H \subseteq J$, then $H \leq J$, and if $|J| < |H|$, then $K_{|J|}$ demonstrates that $H \not\leq J$. Liu, West and Yu [47] investigated the partial order $\leq$ and gave numerous pairs of graphs (H, J) for which $|H| \geq |J|$, but $J \not\leq H$.

A *double star* is a tree with exactly two non-leaf vertices, and $\mathcal{H}_{r,k}$ is the class of graphs with $k+1$ components consisting of a double star of size $r-k$ and k isolated edges. Let $\beta(G)$ be the vertex-cover number of G. The main result of [47] is as follows.

Theorem 8.1 *Let H and J be graphs, where H has order $r+k$ and J has r vertices and m edges. If no graph with $m-k$ edges is contained in both H and J, and if $\beta(J) \geq k+3$, or $\beta(J) = k+2$, and if $J \notin \mathcal{H}_{m,k}$ or $m > 2k+2$, then $\overline{J} + K_{m-1}$ is H-linked, but not J-linked.*

One special case of Theorem 8.1, dealing specifically with the case where H and J have the same order, is presented separately in [47].

Theorem 8.2 *If H and J are graphs of order n, and if $\beta(J) \geq 2$, then $J \leq H$ if and only if $J \subseteq H$.*

If $\mathcal{G}_H$ denotes the set of H-linked graphs, then the poset $\mathcal{P}$ arising from the partial order $\leq$ is simply the containment poset for the families $\mathcal{G}_H$. This poset was introduced by Whalen [75], who also determined the subposet formed by graphs with three edges. Liu, West and Yu determined most of the subposet of $\mathcal{P}$ formed by graphs with four edges, with the exception of five pairs. Here $K'_{1,3}$ denotes the claw with a single subdivided edge.

Problem D *Determine which, if any, of the following hold:*

$$K'_{1,3} \leq C_4,\ C_4 \leq K_2 \cup K_{1,3},\ K_2 \cup K_{1,3} \leq K_2 \cup K_3,\ K_2 \cup K_3 \leq 2P_3,\ C_4 \leq K_2 \cup K_3.$$

All of the other relationships are determined by Theorem 8.1 and the following result from [47]. A *splintering* at a vertex w adds a new vertex w' to G and replaces the edge vw in G with the edge vw'.

Theorem 8.3 *If H' is obtained from H by a splinter operation, then $H \leq H'$.*

Recall that $f(k)$ is the smallest integer for which every $f(k)$-connected graph is k-linked. As observed in [47], $f(k)$ can also be considered as the smallest number t for which $kK_2 \leq K_2 + (t-1)K_1$. This gives rise to the following more general version of Problem A.

Problem E *Given H, determine the minimum integer t for which $K_2 \cup (t-1)K_1$ is H-linked. Equivalently, determine the minimum connectivity $g(H)$ that assures that a graph G is H-linked.*

In line with Theorem 2.8, one can show that if H has k vertices and m edges, then $g(H) \leq f(m) + (k - 2\alpha'(H))$, where $\alpha'(H)$ is the edge-independence number of H. It would be of interest to determine stronger results for arbitrary H.

Many of the sufficient conditions for a graph to be H-linked take a degree-based approach. Kostochka and Yu [42] gave minimum-degree conditions that assure that a graph G is H-linked for every graph H with minimum degree at least 2.

Theorem 8.4 *Let H be a graph of order k with $\delta(H) \geq 2$ and no loops. Then every graph G of order $n \geq 5n + 6$ with $\delta(G) \geq \lfloor \frac{1}{2}(n + k - 1) \rfloor$ is H-linked, and this bound is sharp.*

Subsequently, minimum-degree thresholds were determined for a graph G to be H-linked that are dependent upon the graph H itself. Let

$$b(H) = \max |E(A, B)| + |C|,$$

where the maximum is taken over all partitions $\{A, B, C\}$ of $V(H)$ with $|E(A, B)| \geq 1$. If H is connected, it is not difficult to see that $b(H)$ is precisely the maximum number of edges in a bipartite subgraph of H. Ferrara *et al.* [23] proved the following result.

Theorem 8.5 *For any connected multigraph H, possibly containing loops, if G is a graph of sufficiently large order n with $\delta(G) \geq \lceil \frac{1}{2}(n + b(H) - 2) \rceil$, then G is H-linked. Furthermore, every injection $f : V(H) \to V(G)$ can be extended to an H-subdivision in which each edge-path has at most two intermediate vertices.*

The values of n required for Theorem 8.5 are quite large, in the vicinity of $2^{|E(H)|}$. Independently, Kostochka and Yu [43] obtained a similar result for connected loopless multigraphs with minimum degree at least 2; however, their result required much smaller values of n.

Theorem 8.6 *Let H be a loopless connected multigraph of order k with $\delta(H) \geq 2$. If G is a graph of order $n \geq \frac{15}{2}k$ with $\delta(G) \geq \lceil \frac{1}{2}(n + b(H) - 2) \rceil$, then G is H-linked.*

The stronger aspects of Theorems 8.5 and 8.6 were subsequently combined by Gould, Kostochka and Yu [29].

Theorem 8.7 *Let H be a multigraph of size l, possibly containing loops, and let $k_1 = k_1(H) = l + u(H)$. If G is a graph of order $n \geq \frac{19}{2}(k_1 + 1)$ with minimum degree $\delta(G) \geq \lceil \frac{1}{2}(n + b(H) - 2) \rceil$, then G is H-linked. Furthermore, every injection f : $V(H) \to V(G)$ can be extended to an H-subdivision with at most $5k_1 + 2$ vertices.*

Because Theorem 8.7 permits H to be a disconnected multigraph, the minimum-degree threshold for being k-linked from [39] can be obtained as a consequence of Theorem 8.7 for appropriately large values of n.

Note that $\lceil \frac{1}{2}(n + b(H) - 2) \rceil$ is the minimum-degree threshold for a graph G to be $b(H)$-connected, which is necessary for the H-linked property.

To see that the degree conditions in Theorems 8.5–8.7 are sharp, let $\{A, B, C\}$ be a partition of $V(H)$ for which $|E(A, B)| + |C| = b(H)$. Create G by first adding $|E(A, B)| - 1$ vertices to C to obtain C^*, and then adding vertices to A and B to create sets A^* and B^*, each of size $\frac{1}{2}(n - |C^*|)$. The edges of G are all possible edges in $A^* \cup C^*$ and $B^* \cup C^*$. It is straightforward to see that $\delta(G) \geq \lceil \frac{1}{2}(n + b(H) - 3) \rceil$, yet G is not H-linked, as there are not enough vertices in C^* to create paths representing the edges in $E(A, B)$.

In [26], Ferrara, Jacobson and Pfender obtained minimum semi-degree conditions that assure a directed graph D is H-linked for an arbitrary digraph H. While too complex to cover here in full detail, it is worth noting that the δ^0-threshold for H-linkedness in digraphs depends on a partition of the digraph H into as many as nine parts, providing an interesting contrast to Theorems 8.5–8.7.

Kostochka and Yu obtained Ore-type degree conditions, dependent on k, implying that a graph G is H-linked for *every* graph H with k edges.

Theorem 8.8 *Let G be a graph of order n and let H be a graph with k edges and minimum-degree at least 2. If*

$$\delta_2(G) \geq \begin{cases} n + \frac{1}{2}(3k - 9) & \text{for } n > \frac{5}{2}k - \frac{11}{2}, \\ n + \frac{1}{2}(3k - 8) & \text{for } 2k \leq n \leq \frac{5}{2}k - \frac{11}{2}, \\ 2n - 3, & \text{for } k \leq n \leq \frac{5}{2}k - 1, \end{cases}$$

then G is H-linked.

Recently, Theorem 8.8 was refined by Ferrara *et al.* [22], who gave sharp δ_2-conditions which assure that a graph G is H-linked for general H. Note that, for arbitrary H, this δ_2-threshold for H-linkedness is not necessarily twice the minimum degree given in Theorem 8.7, as results of Faudree *et al.* [19] on k-ordered graphs demonstrate that this is not the case for $H = C_k$, when n is sufficiently large.

The idea of an H-subdivision can be relaxed somewhat if the edge-paths may share vertices but not edges. If the paths are required only to be edge-disjoint, with no branch-vertex appearing internally on any of the paths, then the result is called a *strong immersion* of H, or an *H-immersion*. The branch-vertices and edge-paths of an H-immersion are defined as in an H-subdivision.

The edge-paths in an H-immersion may overlap significantly. Let $\mathcal{I}$ be an H-immersion in a graph G and let $S = f(V(H))$. For each vertex $v \notin S$, define the *repetition number* $r(v, \mathcal{I})$ to be 1 less than the number of paths in $g(E(H))$ containing v if v lies on a path of $\mathcal{I}$, and 0 otherwise. We then define the repetition number

$r(\mathcal{I})$ of an immersion $\mathcal{I}$ to be the sum of the repetition numbers of the vertices of $G - S$.

The problem of determining when a graph G has an H-immersion for a given set of ground vertices relates to the notion of H-linkage in much the same way that the class of weakly k-linked graphs relates to the class of k-linked graphs. Specifically, G is *weakly k-linked* if and only if every injection $f : V(kK_2) \to V(G)$ extends to a kK_2-immersion.

Ferrara *et al.* [24] proved the following result.

Theorem 8.9 *Let H be a loopless non-trivial multigraph with k vertices and m edges, and let G be a graph of order*

$$n \geq \max\{8m^2 + 8mk + 2k^2 - m + 2,\ 34m + 4k + 5\}.$$

If $0 \leq \lambda \leq b(H) - k + 1$ and $\delta(G) \geq \frac{1}{2}(n + b(H) - \lambda - 2)$, then any injective map $f : V(H) \to V(G)$ can be extended to an immersion $\mathcal{I}$ with branch vertices $S = f(V(G))$ and $r(\mathcal{I}) \leq \lambda$.

An H-immersion with repetition number 0 is an H-subdivision. Consequently, the minimum degree condition for H-linkage from Theorem 8.7 follows from Theorem 8.9.

9. *H*-extendible graphs

We now consider several results concerned with the existence of subdivisions of various orders in a graph. An H-subdivision $\mathcal{H}$ is *fully extendible* if G contains an H-subdivision $\mathcal{H}^*$ with the same branch vertices as $\mathcal{H}$ (associated with $V(H)$ in the same manner) and the same order as G. If every H-subdivision in G is fully extendible, we say that G is *H-extendible*. Gould and Whalen [31] proved the following result.

Theorem 9.1 *Let H be a multigraph with k vertices and l edges with n_1 end-vertices and n_0 isolated vertices, and let G be a $(\max\{\alpha(H), \alpha'(H)\} + 1)$-connected graph of order $n \geq 11l + 7(k - n_1)$ for which $\delta_2(G) \geq n + l - k + n_1 + 2n_0$. Then G is H-extendible.*

Theorem 9.1 demonstrates that a number of results on paths and cycles in graphs can be viewed as special cases of broader results on graph subdivisions. For instance, if H is a cycle, then $n + m - k + n_1(H) + 2n_0(H) = n$. As such, for $n \geq 19$, Ore's theorem [59] on Hamiltonian graphs can be obtained as a corollary to Theorem 9.1. Dirac's theorem [14] follows as well. Other easy corollaries to Theorem 9.1 include degree-sum conditions ensuring the existence of 2-factors with a prescribed number of components [6], degree-sum conditions for k-ordered Hamiltonicity (see [11], [19] and [54]), and degree-sum conditions that ensure the existence of a Hamiltonian cycle that visits the components of certain linear forests in a given order [20].

The problem of determining when a graph G has subdivisions of many sizes has also received attention. An H-subdivision $\mathcal{H}$ in a graph G is 1-*extendible* if G contains an H-subdivision $\mathcal{H}^*$ on the same ground set as $\mathcal{H}$ with $|\mathcal{H}^*| = |\mathcal{H}| + 1$. If every non-spanning H-subdivision in G is 1-extendible, then G is called *pan-H-linked*.

Gould *et al.* [30] gave a minimum-degree condition for 1-extendible k-linkages. With the equivalence of the kK_2-linkage and k-linkage properties in mind, we define a graph G to be *pan-k-linked* if every kK_2-subdivision is 1-extendible.

Theorem 9.2 *If $k \geq 2$, then every graph G of order $n \geq 5k - 1$ and minimum degree $\delta(G) \geq \frac{1}{2}(n + 2k - 1)$ is pan-k-linked. This bound is sharp.*

Bondy [6] extended Ore's theorem by showing that the assumption $\sigma_2(G) \geq n$ is almost always sufficient for a graph to contain cycles of all lengths.

Theorem 9.3 *If G is a graph of order $n \geq 3$ with $\sigma_2(G) \geq n$, then either G is pancyclic or G is isomorphic to $K_{n/2,n/2}$.*

Ferrara, Powell and Magnant [25] extended Theorems 9.1 and 9.2 by showing that, for almost all H, there is a similar connection between the minimum-degree condition implied by Theorem 9.1 and the existence of H-subdivisions of many different orders.

Theorem 9.4 *Let H be a multigraph with k vertices and m edges for which $m \geq \max\{(k - n_1(H)) + \alpha(H), 2k - n_1(H) - 1\}$ and no component of H is a cycle. If G is a graph of sufficiently large order n with $\delta(G) \geq \frac{1}{2}(n + m - k + n_1(H) + 2n_0(H))$, then G is pan-H-linked. This bound is sharp.*

As demonstrated by a number of examples, the requirement that no component of H is a cycle is necessary, although we should note that this restriction is subsumed by the restrictions on the number of edges in H when H is connected. Additionally, the alternative condition that $m \geq 2k - n_1(H) - 1$ is sharp, and the alternative condition that $m \geq (k - n_1(H)) + \alpha(H)$ is within 1 of being best possible.

Observe also that it is not feasible to mimic Theorem 9.3 strictly by assuring the existence of an H-subdivision in G with each order from k up to n. The classical extremal theorems of Erdős, Stone and Simonovits [17], [18] imply that the necessary minimum degree to ensure that a copy of H in G would be considerably greater than $\frac{1}{2}n$ for non-bipartite H. As such, it is more logical to focus on the problem of 1-extending given H-subdivisions, while preserving their ground sets, which is in line with the H-extendible property.

Much as Ore's theorem was obtained as a consequence of Theorem 9.1, several known results can be obtained as corollaries to Theorem 9.4, most notably the minimum-degree thresholds for vertex-pancyclicity and edge-pancyclicity, first obtained by Hendry [32]. We conclude this section with a problem in the spirit of Theorem 9.3.

Problem F *Let H be a multigraph with k vertices and*

$$m < \max\{(k - n_1(H)) + \alpha(H), 2k - n_1(H) - 1\}$$

edges. Determine a minimum-degree threshold for a graph G of sufficiently large order n to be pan-H-linked. Furthermore, describe the structure of those graphs G with $\delta(G) \geq \frac{1}{2}(n + m - k + n_1(H) + 2n_0(H))$ that are not pan-H-linked.

References

1. J. Bang-Jensen, On the 2-linkage problem for semi-complete digraphs, Graph theory in memory of G. A. Dirac, Sandbjerg, 1985, *Ann. Discrete Math.* **41** (1989), 23–37.
2. J. Bang-Jensen and G. Gutin, *Digraphs: Theory, Algorithms and Applications*, Springer-Verlag, 2001.
3. J.-C. Bermond and L. Lovász: Problem 3, *Recent Advances in Graph Theory*, (ed. M. Fiedler), Academia Prague (1975), 541.
4. B. Bollobás and A. Thomason, Highly linked graphs, *Combinatorica* **16** (1996), 313–320.
5. J. A. Bondy, Pancyclic graphs, *J. Combin. Theory (B)* **11** (1971), 80–84.
6. S. Brandt, G. Chen, R. Faudree, R. J. Gould and L. Lesniak, Degree conditions for 2-factors, *J. Graph Theory* **24** (1997), 165–173.
7. A. Broder, A. Frieze, S. Suen and E. Upfal, An efficient algorithm for the vertex-disjoint paths problem in random graphs, *Proc. Seventh Annual ACM–SIAM Symposium on Discrete Algorithms,* ACM (1996), 261–268.
8. G. Chartrand and L. Lesniak, *Graphs & Digraphs* (4th edn), Chapman & Hall, 2005.
9. G. Chen, Y. Chen, S. Gao and Z. Hu, Linked graphs with restricted lengths, *J. Combin. Theory (B)* **98** (2008), 735–751.
10. G. Chen, R. J. Gould, K. Kawarabayashi, F. Pfender and B. Wei, Graph minors and linkages, *J. Graph Theory* **49** (2005), 75–91.
11. G. Chen, R. J. Gould and F. Pfender, New conditions for k-ordered hamiltonian graphs, *Ars Combin.* **70** (2004), 245–255.
12. Y. Chen, Y. Mao and Q. Zhang, On modulo linked graphs, *Frontiers in Algorithmics*, Lecture Notes in Computer Science **5598** Springer (2009), 173–180.
13. A. Cypher, An approach to k paths problem, *Proc. 12th Annual ACM Symposium on Theory of Computing* (1980), 211–217.
14. G. A. Dirac, Some theorems on abstract graphs, *Proc. London Math. Soc.* **2** (1952), 69–81.
15. G. A. Dirac, Généralisations du thèoréme de Menger, *C. R. Acad. Sci. Paris* **250** (1960), 4252–4253.
16. G. A. Dirac, In abstraken Graphen vorhandene vollständige 4-Graphen und ihre Unterteilungen, *Math. Nachr.* **22** (1960), 61–85.
17. P. Erdős and A. Stone, On the structure of linear graphs, *Bull. Amer. Math. Soc.* **52** (1946), 1087–1091.
18. P. Erdős and M. Simonovits, A limit theorem in graph theory, *Studia Sci. Math. Hungar.* **1** (1966), 51–57.
19. J. Faudree, R. Faudree, R. J. Gould, M. Jacobson and L. Lesniak, On k-ordered graphs, *J. Graph Theory* **35** (2000), 69–82.
20. J. Faudree, R. Faudree, R. J. Gould, M. Jacobson, L. Lesniak and F. Pfender, Linear forests and ordered cycles, *Discuss. Math. Graph Theory* **24** (2009), 359–372.

21. R. J. Faudree, R. J. Gould, T. Lindquester and R. Schelp, On k-linked graphs, *Combinatorics, Graph Theory, and Algorithms*, Vol. II (Kalamazoo, 1996), New Issues Press (1999), 387–400.
22. M. Ferrara, R. J. Gould, M. Jacobson, F. Pfender, J. Powell and T. Whalen, New Ore-type degree conditions for H-linked graphs, *J. Graph Theory* **71** (2012), 69–77.
23. M. Ferrara, R. J. Gould, G. Tansey and T. Whalen, On H-linked graphs, *Graphs Combin.* **22** (2006), 217–224.
24. M. Ferrara, R. J. Gould, G. Tansey and T. Whalen, On H-immersions, *J. Graph Theory* **57** (2008), 245–254.
25. M. Ferrara, C. Magnant and J. Powell, Pan-H-linked graphs, *Graphs Combin.* **26** (2010), 225–242.
26. M. Ferrara, M. Jacobson and F. Pfender, Degree conditions for H-linked digraphs, submitted.
27. S. Fortune, J. E. Hopcroft and J. Wyllie, The directed subgraph homeomorphism problem, *J. Theoret. Comput. Sci.* **10** (1980), 111–121.
28. W. Goddard, 4-connected maximal planar graphs are 4-ordered, *Discrete Math.* **257** (2002), 405–410.
29. R. J. Gould, A. Kostochka and G. Yu, On minimum degree implying that a graph is H-linked, *SIAM J. Discrete Math.* **20** (2006), 829–840.
30. R. J. Gould, J. Powell, B. Wagner and T. Whalen, Minimum degree and pan-k-linked graphs, *Discrete Math.* **309** (2009), 3013–3022.
31. R. J. Gould and T. Whalen, Subdivision extendibility, *Graphs Combin.* **23** (2007), 165–182.
32. G. R. T. Hendry, Extending cycles in graphs, *Discrete Math.* **85** (1990), 59–72.
33. M. C. Heydemann and D. Sotteau, About some cyclic properties in digraphs, *J. Combin. Theory (B)* **38** (1985), 261–278.
34. T. Hirata, K. Kubota and O. Saito, A sufficient condition for a graph to be weakly k-linked, *J. Combin. Theory (B)* **36** (1984), 85–94.
35. A. Huck, A sufficient condition for graphs to be weakly k-linked, *Graphs Combin.* **7** (1991), 323–351.
36. H. A. Jung, Eine Verallgemeinerung des n-fachen zusammenhangs für Graphen, *Math. Ann.* **187** (1970), 95–103.
37. R. M. Karp, On the complexity of combinatorial problems, *Networks* **5** (1975), 45–68.
38. K. Kawarabayashi, k-linked graphs with girth condition, *J. Graph Theory* **45** (2003), 48–50.
39. K. Karawabayashi, A. Kostochka and G. Yu, On sufficient degree conditions for a graph to be k-linked, *Combin. Probab. Comput.* **15** (2006), 685–694.
40. K. Kawarabayashi and B. Reed, Highly parity-linked graphs, *Combinatorica* **29** (2009), 215–225.
41. A. Kostochka, A lower bound for the Hadwiger number of a graph as a function of the average degree of its vertices, *Discret. Analiz* **38** (1982), 37–58.
42. A. Kostochka and G. Yu, An extremal problem for H-linked graphs, *J. Graph Theory* **50** (2005), 321–329.
43. A. Kostochka and G. Yu, Minimum degree conditions for H-linked graphs, *Discrete Appl. Math.* **156** (2008), 1542–1548.
44. A. Kostochka and G. Yu, Ore-type degree conditions for H-linked graphs, *J. Graph Theory* **58** (2008), 14–26.
45. D. Kühn and D. Osthus, Linkedness and ordered cycles in digraphs, *Combin. Probab. Comput.* **17** (2008), 411–422.

46. D. G. Larman and P. Mani, On the existence of certain configurations within graphs and the 1-skeletons of polytopes, *Proc. London Math. Soc.* **20** (1970), 144–160.
47. Q. Liu, D. B. West and G. Yu, Implications among linkage properties in graphs, *J. Graph Theory* **60** (2009), 327–337.
48. W. Mader, Homomorphiceigenshaften und mittlere Kantendichte von Graphen, *Math. Ann.* **174** (1967), 265–268.
49. W. Mader, Topological minors in graphs of minimum degree n, *Contemporary Trends in Discrete Mathematics*, DIMACS Ser. Discrete Math. Theoret. Comput. Sci. **49** (1999), 199–211.
50. W. Mader, Topological subgraphs in graphs of large girth, *Combinatorica* **18** (1998), 405–412.
51. Y. Manoussakis, k-linked and k-cyclic digraphs, *J. Combin. Theory (B)* **48** (1990), 216–226.
52. K. Menger, Zur allgemeinen Kurventheorie, *Fund. Math.* **10** (1927), 95–115.
53. R. Mori, (3, 3)-linked planar graphs, *Discrete Math.* **308** (2008), 5280–5283.
54. L. Ng and M. Schultz, k-ordered hamiltonian graphs, *J. Graph Theory* **24** (1997), 45–57.
55. H. Okamura, Multicommodity flows in graphs II, *Japan J. Math. (N.S.)* **10** (1984), 99–116.
56. H. Okamura, Paths and edge-connectivity in graphs, *J. Combin. Theory (B)* **37** (1984), 151–172.
57. H. Okamura, Paths in k-edge-connected graphs, *J. Combin. Theory (B)* **45** (1988), 345–355.
58. H. Okamura, Every $4k$-edge connected graph is weakly $3k$-linked, *Graphs Combin.* **6** (1990), 179–185.
59. O. Ore, A note on hamilton circuits, *Amer. Math. Monthly* **67** (1960), 55.
60. F. Pfender, A result about dense k-linked graphs, *Congr. Numer.* **140** (1999), 209–211.
61. N. Robertson and P. Seymour, Graph minors XIII. The disjoint paths problem, *J. Combin. Theory (B)* **63** (1995), 65–110.
62. A. Schrijver, Finding k disjoint paths in a directed planar graph, *SIAM J. Comput.* **23** (1994), 780–788.
63. P. Seymour, Disjoint paths in graphs, *Discrete Math.* **29** (1980), 293–309.
64. Y. Shiloach, A polynomial solution to the undirected two paths problem, *J. Assoc. Comput. Mach.* **27** (1980), 445–456.
65. T. Tholey, Solving the 2-disjoint paths problem in nearly linear time, *Theory Comput. Syst.* **39** (2006), 51–78.
66. R. Thomas and P. Wollan, An improved linear edge bound for graph linkage, *Europ. J. Combin.* **26** (2005), 309–324.
67. A. Thomason, An extremal function for complete subgraphs, *Math. Proc. Camb. Philos. Soc.* **95** (1984), 261–265.
68. C. Thomassen, 2-linked graphs, *Europ. J. Combin.* **1** (1980), 371–378.
69. C. Thomassen, Connectivity in tournaments, *Graph Theory and Combinatorics* (ed. B. Bollobás), Academic Press (1984), 305–313.
70. C. Thomassen, Highly connected non-2-linked digraphs, *Combinatorica* **11** (1991), 393–395.
71. C. Thomassen, Paths, circuits and subdivisions, *Selected Topics in Graph Theory, 3* (eds. L. W. Beineke and R. J. Wilson), Academic Press, 1988.
72. C. Thomassen, Graph decomposition with applications to subdivisions and path systems modulo k, *J. Graph Theory* **7** (1983), 261–271.

73. C. Thomassen, The Erdős-Pósa property for odd cycles in graphs of high connectivity, *Combinatorica* **21** (2001), 321–333.
74. C. Thomassen, Totally odd K_4-subdivisions in 4-chromatic graphs, *Combinatorica* **21** (2001), 417–443.
75. T. Whalen, *Degree Conditions and Relations to Distance, Extendability, and Levels of Connectivity in Graphs*, Ph.D. thesis, Emory University, 2003.
76. W. Zang, Proof of Toft's conjecture: Every graph containing no fully odd K_4 is 3-colorable, *J. Comb. Optim.* **2** (1998), 117–188.

7

Tree-width and graph minors

DIETER RAUTENBACH and BRUCE REED

In this chapter, we discuss tree decompositions, motivating our treatment by discussing how they are used in the characterization of H-minor-free graphs. We focus in particular on one invariant of tree decompositions, their width, which is of central importance. We also outline the proof of Wagner's conjecture as one of the main results of Robertson and Seymour's Graph Minor Project.

1. Introduction

One of the most important bodies of work in the field of graph theory is the Graph Minor Project of Neil Robertson and Paul Seymour, which extends over a series of more than twenty papers (see [20]–[27]). The central result of this collection is a good characterization of the class of graphs that do not contain a fixed graph H as a minor.

The previous sentence is intentionally ambiguous. On the one hand, following Edmonds [12], we use *good characterization* to imply a polynomial-time recognition algorithm for the class of H-minor-free graphs. More importantly, from the perspective of this chapter, Robertson and Seymour's characterization of H-minor-free graphs is good in that it allowed them to prove the following conjecture, which was the primary motivation for their work.

Wagner's conjecture *For any infinite sequence of graphs $G_1, G_2, G_3, \ldots$, there exist indices $i < j$ for which G_i is a minor of G_j.*

The approach that Robertson and Seymour took to characterizing H-minor-free graphs was to extend Wagner's characterization of graphs with no K_5-minor. Wagner [33] showed that K_5-minor-free graphs can be decomposed in a 'tree-like' way into pieces, each of which is either planar or isomorphic to one exceptional graph. Tree decompositions also play an important role in Robertson and Seymour's characterization.

In this chapter, we discuss tree decompositions, motivating our treatment by describing how they are used in the characterization of H-minor-free graphs. We focus in particular on one invariant of tree decompositions, the *tree-width*, which is of central importance.

In Sections 2, 3 and 4, we present the basic definitions and properties of tree decompositions. In Section 5 we relate tree-width to forbidden planar minors, and in Section 6 we begin our exposition of Wagner's conjecture. In Section 7 we develop a dual to tree-width, and in Section 8 we show that every (labelled) graph has a canonical tree decomposition that splits it into its most connected pieces using the smallest possible cutsets. These results are needed to complete our exposition of the proof of Wagner's conjecture in Section 9.

In Section 10 we turn to algorithmic implications of the Graph Minors Project. We begin the section with an algorithm for testing whether an input graph has a fixed graph H as a minor, and then proceed to describe how the same techniques can be used to solve the 'k disjoint rooted paths problem': *Given a graph G, a positive integer k, and two sets $X = \{x_1, x_2, \ldots, x_k\}$ and $Y = \{y_1, y_2, \ldots, y_k\}$ of k vertices in G, determine whether there exist k disjoint paths P_i such that P_i is an x_i–y_i path.* (For an exact statement of the problem, see Section 10.)

This problem differs from the well-known Network Flow problem, which asks only for k disjoint paths from X to Y. Specifying the pairing of the endpoints, as is done here, makes the problem much harder. Indeed, it is NP-complete if k is part of the input (see [17]).

For surveys that focus on different aspects of the topic of this chapter, see Bodlaender [6] and Reed [19].

2. Subtree intersection representation

A graph is *chordal* if it has no induced cycle of length greater than 3. We consider chordal graphs at this point, because there is a structural characterization of this family of graphs that is closely related to tree decompositions.

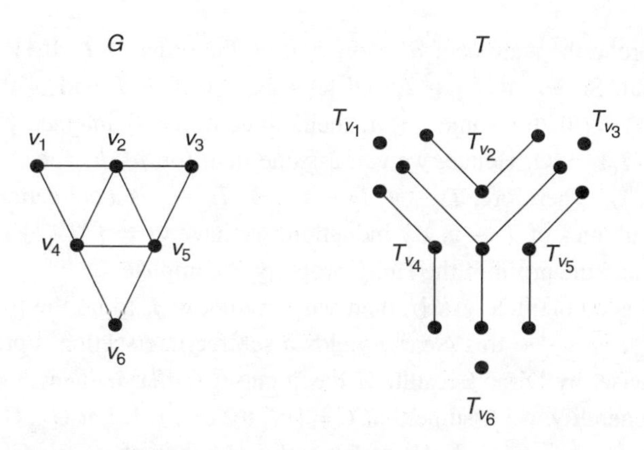

Fig. 1. A subtree intersection representation $\mathcal{T}$ of a graph G

A *subtree intersection representation* of a graph G is a collection $\mathcal{T} = (T_v : v \in V(G))$ of subtrees of a tree T such that, for every pair v and w of vertices of G, vw is an edge of G if and only if the trees T_v and T_w intersect. For example, Fig. 1 shows a graph G and a subtree intersection representation $\mathcal{T}$ of G.

We have the following result of Gavril [13].

Theorem 2.1 *A graph is chordal if and only if it has a subtree intersection representation.*

Proof The key to the proof is a classical result of Dirac [11] that states that a graph is chordal if and only if each of its induced subgraphs is a clique, or has a cutset that induces a clique. With this result at hand, we turn to the proof of the theorem.

First, we show that if a graph G with vertices $v_1, v_2, \ldots, v_n$ has a subtree intersection representation $\mathcal{T} = (T_{v_1}, T_{v_2}, \ldots, T_{v_n})$, then it is chordal. To this end, let H be an induced subgraph of G. If every pair of subtrees in H intersect, then H is a clique. Otherwise, we let S_1 and S_2 be a pair of non-intersecting trees in $\mathcal{T}' = (T_v : v \in V(H))$. This implies that there is an edge wx in T for which S_1 and S_2 are contained in different components of $T - wx$. Note that if T_{v_i} and T_{v_j} are contained in different components of $T - wx$, then they are disjoint, so there is no edge joining v_i and v_j in H. Now, the set $C = \{v_i \in V(H) : \{w, x\} \subseteq V(T_{v_i})\}$ induces a clique, and the preceding remark shows that it is a cutset of H separating the vertex corresponding to S_1 from that corresponding to S_2. Applying Dirac's characterization, we see that G is chordal.

Now, we prove by induction that every chordal graph has a subtree intersection representation. We use the following well-known fact in the proof.

Helly property of trees *If $\mathcal{T} = (T_1, T_2, \ldots, T_h)$ is a collection of pairwise intersecting subtrees of some tree T, then they have a vertex in common – that is, $\bigcap_{i=1}^{h} V(T_i) \neq \varnothing$.*

Proof We prove the statement by induction on the order of T. If $|V(T)|=1$, the claim is trivial. So let $|V(T)|\geq 2$, and let v be a leaf of T and w the neighbour of v. If $V(T_j)=\{v\}$, for some $j\leq h$, then, since every T_i intersects T_j, we have $v\in\bigcap_{i=1}^{h}V(T_i)\neq\varnothing$. Hence, we can assume that, for $j\leq h$, $v\in V(T_j)$ implies that $w\in V(T_j)$. Therefore, $T_1-v, T_2-v,\ldots,T_h-v$ is a collection of pairwise intersecting subtrees of $T-v$. By induction, we have $\varnothing\neq\bigcap_{i=1}^{h}(V(T_i)\setminus\{v\})\subseteq\bigcap_{i=1}^{h}V(T_i)$, and the proof of the Helly property is complete.

Now, if G is a complete graph, then we can choose T to be the trivial tree and set $T_{v_i}=T$ for all $i\leq n$: this clearly yields a subtree intersection representation of G, for otherwise, by Dirac's result, G has a cutset C that induces a clique. Without loss of generality, we assume that $C=\{v_1,v_2,\ldots,v_h\}$. Let $G'_1,G'_2,\ldots,G'_l$ be the components of $G-C$. Each of the induced subgraphs $G_i=G[V(G'_i)\cup C]$ for $i\leq l$ is chordal, and so by induction has a subtree intersection representation $\mathcal{T}^i=(T^i_v : v\in V(G_i))$ of subtrees of some trees T^i. Since C is a clique in G, each pair of the sets $V(T^i_{v_1}),V(T^i_{v_2}),\ldots,V(T^i_{v_h})$ intersects for each $1\leq i\leq l$. Thus, by the Helly property, there is a vertex $w_i\in\bigcap_{j=1}^{h}V(T^i_{v_j})$. Let the tree T arise from the trees $T^1,T^2,\ldots,T^l$ by adding a new vertex w joined to $w_1,w_2,\ldots,w_l$. Furthermore, for $1\leq i\leq h$, let $T_{v_i}=T[\{w\}\cup\bigcup_{j=1}^{l}V(T^j_{v_i})]$, and for $h+1\leq i\leq n$, let j^i be such that $v_i\in V(G_{j^i})$ and set $T_{v_i}=T^{j^i}_{v_i}$. It is easy to see that the collection $\mathcal{T}=(T_1,T_2,\ldots,T_n)$ of subtrees of T defines a subtree intersection representation of T, and this completes the proof of the theorem. ■

Given a subtree intersection representation of a chordal graph G, for each vertex t of T, let W_t be the set $\{v_i\in V(G) : t\in V(T_{v_i})\}$. Clearly, each W_t induces a clique. On the other hand, if C is a clique of G, then by the Helly property applied to the subtrees corresponding to the vertices of C, there exists a vertex t which is in all of these subtrees – that is, there is a vertex t with $C\subseteq W_t$. We have thus shown the following result.

Theorem 2.2 *Let $\mathcal{T}$ be a subtree intersection representation of a chordal graph G. Then G has no clique of size k or greater if and only if no vertex is contained in k or more of the trees in $\mathcal{T}$.*

3. Tree decomposition and tree-width

A *tree decomposition* of a graph G consists of a tree T and an assignment of a subset W_t of the vertex-set of G to each vertex t of T so that

(a) for each vertex v of G, the set $S_v=\{t\in V(T) : v\in W_t\}$ induces a subtree of T;
(b) for every edge vw of G, $S_v\cap S_w\neq\varnothing$.

Let $\mathcal{W}$ denote the collection of all sets W_t – that is, $\mathcal{W}=(W_t : t\in V(T))$ – and let $[T,\mathcal{W}]$ denote the corresponding tree decomposition.

It is obvious from the definition that every subtree intersection representation is a tree decomposition; more strongly, $\mathcal{T} = (T[S_v] : v \in V(G))$ is the subtree intersection representation of a chordal supergraph of G. In view of property (b), we can assign each edge vw of G to some vertex t of T so that v and w are in W_t. Doing this for all edges of G defines a collection of edge-disjoint subgraphs $\mathcal{X} = (X_t : t \in V(T))$ of G for which $V(X_t) = W_t$ and the union of the graphs in $\mathcal{X}$ is G. This observation allows a different definition of a tree decomposition (see [21]), where the sets W_t are replaced by the subgraphs X_t. We do not need this level of detail, except in Section 8.

As we mentioned in the introduction, tree decompositions have one very important invariant, their 'width'. The *width* of a tree decomposition $[T, \mathcal{W}]$ is $\max\{|W_t| - 1 : t \in V(T)\}$, and the *tree-width* $\text{tw}(G)$ of a graph G is the minimum width of a tree decomposition of G.

Here are some examples that illustrate this concept.

- Trivially, every graph G has a tree decomposition $[T, \mathcal{W}]$, where T is the tree on just one vertex t and W_t consists of the n vertices of G. The width of this decomposition is $n - 1$, and hence the tree-width of G is at most this value.
- If G has order n and has non-adjacent vertices v and w, then $[T, \mathcal{W}]$ with $V(T) = \{s, t\}$, $E(T) = \{st\}$, $W_s = V(G) \setminus \{v\}$ and $W_t = V(G) \setminus \{w\}$ is a tree decomposition of G of width $n - 2$. Together with the first remark, this implies that the complete graph K_n is the only graph of order n with tree-width $n - 1$.
- Let G be a tree, rooted at vertex v_1. If vw is an edge of G with v closer to v_1 than w, then v is called the *parent* of w and w is called a *child* of v. We obtain a tree decomposition $[T, \mathcal{W}]$ of G with $T \cong G$ by taking $W_{v_1} = \{v_1\}$ and, for $2 \le i \le n$, the set W_{v_i} to consist of v_i and its parent. For each i, the set $S_{v_i} = \{v \in V(T) : v_i \in W_v\}$ induces a star in G with centre v_i that contains all children of v_i. Since each $|W_{v_i}| \le 2$, the tree-width of G is at most 1 (see Fig. 2 for an example). If some graph G has tree width 0, then there is some tree decomposition $[T, \mathcal{W}]$ of G for which $|W_r| \le 1$ for each $r \in V(T)$. This implies that the sets S_v and S_w cannot intersect for different vertices v and w, and hence G has no edges. Thus, the tree-width of every non-trivial tree is 1.

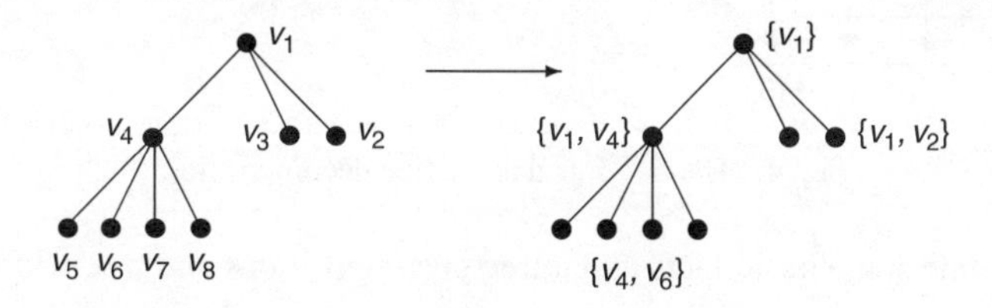

Fig. 2. A tree decomposition of a tree

- Fig. 3 contains a graph G and a tree decomposition $[T, \mathcal{W}]$ of G of width 2. Since G is chordal and the maximum order of clique in G is 3, Theorem 2.2 and the remark relating subtree intersection representations to tree decompositions imply that $\text{tw}(G) = 2$.

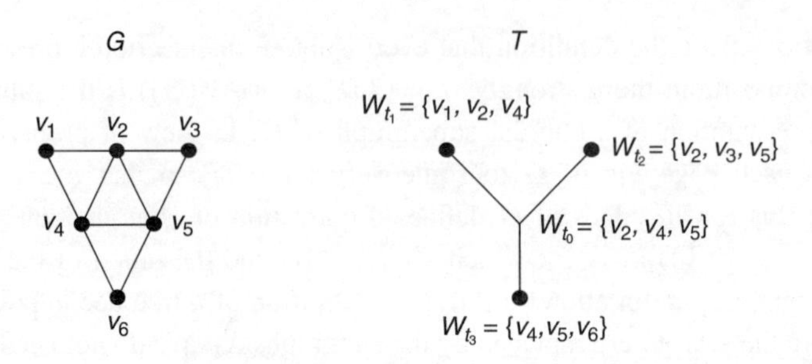

Fig. 3. A tree decomposition of a graph

- Our final example, the square $k \times k$ grid $P_k \times P_k$, is of central importance for the theory of tree decompositions. (See Fig. 4 for $P_5 \times P_5$.) Assume that its vertices are labelled $v_1, v_2, \ldots, v_{k^2}$ with $v_l = (i-1)k + j$, for $1 \leq i, j \leq k$, and its edges are $(i, j)(i', j')$ whenever $|i - i'| + |j - j'| = 1$, for $1 \leq i, i', j, j' \leq k$.

  To obtain a tree decomposition $[T, \mathcal{W}]$ of the $k \times k$ grid, let T be a path $t_1, t_2, \ldots, t_{k^2-k}$ of order $k^2 - k$, and let $W_{t_i} = \{v_i, v_{i+1}, v_{i+2}, \ldots, v_{i+k}\}$ for $1 \leq i \leq k^2 - k$. For each vertex v_i, the set $S_{v_i} = \{t_j : \max\{1, i-k\} \leq j \leq \min\{k^2 - k, i\}\}$ induces a non-trivial subpath in T. If $v_i v_j$ is an edge of the $k \times k$ grid with $j > i$, then $j - i \leq k$ and $v_i, v_j \in W_{t_i}$. This implies that t_i is in both S_{v_i} and S_{v_j}, and so their intersection is non-empty.

  The width of this decomposition is k. As we shall see later, it is in fact an optimal tree decomposition – that is, the tree - width of the $k \times k$ grid $P_k \times P_k$ is k. Fig. 4 shows the case $k = 5$.

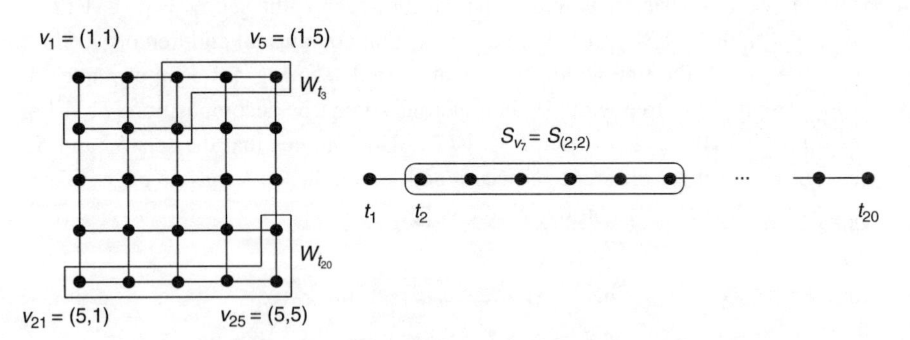

Fig. 4. The 5×5 grid and a tree decomposition

We now state some basic facts about tree decompositions and tree-width.

Theorem 3.1 *If H is a subgraph of G, then* $\text{tw}(H) \leq \text{tw}(G)$.

Proof If $[T, \mathcal{W}]$ is a tree decomposition of G of width k, then we obtain a tree decomposition $[T', \mathcal{W}']$ of H of width at most k by setting $T' = T$ and putting $W'_t = W_t \cap V(H)$ for all $t \in V(T')$. ■

Note that this implies that if a graph contains either the complete graph of order $k+1$ or the $k \times k$ grid, then it has tree-width at least k.

Theorem 3.2 *For any vertex v in a graph G,* $\text{tw}(G) \le \text{tw}(G - v) + 1$.

Proof This follows from the fact that a tree decomposition of G is obtained from a tree decomposition $[T, \mathcal{W}]$ of $G - v$ by adding v to each set W_t. ■

Note that this implies that cycles have tree-width at most 2, which we could also have shown directly.

Theorem 3.3 *If C is a clique cutset of a graph G, and if the components of $G - C$ are $G_1, G_2, \ldots, G_k$, then the tree-width of G is the maximum of the tree-widths of the induced subgraphs $G[C \cup V(G_1)], G[C \cup V(G_2)], \ldots, G[C \cup V(G_k)]$.*

Proof In order to prove this result, we mimic the proof of Theorem 2.1; the only difference is that we use tree decompositions (rather than subtree intersection representations) of the pieces to construct a tree decomposition (rather than a subtree intersection representation) of G. ■

In particular, this result implies that the tree-width of any graph is the maximum of the tree-widths of its blocks.

Theorem 3.4 *If G is a graph with n vertices and m edges, then* $\text{tw}(G) \ge m/n$.

Theorem 3.5 *Every graph G has a tree decomposition* $[T, \mathcal{W}]$ *of width* $\text{tw}(G)$, *with* $|V(T)| \le |V(G)|$.

We can prove these last two results by considering a special kind of tree decomposition. A tree decomposition $[T, \mathcal{W}]$ is called *nice* if there do not exist two adjacent vertices v and w of the tree T for which $W_v \subseteq W_w$. Let $[T, \mathcal{W}]$ be a tree decomposition of a graph G. If v and w are adjacent vertices of T for which $W_v \subseteq W_w$, then we can get a new tree decomposition $[T', \mathcal{W}']$ of G having the same width as $[T, \mathcal{W}]$ by contracting vw – that is, by setting $V(T') = V(T) \setminus \{v\}$, $E(T') = (E(T) \setminus \{vx : x \in N_T(v)\}) \cup \{wx : x \in N_T(v) \setminus \{w\}\}$ and $W'_x = W_x$ for all $x \in V(T')$. Possibly repeating this process, we eventually obtain a nice tree decomposition with the same width as $[T, \mathcal{W}]$, so every graph G has a nice tree decomposition of width $\text{tw}(G)$. Repeating this process as often as necessary, we eventually obtain a nice tree decomposition with the same width as $[T, \mathcal{W}]$, so every graph G has a nice tree decomposition of width $\text{tw}(G)$.

It is straightforward to prove that such a tree decomposition has at most $|V(G)|$ vertices, thereby proving Theorem 3.5. By considering the leaves of the tree in this decomposition, we can also show that the minimum vertex-degree in a graph of tree-width k is at most k. This implies Theorem 3.4.

The proof of the next result is left as an exercise.

Theorem 3.6 *If H is a connected subgraph of G, then for any tree decomposition $[T, \mathcal{W}]$ of G, $\bigcup\{S_v : v \in V(H)\}$ induces a subtree of T.*

We have already mentioned that tree decompositions correspond to subtree intersection representations of chordal supergraphs. This fact, combined with Theorem 2.2, yields a second characterization of tree-width using chordal graphs (see [22]).

Theorem 3.7 *The tree-width of a graph G is 1 less than the minimum order of a largest clique in any chordal supergraph of G.*

Halin [14] was the first to define tree-width, but it did not receive much attention until it was rediscovered by Robertson and Seymour [21]. An equivalent definition comes through another family of graphs.

The graphs called *k-trees* were first defined by Beineke and Pippert [5] recursively:

- The complete graph K_k is a k-tree,
- every graph that is obtained from a k-tree G by adding a new vertex adjacent to k mutually adjacent vertices is also a k-tree.

Thus, every tree is a 1-tree; an example of a 2-tree is the graph G in Fig. 1. A *partial k-tree* is a subgraph of a k-tree. This definition relates to tree-width as follows.

Theorem 3.8 *The tree-width of a graph G is the minimum k for which there is a k-tree that contains G.*

Proof We show that k-trees of order greater than k are precisely those chordal graphs for which the maximum order of a clique is $k+1$. In view of Theorem 3.7, the desired result follows.

First, let G be a k-tree, and let C be a cycle of G of length at least 4. If v is the last vertex of C added to G, then the two neighbours of v in C are adjacent in G, and so C has a chord. This implies that G is chordal. The definition implies that the maximum order of a clique in G is $k+1$.

Now, let G be a chordal graph for which the maximum order of a clique is $k+1$. It is enough to show that G contains a vertex whose neighbours induce a clique, since we can delete this vertex and apply induction. To do so, we consider a subtree intersection representation of G. Applying our procedure for constructing nice tree decompositions, we obtain a nice subtree intersection representation $\mathcal{T}$ – that is, the corresponding tree decomposition is nice. Let s be a leaf of the corresponding tree. Since we can assume that G is not a clique, s has a neighbour t in the tree. Since $\mathcal{T}$ is nice, there is a vertex v in G for which $\{s\} = V(T_v)$. Note that the neighbours of v are precisely the vertices w of G such that $s \in V(T_w)$, and these vertices induce a clique. This completes the proof. ∎

4. Tree decompositions decompose

We now show that the term 'tree decomposition' is justified, since a tree decomposition of a graph indicates a way of decomposing the graph in a tree-like fashion. If $[T, \mathcal{W}]$ is a tree decomposition of a graph G and S is a subtree of T, then we denote $\bigcup\{W_s : s \in V(S)\}$ by V_S.

Theorem 4.1 *Let $[T, \mathcal{W}]$ be a tree decomposition of a graph G, let rs be an edge of T and let R and S be the components of $T - rs$ containing r and s, respectively. Then $(V_R \setminus W_s, V_S \setminus W_r)$ is a partition of $V(G) \setminus (W_r \cap W_s)$, and there is no edge of G between $V_R \setminus W_s$ and $V_S \setminus W_r$.*

Proof For every vertex v of G not in both W_r and W_s, S_v is totally contained in either R or S, but not both – that is, v is in either $V_R \setminus W_s$ or $V_S \setminus W_r$, but not both. Hence, $(V_R \setminus W_s, V_S \setminus W_r)$ is a partition of $V(G) \setminus (W_r \cap W_s)$ (see Fig. 5). Furthermore, if $v \in V_R \setminus W_s$ and $w \in V_S \setminus W_r$, then $S_v \subseteq V(R)$ and $S_w \subseteq V(S)$. Thus $S_v \cap S_w = \varnothing$, and so the edge vw is not in G. ■

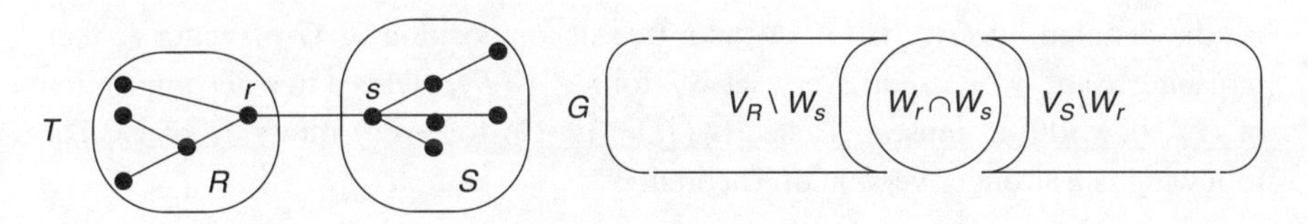

Fig. 5. $W_r \cap W_s$ decomposes the graph G

Let $[T, \mathcal{W}]$ be a tree decomposition of a graph G. By Theorem 4.1, the edges of T correspond to cutsets of G, unless $V_S \subseteq W_s \subseteq W_r$ or $V_R \subseteq W_r \subseteq W_s$; note that no such exceptional edge exists in a nice tree decomposition. In the same way, for each vertex t of T, the removal of W_t decomposes G into pieces that correspond to the components of $T - t$; typically, there are at least two such pieces, so W_t is a cutset. In Section 8, we discuss how these cutsets fit together in a tree-like way. For the moment, we content ourselves with proving that, just as any tree can be decomposed into two pieces of approximately equal sizes by removing a vertex or edge, for a graph G of bounded tree-width, one of these cutsets splits G into balanced pieces.

Theorem 4.2 *For each set S of vertices in a graph G, there exists a set C with $|C| \leq \text{tw}(G) + 1$, for which no component of $G - C$ has more than $\frac{1}{2}|S|$ vertices of S.*

Proof Assume the contrary. Let $[T, \mathcal{W}]$ be a tree decomposition of G of minimum width, and let rs be an edge of T. By Theorem 4.1, either $V_R \setminus W_s$ or $V_S \setminus W_r$ contains more than half of the vertices of S. If $V_R \setminus W_s$ contains more than half of the vertices of S, then we direct rs from s to r; otherwise, we direct rs from r to s. The directed tree that arises from T in this way is acyclic, and hence has a sink (that is, a vertex r for which every edge at r is directed to r). This implies that no component of $G - W_r$ contains more than $\frac{1}{2}|S|$ vertices of S, which is the desired contradiction. ■

This result is a useful tool for establishing lower bounds on the tree-width of a graph. We illustrate how to proceed by proving that the tree-width of the $2k \times 2k$ grid is at least k. (In Section 7, we show that it is in fact $2k$.) If we delete a set C of at most k vertices from this grid, then C misses at least k rows and one column. Since these rows all intersect this column, they are all in the same component, which therefore contains more than half the vertices of the grid. So, applying Theorem 4.2 with $S = V(G)$, we obtain the desired result.

5. Excluding planar minors

Minors appear in many natural ways in graph theory. For example, if a graph is embeddable in a given surface, then so are all of its minors. The simplest observation relating minors to tree decompositions is the following. Let G be a graph, let xy be an edge of G, and let the graph obtained by contracting the edge xy to a single vertex v_{xy} be denoted by G_{xy}. If $[T, \mathcal{W}]$ is a tree decomposition of G of width k, then replacing x and y by v_{xy} in every set W_t, for $t \in V(T)$, yields a tree decomposition of G_{xy} of width at most k. Hence $\mathrm{tw}(H) \leq \mathrm{tw}(G)$ for every minor H of G. The following is a stronger version of Theorem 3.1.

Theorem 5.1 *If H is a minor of graph G, then* $\mathrm{tw}(H) \leq \mathrm{tw}(G)$.

There is a simple relation between tree-width and the existence of small clique minors.

Theorem 5.2 *For $k \in \{0, 1, 2\}$, a graph has tree-width at most k if and only if it has no K_{k+2}-minor.*

Proof We consider only the case $k = 2$, as the other two cases are straightforward. If G contains a K_4-minor, then $\mathrm{tw}(G) \geq \mathrm{tw}(K_4) = 3$. So we need to show that if G contains no K_4-minor, then $\mathrm{tw}(G) \leq 2$. We proceed by induction on the order of G, and assume that G is connected.

If G has no cycle, then $\mathrm{tw}(G) \leq 1$ and we are done. Assume that G contains a chordless cycle C. If C spans G, then $\mathrm{tw}(G) = 2$, so we assume that there is a vertex v not on C. Since G has no K_4-minor, there do not exist three paths from v to C that are vertex-disjoint except for v. Thus, by Menger's theorem, G has a cutset of order 1 or 2. If G has a cut-vertex or a cutset of two adjacent vertices, then we can apply induction and Theorem 3.3 to obtain the desired result. So we assume that G has a minimum cutset X consisting of two non-adjacent vertices x and y. By the minimality of X, for each component F of $G - X$, there is an x–y path in $G[X \cup V(F)]$, and hence such a path in $G - F$ itself. Thus, the graph obtained from $G[X \cup V(F)]$ by adding the edge xy is a minor of G, and so has no K_4-minor. We can thus apply induction and Theorem 3.3 to these auxiliary graphs to obtain a tree decomposition of $G + xy$, and hence of G, of width 2. ■

It is not possible to extend Theorem 5.2 immediately to larger values of k, for the simple reason that there exist graphs with no K_5-minor but with arbitrarily large tree-width. In particular, since the $k \times k$ grid is planar, by Kuratowski's theorem [16], it contains no K_5-minor, but its tree-width is k (we showed earlier that it is at least $\frac{1}{2}k$). In Section 9 we give a characterization of K_l-minor-free graphs, for $l \geq 5$, in terms of tree decompositions. To close the present section, we show how to characterize graphs with bounded tree-width by forbidding minors (see [10], [28]).

Theorem 5.3 *For $k \geq 1$, the tree-width of every graph with no minor isomorphic to the $k \times k$ grid is at most 20^{2k^5}.*

Since a graph with a $k \times k$ grid minor has tree-width at least k, this theorem shows that the tree-width of a graph is large, precisely when it has a large grid minor.

6. Wagner's conjecture

In this section we begin our exposition of Robertson and Seymour's proof of Wagner's conjecture. In 1960 Kruskal [15] proved that the conjecture holds for trees. An alternative proof was later given by Nash-Williams [18], and by mimicking his argument (and proving many auxiliary results), Robertson and Seymour [23] proved the conjecture for graphs of bounded tree-width. As we show now, it follows easily from this result and Theorem 5.3 that the conjecture holds for all planar graphs.

Theorem 6.1 *Wagner's conjecture holds for planar graphs.*

Proof Let $G_1, G_2, \ldots$ be an infinite sequence of planar graphs. We may assume that G_1 is not a minor of any G_i, for $i \geq 2$, since otherwise we are done. Since G_1 is planar, there is some integer $k \geq 1$ for which it is a minor of the $k \times k$ grid. This can most easily be seen by superposing a very fine grid on a plane embedding of G_1. The vertices and edges of G_1 can obviously be 'approximated' by cycles and paths (respectively) in the grid (see Fig. 6). Hence, for $i \geq 2$, the $k \times k$-grid is not a minor of G_i, and so Theorem 5.3 implies that $G_2, G_3, \ldots$ is a sequence of graphs of bounded tree width. Since Wagner's conjecture holds for such graphs, the proof is complete. ■

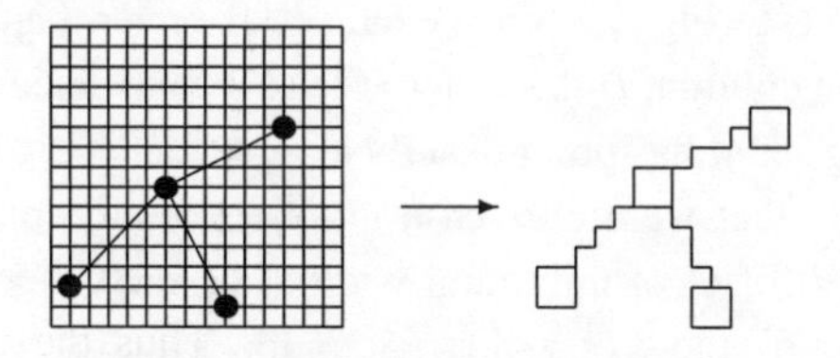

Fig. 6. G_1 is a minor of some grid

Robertson and Seymour [24] later extended this result to graphs that are embeddable on an arbitrary surface.

Now, given a sequence $G_1, G_2, \ldots$ of graphs that is supposedly a counter-example to Wagner's conjecture, we would like to imitate the proof of Theorem 6.1 to obtain a contradiction. As before, we use the fact that, for $i \geq 2$, G_1 is not a minor of G_i. Since, if G_1 has order l, it is a minor of the complete graph K_l, we deduce that K_l cannot be a minor of G_i, for $i \geq 2$. In order to complete the proof, we need a structural characterization of graphs with no K_l-minor to play the role of Theorem 5.3. Robertson and Seymour developed such a characterization, but it is too complicated even to state in its entirety here. Over the next three sections, we present some of the ideas used in their characterization.

7. The dual of tree-width

By Theorem 3.7, the tree-width of a chordal graph G is precisely one less than the order of a largest clique of G. Thus, to certify that a tree decomposition of G is optimal, we need only provide a clique of the appropriate size. For an arbitrary graph G, we can use the largest order of a clique-minor or the largest k for which G contains a $k \times k$ grid minor as a lower bound on its tree-width. However, these bounds can be far from exact. There are graphs (grids, in particular) with arbitrarily large tree-width but for which the first bound is 4. And, although Theorem 5.3 shows that the second bound is more closely tied to tree-width, it implies only that if this bound is k, then the tree-width is at most 20^{2k^5}. A tool that certifies an exact lower bound on the tree-width of G is a dual concept. This is the topic of the present section.

We say that two connected subgraphs of a graph *touch* if they either share a vertex or are joined by an edge. A *bramble* in a graph G is a set $\mathcal{B}$ of connected subgraphs of G for which any two graphs in $\mathcal{B}$ touch. A *hitting set* of a bramble $\mathcal{B}$ is a set X of vertices of G that intersects each element of $\mathcal{B}$, and the *order* $\mathrm{ord}(\mathcal{B})$ of a bramble $\mathcal{B}$ is the minimum cardinality of a hitting set of $\mathcal{B}$. The *bramble number* $\mathrm{bn}(G)$ of G is the maximum order of a bramble in G.

In the complete graph K_n, the set of one-vertex subgraphs is obviously a bramble of order n. Furthermore, as we can easily verify, a clique minor of order k is precisely a bramble of order k whose elements are disjoint. In a $k \times k$ grid, the set of all subgraphs that consist of one row and one column – that is, that are induced by the set of vertices $\{(i, h) : 1 \leq h \leq k\} \cup \{(h, j) : 1 \leq h \leq k\}$, for some $1 \leq i, j \leq k$ – forms a bramble $\mathcal{B}$ (see Fig. 7). Since each set of fewer than k vertices misses at least one row and one column, $\mathcal{B}$ has order at least k, and since the vertices of a row clearly form a hitting set, it has order exactly k.

Now let us assume that we are given a graph G, a bramble $\mathcal{B}$ of G and a tree decomposition $[T, \mathcal{W}]$ of G of minimum width. For each H in $\mathcal{B}$, since H is connected (by Theorem 3.6), the set $\bigcup_{v \in V(H)} S_v$ induces a non-trivial subtree T_H of T. Furthermore, since every two graphs H_1 and H_2 in $\mathcal{B}$ touch, the two subtrees T_{H_1}

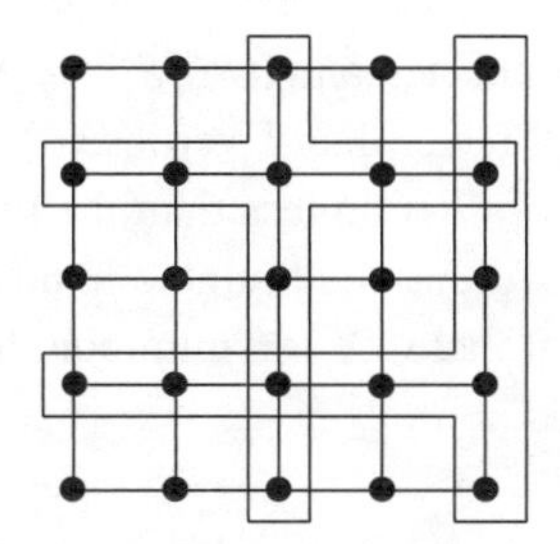

Fig. 7. Two elements of a bramble in the 5×5 grid

and T_{H_2} of T share a vertex. This implies that $\{T_H : H \in \mathcal{B}\}$ is a set of pairwise intersecting subtrees of a tree. By the Helly property, some vertex t of T is in all of the trees T_H for $H \in \mathcal{B}$. The set W_t is clearly a hitting set of $\mathcal{B}$, and so the order of $\mathcal{B}$ is at most $|W_t| \leq \text{tw}(G) + 1$. We have just proved the first (and simpler) half of the following duality result [32].

Theorem 7.1 *The tree-width of a graph is exactly one less than its bramble number.*

In [32] Seymour and Thomas stated this theorem in a different way. They considered a game played on a graph by a 'robber' and some 'cops' who want to catch the robber. They proved that a graph has tree-width at least $k-1$ if and only if the robber can escape fewer than k cops. Their proof relies on the duality of tree decompositions and brambles. If the tree-width of the graph is small, then the cops can use a tree decomposition of small width to corner, and then eventually to catch, the robber. Conversely, if the tree-width of the graph is large, then it has a bramble of large order, and the robber can evade the cops by running between the subgraphs in the bramble (note that all of these subgraphs touch).

With the help of Theorem 7.1, we can now easily establish lower bounds on the tree-width of a graph by exhibiting a bramble of large order. For example, we showed above that the $k \times k$ grid has a bramble of order k, and hence has tree-width at least $k - 1$. It is a straightforward exercise to prove that this grid actually contains a bramble of order $k + 1$ and hence has tree-width at least k; in fact, it is exactly k, as was shown in Section 3.

Small tree-width implies that a graph can be decomposed using small cutsets, and the existence of a bramble of large order implies that the graph has an indecomposable piece. Thus, the bramble number is a measure of graph connectivity. We now give another, more natural, measure that points to an indecomposable piece of a graph, and which is also an approximate dual of tree-width. For a discussion of this and other connectivity measures that are duals of tree-width, see [19].

Let G be a graph, and let k be a positive integer. A set S of vertices of G is *k-linked* if, for every set X of fewer than k vertices, some component of $G - X$ contains more than half the vertices of S. The *linkedness* link(G) of a graph G is the largest k for which the graph contains a k-linked set. Now, for any set S, the set

$\mathcal{B}_S = \{\mathcal{U} \subseteq \mathcal{V}(\mathcal{G}) : \mathcal{G}[\mathcal{U}]$ is connected and $|U \cap S| > \frac{1}{2}|S|\}$ is clearly a bramble of G. If S is k-linked, then no set of fewer than k vertices of G can be a hitting set of $\mathcal{B}_S$. Hence, the bramble number of G is no less than the linkedness of G. On the other hand, it is not difficult to prove that the bramble number of G is at most twice the linkedness; in fact, it turns out that if X is a minimum hitting set of a bramble in G, then G must be $\lceil\frac{1}{2}|X|\rceil$-linked.

8. A canonical tree decomposition

In Section 3 we saw that the edges of a tree in a tree decomposition correspond to cutsets of the graph. In fact, they define a very special collection of 'separations' of the graph. A *separation* of a graph G is a pair (A, B) of edge-disjoint subgraphs of G whose union is G. The *order* of the separation (A, B) is the number of vertices in both A and B.

Now let $[T, \mathcal{W}]$ be a tree decomposition of a graph G. As noted in Section 3, for each $t \in V(T)$, we can choose edge-disjoint subgraphs X_t with vertex-set W_t so that $G = \bigcup_{t\in V(T)} X_t$. For every edge rs of T, let R and S be the components of $T - rs$ that contain r and s, respectively. By Theorem 4.1, the graphs $A = \bigcup_{t\in V(R)} X_t$ and $B = \bigcup_{t\in V(S)} X_t$ form a separation (A, B) of G of order $|W_r \cap W_s| = |V(A) \cap V(B)|$. Hence, every edge rs of a tree in a tree decomposition defines exactly two separations, (A, B) and (B, A).

If (A, B) and (A', B') are separations defined by two edges of the same tree, then $A \subseteq A'$ and $B' \subseteq B$, or $A \subseteq B'$ and $A' \subseteq B$, or $A' \subseteq A$ and $B \subseteq B'$, or $B' \subseteq A$ and $B \subseteq A'$. Any pair of separations with this property is called *laminar*. With this definition, we can summarize our argument by saying that the edges in every tree decomposition $[T, \mathcal{X}]$ of a graph G define a set of $|E(T)|$ pairwise laminar separations. We now show that the converse is also true (see [25]).

Theorem 8.1 *Let G be a graph, and let $\mathcal{S} = \{(A_1, B_1), (A_2, B_2), \ldots, (A_k, B_k)\}$ be a set of pairwise laminar separations of G. Then there is a tree decomposition $[T, \mathcal{X}]$ of G for which*

- *every separation (A_i, B_i), for some $1 \le i \le k$, is defined by exactly one edge of T;*
- *for each edge rs of T, at least one of the two separations defined by rs belongs to $\{(A_i, B_i) : 1 \le i \le k\}$.*

Proof We prove the result by induction on k. For $k = 1$, let $[T, \mathcal{X}]$ be such that $V(T) = \{r, s\}$, $E(T) = \{rs\}$, $X_r = A_1$ and $X_s = B_1$. Then $[T, \mathcal{X}]$ has the desired properties.

Now let $k \ge 2$. Among the separations in $\mathcal{S}$, we choose a separation (A, B) for which no A_i or B_i is a proper subgraph of A. The laminarity of $\mathcal{S}$ implies that, for every $(A_i, B_i) \in \mathcal{S}$, either $A \subseteq A_i$ or $A \subseteq B_i$, since if (for example) $A_i \subseteq A$, then

$A_i = A$. Let $[T, \mathcal{X}]$ be a tree decomposition of G for which the two properties are satisfied for the set of separations $\mathcal{S} \setminus \{(\mathcal{A}, \mathcal{B}), (\mathcal{B}, \mathcal{A})\}$. Each edge rs of T corresponds to some separation (C, D) with $C = \bigcup_{t \in V(R)} X_t$ and $D = \bigcup_{t \in V(S)} X_t$ (see Theorem 4.1). If $A \subseteq C$ we direct the edge rs from s to r, while if $A \subseteq D$ we direct rs from r to s. The directed tree that arises in this way has exactly one sink t. Let T' be such that $V(T') = V(T) \cup \{t'\}$, $E(T') = E(T) \cup \{tt'\}$ and let $X'_t = X_t \setminus (V(A) \setminus V(B))$, $X'_{t'} = A$ and $X'_s = X_s$, for all $s \in V(T') \setminus \{t, t'\}$. It is straightforward to verify that $[T', \mathcal{X}']$ with $\mathcal{X}' = (X'_t : t \in V(T'))$ is a tree decomposition of G that has the desired properties. ■

We have just seen that a tree decomposition is simply a set of laminar separations. Our next theorem shows that every graph has a tree decomposition, called its *canonical tree decomposition*, that cuts it into its most highly connected pieces, using cutsets that are as small as possible. In order to define it, we have to consider a special class of brambles called 'tangles'.

A set $\mathcal{T}$ of connected subgraphs of G is a *tangle* if, for any triple T_1, T_2, T_3 of elements of $\mathcal{T}$, either there is a vertex in all three, or there is an edge e for which each $V(T_i)$ contains an endpoint of e. By taking $T_2 = T_3$, we see that every tangle is a bramble.

The *tangle number* $\text{tn}(G)$ of G is the maximum order of a tangle in G, where the order of a tangle is its order as a bramble. Clearly, the tangle number of a graph does not exceed its bramble number. Robertson and Seymour [25] defined a *branch decomposition* and the *branch-width* of a graph, and used the duality of branch decompositions and tangles to prove that the bramble number of G is at most $\frac{3}{2}$ of its tangle number. They also observed that this bound is sharp, in view of the complete graph. (For a simple proof that the bramble number is at most three times the tangle number, see Reed [19].)

Let $\mathcal{B}$ be a bramble of a graph G, and let X be a set of vertices for which $|X| < \text{ord}(\mathcal{B})$. Then the graph $G - X$ has exactly one component containing an element of $\mathcal{B}$, which we denote by $f_{\mathcal{B}}(X)$. Two brambles $\mathcal{B}_1$ and $\mathcal{B}_2$ are *distinguishable* if there is a set X of vertices for which $|X| < \min\{\text{ord}(\mathcal{B}_1), \text{ord}(\mathcal{B}_2)\}$ and $f_{\mathcal{B}_1}(X) \neq f_{\mathcal{B}_2}(X)$; such a set X is called a $(\mathcal{B}_1, \mathcal{B}_2)$-*distinguisher*. A bramble (or tangle) is called *maximal* if there is no bramble (or tangle) of higher order that is indistinguishable from it.

Just before the statement of Theorem 7.1 we saw that, for any tangle (or bramble) $\mathcal{B}$ and any tree decomposition $[T, \mathcal{X}]$, there is some vertex t in T for which $W_t = V(X_t)$ is a hitting set of $\mathcal{B}$. The next theorem of Robertson and Seymour [25] sheds more light on this relationship.

Canonical tree decomposition theorem *Every graph G has a tree decomposition $[T, \mathcal{X}]$ with the following properties:*

- *For each maximal tangle $\mathcal{T}$, there is exactly one vertex $t(\mathcal{T})$ in T for which $W_t = V(X_t)$ is a hitting set of $\mathcal{T}$.*

- *If $\mathcal{T}_1$ and $\mathcal{T}_2$ are indistinguishable maximal tangles (and hence of the same order), then $t(\mathcal{T}_1) = t(\mathcal{T}_2)$.*
- *If $\mathcal{T}_1$ and $\mathcal{T}_2$ are distinguishable maximal tangles, then $t(\mathcal{T}_1) \neq t(\mathcal{T}_2)$ and there is an edge st on the path in T from $t(\mathcal{T}_1)$ to $t(\mathcal{T}_2)$ for which $W_s \cap W_t$ is a $(\mathcal{T}_1, \mathcal{T}_2)$-distinguisher of minimum order.*
- *For every vertex s of T, there is a maximal tangle $\mathcal{T}$ with $s = t(\mathcal{T})$.*

The key to proving this theorem lies in showing that the minimal distinguishers under consideration are laminar. By the last two properties above, the canonical tree decomposition is nice and has at most n vertices. Therefore, every graph has at most n indistinguishable maximal tangles, which is not the case for indistinguishable maximal brambles.

9. Wagner's conjecture for arbitrary graphs

In the preceding sections we introduced the theoretical tools needed to complete our description of Robertson and Seymour's proof of Wagner's conjecture. As noted above, we wish to have a structural characterization of graphs with no K_l-minor, for $l \geq 5$. We can formulate such a characterization in terms of canonical tree decompositions.

Theorem 9.1 *Let $l \geq 1$, let G be a graph with no K_l-minor, and let $[T, \mathcal{X}]$ be its canonical tree decomposition. Then, for every vertex t in T, the graph X_t is 'almost embeddable' in some surface in which K_l cannot be embedded.*

We shall not give the exact definition of being 'almost embeddable' in a surface Σ, but here mention only that one case in which a graph G is almost embeddable in a surface Σ is when there is a set $Z \subset V(G)$ of bounded size (depending on l) for which $G - Z$ is embeddable in Σ.

Actually, the characterization theorem of Robertson and Seymour is much stronger than the above result suggests. They proved the same statement when $[T, \mathcal{X}]$ is obtained by taking not all of the laminar separations in a tree decomposition, but only those whose order is less than some bound $f(l)$. (By Theorem 8.1, we do get a tree decomposition from these separations.) We refer to this tree decomposition as the *l-canonical tree decomposition* of G.

Now let $G_1, G_2, \ldots$ be an arbitrary sequence of graphs. As in the proof of Theorem 6.1 we can assume that, if the order of G is l, then K_l is not a minor of any G_i, for $i \geq 2$. This implies that, for $i \geq 2$, an l-canonical tree decomposition splits the graph G_i into pieces that are almost embeddable in some fixed surface. Robertson and Seymour first proved that Wagner's conjecture holds for these pieces, and then completed the proof by extending Kruskal's result again to graphs that are built up in a tree-like manner from such pieces.

We point out an important consequence of Wagner's conjecture, now a theorem.

Theorem 9.2 *Every class $\mathcal{G}$ of graphs that is closed under taking minors has a characterization in terms of finitely many forbidden minors.*

Proof Let Forb($\mathcal{G}$) be the set of all graphs H for which H does not belong to $\mathcal{G}$, but every proper minor of H does. Since $\mathcal{G}$ is closed under the taking of minors, a graph G belongs to $\mathcal{G}$ if and only if G has no minor in Forb($\mathcal{G}$). Furthermore, no graph in Forb($\mathcal{G}$) can be a minor of any other graph in Forb($\mathcal{G}$). By the truth of Wagner's conjecture, Forb($\mathcal{G}$) must be a finite set, and hence $\mathcal{G}$ is characterized in terms of finitely many forbidden minors. ■

Archdeacon and Huneke [2] proved earlier that the class of graphs that are embeddable in some fixed non-orientable surface has a characterization in terms of finitely many forbidden minors; this was a major result that is now just a special case of Robertson and Seymour's theorem.

There are only a few examples of minor-closed classes for which the complete finitely many forbidden minors are explicitly known. The prototype of such results is Kuratowski's characterization [16] of planar graphs. Archdeacon [1] determined all 35 forbidden minors for the graphs that can be embedded in the projective plane.

We give one more example here. A graph G is called *linklessly embeddable* if it can be embedded in three-dimensional space in such a way that every two cycles in G can be pulled apart – that is, no two cycles are joined together like two consecutive links of a chain. The problem of characterizing linklessly embeddable graphs arose in [7], [30] and [31]. In [29], Robertson, Seymour and Thomas determined the finite list of forbidden minors, the 'Petersen family', for this class of graphs.

10. Efficient characterization of H-minor-free graphs

We now turn our attention to the disjoint rooted paths problem, mentioned in Section 1.

***k* Disjoint Rooted Paths Problem**

Instance: *A graph G and two sets of vertices of G: $X=\{x_1, x_2, \ldots, x_k\}$ and $Y=\{y_1, y_2, \ldots, y_k\}$.*

Question: *Are there k disjoint paths $P_1, P_2, \ldots, P_k$ in G for which each P_i is an x_i–y_i path?*

For graphs G with n vertices and m edges, Robertson and Seymour [26] gave an algorithm for this problem that runs in $O(n^3)$ time. The current fastest algorithm, which runs in $O(m + n \cdot \text{polylog } n)$ is due to Kawarabayashi and Reed.

Robertson and Seymour's algorithm relies on a simple key observation. Assume that the graph G for which we want to solve the problem contains a large grid as a minor, and assume further that the k disjoint paths $P_1, P_2, \ldots, P_k$ actually exist. If the grid minor is large enough, then it should always be possible to find detours for the parts of the paths $P_1, P_2, \ldots, P_k$ that pass through the minor, in such a way that some vertex v in the 'middle' of the grid is not used by any of them. Then the vertex v is 'irrelevant' to the existence of the paths, and so it can be deleted and the algorithm started again on $G - v$.

If the tree-width of G is large, then such a grid minor always exists (see Theorem 5.3), and in fact, so does an irrelevant vertex. The proof of this splits into two parts. If G actually has a large clique minor, then it is relatively easy to prove that an irrelevant vertex exists. For graphs with no large clique minor, we need to use Robertson and Seymour's characterization of such graphs to prove that an irrelevant vertex exists. Robertson and Seymour actually showed how to find an irrelevant vertex in a graph with a large grid minor in polynomial time. Repeatedly deleting irrelevant vertices finally results in a graph H of bounded tree-width for which the answer to the k disjoint rooted paths problem is the same as for G. Since H has bounded tree-width, the problem can now be solved by a dynamic programming approach that generalizes dynamic programming on trees. This approach actually permits the solution of many NP-hard problems in polynomial time for graphs of bounded tree-width. We do not go into detail here, but refer the reader to [3], [4], [6], [8] and [19].

As we discuss below, the polynomial-time algorithm for the disjoint rooted paths problem implies the following important result.

Theorem 10.1 *For every fixed graph H, there is a polynomial-time algorithm to determine whether H is a minor of a given graph.*

Combining this theorem with Theorem 9.2, we obtain the following additional result.

Theorem 10.2 *Every class of graphs that is closed under taking minors can be recognized in polynomial time.*

In order to prove Theorem 10.1, it is actually sufficient to consider only subdivisions instead of minors, and these are somewhat simpler objects. The following observation allows us to reduce checking for minors to checking for subdivisions.

Theorem 10.3 *For every graph H, there is a finite family $\mathcal{H}$ of graphs with the property that a graph G has H as a minor if and only if it contains a subdivision of some graph in $\mathcal{H}$.*

Proof Let H be a minor of G. The graph H arises from a subgraph G' of G by repeated contractions of edges. Hence, for each vertex v in H there is a set $\mathrm{Im}(v)$ of vertices of G' that are identified to form v by these contractions. The sets $\mathrm{Im}(v)$ induce connected graphs, and for each edge vw of H there is an edge of G' that joins a vertex in $\mathrm{Im}(v)$ to a vertex in $\mathrm{Im}(w)$.

We can actually assume that, for each vertex v in H, the subgraph $G'[\mathrm{Im}(v)]$ is a tree. Since there are only finitely many non-isomorphic trees with a fixed maximum number of vertices of degree 1 or 2, each of the trees $G'[\mathrm{Im}(v)]$ is a subdivision of one of finitely many trees. Replacing the vertices of H by these trees in all possible ways, we obtain a finite collection of graphs $\mathcal{H}$ with the desired properties. ■

Together with the following result, this implies Theorem 10.1.

Theorem 10.4 *For every fixed graph H, there is a polynomial-time algorithm to determine whether a given graph contains H as a subdivision.*

Proof Consider an injection $f : V(H) \to V(G)$, of which there are only polynomially many, and for each vertex v in H replace $f(v)$ by $\deg_H(v)$ copies of $f(v)$ with the same neighbours as $f(v)$. Now, considering the copies of the vertices $f(v)$ as roots, solve (with $k = |E(H)|$) the k disjoint rooted paths problem, and try to find disjoint paths corresponding to the edges of H. This yields a polynomial time algorithm for the given problem. ■

Since it is often easy to verify that a class of graphs is minor-closed, Theorem 10.2 is easy to apply. As an illustration, we cite again the linklessly embeddable graphs. Prior to the discovery of Theorem 10.2, it was not known whether there is any recognition algorithm for this class, polynomial or otherwise – that is, determining linkless embeddability was not known to be decidable. Since the forbidden minors for this class have been determined, a polynomial-time recognition algorithm is now explicitly known.

References

1. D. Archdeacon, A Kuratowski theorem for the projective plane, *J. Graph Theory* **5** (1981), 243–246.
2. D. Archdeacon and P. Huneke, A Kuratowski theorem for nonorientable surfaces, *J. Combin. Theory (B)* **46** (1989), 173–231.
3. S. Arnborg, J. Lagergren and D. Seese, Easy problems for tree-decomposable graphs, *J. Algorithms* **12** (1991), 308–340.
4. S. Arnborg, B. Courcelle, A. Proskurowski and D. Seese, An algebraic theory of graph reduction, *J. Assoc. Comput. Mach.* **40** (1993), 1134–1164.
5. L. W. Beineke and R. E. Pippert, The number of labeled k-dimensional trees, *J. Combin. Theory* **6** (1969), 200–205.
6. H. L. Bodlaender, Dynamic programming on graphs with bounded treewidth, *Automata, Languages and Programming*, Lecture Notes in Computer Science **317**, Springer (1988), 105–118.
7. T. Böhme, On spatial representations of graphs, *Contemporary Methods in Graph Theory* (1990), 151–167.
8. B. Courcelle, The monadic second-order logic of graphs. I: Recognizable sets of finite graphs, *Inf. Comput.* **85** (1990), 12–75.
9. R. Diestel, *Graph Theory* (3rd edn), Springer, 2006.
10. R. Diestel, T. R. Jensen, K. Y. Gorbunov and C. Thomassen, Highly connected sets and the excluded grid theorem, *J. Combin. Theory (B)* **75** (1999), 61–73.
11. G. A. Dirac, On rigid circuit graphs, *Abh. Math. Sem. Univ. Hamburg* **25** (1961), 71–76.

12. J. Edmonds, Minimum partition of a matroid into independent subsets, *J. Res. Nat. Bur. Standards* **69** (1965), 67–72.
13. F. Gavril, The intersection graphs of subtrees in trees are exactly the chordal graphs, *J. Combin. Theory (B)* **16** (1974), 47–56.
14. R. Halin, S-functions for graphs, *J. Geometry* **8** (1976), 171–186.
15. J. B. Kruskal, Well-quasi ordering, the tree theorem, and Vazsonyi's conjecture, *Trans. Amer. Math. Soc.* **95** (1960), 210–225.
16. C. Kuratowski, Sur le problème des courbes gauches en topologie, *Fund. Math.* **15** (1930), 271–283.
17. J. F. Lynch, The equivalence of theorem proving and the interconnection problem, *ACM SIGMA Newsletter* **5** (1975), 31–65.
18. C. St.J. A. Nash-Williams, On well-quasi-ordering infinite trees, *Proc. Cambridge Philos. Soc.* **61** (1965), 697–720.
19. B. A. Reed, Tree width and tangles: A new connectivity measure and some applications, *Surveys in Combinatorics, Proceedings of the 16th British Combinatorial Conference* (ed. R. A. Bailey), London Math. Soc. Lecture Notes **241**, Cambridge University Press (1997), 87–162.
20. N. Robertson and P. D. Seymour, Graph minors I: Excluding a forest, *J. Combin. Theory (B)* **35** (1983), 39–61.
21. N. Robertson and P. D. Seymour, Graph minors II: Algorithmic aspects of tree-width, *J. Algorithms* **7** (1986), 309–322.
22. N. Robertson and P. D. Seymour, Graph minors, III: Planar tree-width, *J. Combin. Theory (B)* **36** (1984), 49–64.
23. N. Robertson and P. D. Seymour, Graph minors IV: Tree-width and well-quasi-ordering, *J. Combin. Theory (B)* **48** (1990), 227–254.
24. N. Robertson and P. D. Seymour, Graph minors VIII: A Kuratowski theorem for general surfaces, *J. Combin. Theory (B)* **48** (1990), 255–288.
25. N. Robertson and P. D. Seymour, Graph minors X: Obstructions to tree-decomposition, *J. Combin. Theory (B)* **52** (1991), 153–190.
26. N. Robertson and P. D. Seymour, Graph minors XIII: The disjoint paths problem, *J. Combin. Theory (B)* **63** (1995), 65–110.
27. N. Robertson and P. D. Seymour, Graph minors XX: Wagner's conjecture, *J. Combin. Theory (B)* **92** (2004), 162–210.
28. N. Robertson, P. D. Seymour and R. Thomas, Quickly excluding a planar graph, *J. Combin. Theory (B)* **62** (1994), 323–348.
29. N. Robertson, P. D. Seymour and R. Thomas, Sachs' linkless embedding conjecture, *J. Combin. Theory (B)* **64** (1995), 185–227.
30. H. Sachs, On a spatial analogue of Kuratowski's theorem on planar graphs – an open problem, *Graph Theory, Proc. Conf. Lagow, Poland*, 1981, Lecture Notes in Mathematics, Springer **1018** (1983), 230–241.
31. H. Sachs, On spatial representations of finite graphs, *Finite and Infinite Sets, Proc. 6th Hungar. Combin. Colloq.* 1981, Vol. II, Colloq. Math. Soc. János Bolyai **37** (1984), 649–662.
32. P. D. Seymour and R. Thomas, Graph searching and a min-max theorem for tree-width, *J. Combin. Theory (B)* **58** (1993), 22–33.
33. K. Wagner, Über eine Erweiterung eines Satzes von Kuratowski, *Deutsche Math.* **2** (1937), 280–285.

8

Toughness and binding numbers

IAN ANDERSON

The toughness and binding number of a graph are two measures of how its edges are distributed and how well the graph is connected. We survey the main properties of toughness and binding number, and study the conditions upon them that ensure the existence of factors and cycles.

1. Introduction

The concept of toughness was introduced by Václav Chvátal [21] in 1973. As he wrote, toughness 'measures in a simple way how tightly various pieces of graph hold together'.

The origin of the concept can be found in the observation that if a graph G has a Hamiltonian cycle then, whenever a set S of s vertices is removed from the graph, the remaining graph $G - S$ can have at most s connected components. A result of a similar nature is the well-known factor theorem of Tutte, which states that a graph G of even order has a perfect matching if and only if, for every subset S of $V(G)$, the number of components of $G - S$ of *odd* order is at most $|S|$. In both of these situations, the number of components of $G - S$ is critical.

Now let S be a cutset of a graph G, and let $\omega(G - S)$ denote the number of components of $G - S$. Since S is a cutset, $\omega(G - S) \geq 2$. If G is not complete, we say that G is *t-tough* if $t \leq |S|/\omega(G - S)$ for every cutset S, and we define the *toughness* $\tau(G)$ of G to be the largest value of t for which G is t-tough. Thus

$$\tau(G) = \min\{|S|/\omega(G - S)\},$$

where the minimum is taken over all cutsets S of G. Sometimes a 1-tough graph is called simply a *tough graph*. The toughness of a complete graph is usually taken to be ∞.

Much of the subsequent interest in toughness was inspired by several conjectures made by Chvátal in [21]. One of the most challenging of these conjectures is still an open question:
Is there a value of t such that if $\tau(G) \geq t$, then G contains a Hamiltonian cycle?
Other conjectures relate to the existence of factors of various types, and we shall look at these later.

Our first theorem brings together some of the basic results on toughness, established in [21] and [50].

Theorem 1.1 *Let G be a graph. Then*

- $\tau(G) = 0$ *if and only if G is disconnected;*
- $\tau(C_n) = 1$ *for all $n \geq 4$;*
- $\tau(K_{r,s}) = r/s$ *for $r \leq s$;*
- $\tau(G) \leq 1$ *if G is bipartite;*
- $\tau(H) \leq \tau(G)$ *if H is a spanning subgraph of G.*

About the same time as Chvátal introduced toughness, Douglas Woodall [61] introduced the concept of binding number. For S a subset of the vertices of a graph G, let $N(S)$ denote the set of neighbours of the vertices in S. We say that G is *b-bound* if $b \leq |N(S)| \,/\, |S|$ for all non-empty subsets S with $N(S) \neq V(G)$. The *binding number* $\beta(G)$ is the greatest number b for which G is b-bound – that is,

$$\beta(G) = \min\{|N(S)|/|S| : \varnothing \neq S \subseteq V(G), N(S) \neq V(G)\}.$$

Any set S with $|N(S)| = \beta(G)|S|$ is called a *binding set* of G.

Like toughness, the binding number measures how well connected the graph is, and, not surprisingly, there are some general relationships between binding number, toughness and connectivity, which we shall explore.

Woodall was led to his definition by a result of Anderson [2], which, in binding number terms, asserts that if the order of G is even and $\beta(G) \geq \frac{4}{3}$, then G has a 1-factor. Anderson had been trying to find a result for graphs in general, similar to the classical marriage theorem, which asserts that for bipartite graphs G of even order the condition $|N(X)| \geq |X|$ for all subsets X of $V(G)$ is both necessary and sufficient for the existence of a perfect matching. Woodall abstracted the concept of binding number from this work, and laid the foundations for future study in [61]. Remarkably, in that first paper he was able to prove the major result that *every $\frac{3}{2}$-bound graph is*

Hamiltonian. This achievement contrasts with the lack (so far) of a similar result for toughness.

Our second theorem is due to Woodall [61] and brings together results for binding number similar to those for toughness in the previous theorem.

Theorem 1.2 *Let G be a graph. Then*

- $\beta(K_n) = n - 1$ *for* $n \geq 2$;
- $\beta(C_n) = 1$ *if n is even, and* $(n-1)/(n-2)$ *if n is odd;*
- $\beta(K_{r,s}) = r/s$ *for* $r \leq s$;
- $\beta(G) \leq 1$ *if G is bipartite;*
- $\beta(H) \leq \beta(G)$ *if H is a spanning subgraph of G.*

In this chapter we survey the main results concerning toughness and binding number, in particular relating them to the existence of k-factors and cycles of different lengths. We make no claim to be exhaustive; rather, it is our aim to give a flavour of the results that have been obtained.

2. Toughness and connectivity

By definition, every graph G of connectivity κ has a set S of order κ for which $\omega(G - S) \geq 2$, and so

$$\tau(G) \leq |S|/\omega(G - S) \leq \tfrac{1}{2}\kappa(G).$$

Thus we have the following result of Chvátal [21].

Theorem 2.1 *If G is not a complete graph, then* $\kappa(G) \geq 2\tau(G)$.

Equality does not hold in general, but it does hold for certain families of graphs; for example, Matthews and Sumner [47] showed that if G is not complete and is claw-free (that is, it does not contain $K_{1,3}$ as an induced subgraph), then $\kappa(G) = 2\tau(G)$. It follows from Theorem 2.1 that large toughness implies large connectivity. The converse is, however, not true; for example, $\kappa(K_{r,r}) = r$, while $\tau(K_{r,r}) = 1$.

Note that if G is t-tough, then G is $2t$-connected. As an illustration, consider the implications of being 1-tough and being 2-connected. If G is 2-connected, the removal of two vertices may create more than one component – indeed, it may create many components. If G is 1-tough, then G is indeed 2-connected, but the removal of two vertices can create at most two components.

Since $\kappa(G) \leq \delta(G)$, the minimum vertex degree in G, it follows from Theorem 2.1 that $\tau(G) \leq \frac{1}{2}\delta(G)$. We now consider the connection between $\tau(G)$ and $\Delta(G)$, the maximum vertex degree. If $\alpha(G)$ is the independence number of G, and if S is any cutset of G, then $|S| \geq \kappa(G)$ and $\omega(G - S) \leq \alpha(G)$, so that

$$\alpha(G)\tau(G) \geq \kappa(G).$$

Further, it was shown by Goddard and Swart [32] that

$$\psi(G)\tau(G) \geq \kappa(G),$$

where $\psi(G)$ denotes the largest value of s for which G contains $K_{1,s}$ as an induced subgraph. This result generalizes that of Matthews and Sumner above, and yields as a corollary the following theorem.

Theorem 2.2 *If G is not complete, then*

$$\kappa(G)/\Delta(G) \leq \tau(G) \leq \tfrac{1}{2}\delta(G).$$

As an immediate corollary, we can determine the toughness of any tree.

Theorem 2.3 *If T is any non-trivial tree, then* $\tau(T) = 1/\Delta(T)$.

If G has toughness $\tau(G)$, then any cutset S for which $|S| = \tau(G)\ \omega(G - S)$ is called a *tough set* and any component of $G - S$ is a *tough component*. How tough is such a tough component C? Plummer [52] showed that if $\tau(G) \geq 1$ and if C is not complete, then $\tau(C) \geq \frac{1}{2}\tau(G)$.

We close this section by considering planar graphs. The following result is essentially due to Goddard, Plummer and Swart [31].

Theorem 2.4

- *If G is a planar graph, then* $\frac{1}{2}\kappa(G) - 1 < \tau(G) \leq \frac{1}{2}\kappa(G)$.
- *If G is a non-planar graph with genus* $\gamma(G)$, *then* $\tau(G) \geq \frac{1}{2}\kappa(G) - \gamma(G)$.

3. Toughness and cycles

Chvátal [21] observed that, since every cycle is 1-tough, every Hamiltonian graph must also be 1-tough. He further observed that the converse is not true; for example, the Petersen graph is 1-tough (in fact, it is $\frac{4}{3}$-tough) but it is not Hamiltonian. Chvátal made the following conjecture.

Conjecture A *There exists* $t > 1$ *such that every t-tough graph is Hamiltonian.*

Having constructed $\frac{3}{2}$-tough graphs with no Hamiltonian cycles, Chvátal originally conjectured that any $t > \frac{3}{2}$ would work, but this was disproved by Thomassen. For many years it was thought that $t = 2$ would be sufficient, but it is now known, through the work of Bauer, Broersma and Veldman [6], that if such a number t exists, then $t \geq \frac{9}{4}$.

Theorem 3.1 *For every* $\varepsilon > 0$, *there exists a* $(\frac{9}{4} - \varepsilon)$*-tough graph that is not Hamiltonian.*

Conjecture A still remains open. If it is true, it illustrates another difference between toughness and connectivity; for high connectivity does not ensure Hamiltonicity, as is shown by $K_{r,r+1}$ which is not Hamiltonian but has connectivity r.

Some authors have sought to obtain sufficient conditions for Hamiltonicity by combining toughness with adaptations of known sufficient conditions. Let $\delta_r(G)$ denote the minimum sum of the degrees of r independent vertices in G; then Ore's well-known sufficient condition for Hamiltonicity gives $\delta_2(G) \geq n$. The following result is due to Jung [36].

Theorem 3.2 *Every 1-tough graph G of order $n \geq 11$ with $\delta_2(G) \geq n - 4$ is Hamiltonian.*

It follows as a corollary that, if G is 1-tough with order $n \geq 11$ and if $\delta(G) \geq \frac{1}{2}(n-4)$, then G is Hamiltonian. This should be compared with the classic result of Dirac (a special case of Ore's theorem), that G is Hamiltonian if $\delta(G) \geq \frac{1}{2}n$. We note in passing that it was later shown by Faudree *et al.* [29] that being 1-tough with $\delta(G) \geq \frac{1}{2}(n-4)$ also ensures the existence of 2-factors with exactly k cycles, for all $k \leq \frac{1}{4}(n-16)$.

The next result, due to Bauer *et al.* [9], is along similar lines to Theorem 3.2.

Theorem 3.3 *Every 2-tough graph G of order n with $\delta_3(G) \geq n$ is Hamiltonian.*

Again, there is an immediate corollary that, if G is 2-tough and $\delta(G) \geq \frac{1}{3}n$, then G is Hamiltonian. The natural question now is:
Is there a similar result for t-tough graphs with $t \geq 3$?
The following general result has been obtained by Bauer *et al.* [5].

Theorem 3.4 *Every t-tough graph G of order $n \geq 3$ with $\delta(G) > (n-t-1)/(t+1)$ is Hamiltonian.*

Other results involve both the minimum degree δ and the connectivity κ. For example, the following is a result of Bauer and Schmeichel [10].

Theorem 3.5 *Every 1-tough graph G of order $n \geq 3$ with $\delta(G) \geq \frac{1}{3}(n+\kappa-2)$ is Hamiltonian.*

Now recall Chvátal's result that if the vertex-degrees $d_1 \leq d_2 \leq \ldots \leq d_n$ of G are such that $d_{n-i} \geq n-i$ whenever $d_i \leq i < \frac{1}{2}n$, then G is Hamiltonian. Let $P(t)$ denote the condition

$$P(t): \textit{if } t \leq i < \tfrac{1}{2}n, \textit{ then } d_i \leq i \Rightarrow d_{n-i+t} \geq n-i.$$

Hoàng [34] established the following result.

Theorem 3.6

- *For $t \leq 3$, every t-tough graph that satisfies $P(t)$ is Hamiltonian.*
- *For $t \geq 4$, every $(\frac{1}{4}t^2 + t + 1)$-tough graph that satisfies $P(t)$ is Hamiltonian.*

Hoàng conjectured that the restriction $t \leq 3$ in the first result is unnecessary.

Some stronger results relating toughness and Hamiltonicity can be obtained for restricted families of graphs. For example, Chen *et al.* [20] showed that Conjecture A is true for *chordal graphs* (graphs containing no induced cycle of length greater than 3).

Theorem 3.7 *Every* 18*-tough chordal graph is Hamiltonian.*

Presumably, 18 is not the best possible number here. Whatever the best number is, it cannot be less than $\frac{7}{4}$, because Brauer, Broersma and Veldman [6] have shown that, for every $\varepsilon > 0$, there exists a $(\frac{7}{4} - \varepsilon)$-tough non-Hamiltonian chordal graph. It remains an open question whether or not all 2-tough chordal graphs are Hamiltonian.

Various authors have studied subclasses of chordal graphs for which Hamiltonicity can be guaranteed under less tough restrictions. One such class of graphs is the class of k-trees, defined in the Preliminaries to this book. (There is some confusion in the literature since 'k-tree' has also been taken to mean a spanning tree with maximum degree at most k.) By our definition, 1-trees are trees, and all k-trees are chordal. The following results are due to Broersma, Xiong and Yoshimoto [15].

Theorem 3.8

- *Every* 1*-tough* 2*-tree with* $n \geq 3$ *is Hamiltonian.*
- *For* $k \geq 2$*, every* $\frac{1}{3}(k+1)$*-tough* k*-tree is Hamiltonian.*

The authors of this theorem also presented infinitely many non-Hamiltonian 1-tough k-trees, for each $k \geq 3$.

Turning to planar graphs, we recall Tutte's result that every 4-connected planar graph is Hamiltonian [58]. Thus, if G is planar but not Hamiltonian, then it must contain a set S with at most three vertices for which $\omega(G-S) \geq 2$, whence $t(G) \leq \frac{3}{2}$. This establishes the following result.

Theorem 3.9 *For* $t > \frac{3}{2}$*, every* t*-tough planar graph is Hamiltonian.*

If planarity is combined with chordality, then the bound in Theorem 3.9 can be considerably reduced, as in this result of Böhme, Harant and Tkáč [12].

Theorem 3.10 *For* $t > 1$*, every* t*-tough chordal planar graph is Hamiltonian.*

This result is sharp, in the sense that not all 1-tough chordal planar graphs are Hamiltonian [12].

With the lack of a proof of Chvátal's conjecture, efforts have been made, some successfully, to prove similar results. One is the following theorem of Ellingham and Zha [24].

Theorem 3.11 *Every* 4*-tough graph has a closed walk that visits every vertex either once or twice.*

As a generalization of the conclusion in this theorem, we call a closed walk that visits every vertex at least once and at most k times a $[1, k]$-*walk*; thus, a Hamiltonian cycle is a $[1, 1]$-walk, and Ellingham and Zha's theorem says that every 4-tough graph has a $[1, 2]$-walk. An earlier result on the existence of $[1, k]$-walks, for $k \geq 3$, was found by Jackson and Wormald [35].

Theorem 3.12 *For $k \geq 3$, every $(1/(k-2))$-tough graph has a $[1, k]$-walk.*

Ellingham and Zha also used the methods of [6] to construct graphs of toughness approximately $(8k+1)/4k(2k-1)$ that do not possess a $[1, k]$-walk. Note that $k=1$ gives $\frac{9}{4}$ (see Theorem 3.1), but for $k=2$ note the gap between the toughness $\frac{17}{24}$ here and the value 4 in Theorem 3.11. Jackson and Wormald [35] conjectured the stronger result that, for $k \geq 2$, every $1/(k-1)$-tough graph has a $[1, k]$-walk. The bound of Ellingham and Zha gets close to this, for large k.

Cycles other than Hamiltonian cycles have also been studied in the context of toughness. One of the many results is the following, due to Jung and Wittman [37], which has Theorem 3.4 as a special case.

Theorem 3.13 *If G is 2-connected and t-tough, with minimum vertex-degree δ, then G contains a cycle of length at least* $\min\{n, (t+1)\delta + t\}$.

Chvátal made two further conjectures involving toughness and cycles, but both turned out to be false. The first of these was that every sufficiently tough graph has a 3-cycle. It was shown by Bauer, van den Heuvel and Schmeichel [7] that, for all t, there exists a triangle-free graph G with $\tau(G) > t$, a result that was extended to large girth by Alon [1].

Theorem 3.14 *For all t and g, there exists a t-tough graph with girth greater than g.*

Interesting examples of triangle-free graphs of arbitrarily large order were also given by Vinh and Dung [59], who used finite fields to construct, for each prime $q \equiv 7 \pmod{12}$, a triangle-free graph of order $n = q^2$ with toughness at least $\frac{1}{6}\sqrt{q}$.

It is worth mentioning that, since $\chi(G) \geq n/\alpha(G) \geq \tau(G) + 1$ (as is easily shown), Theorem 3.14 is a strengthening of the previously known fact that there are graphs of arbitrarily large girth and with arbitrarily large chromatic number.

The second Chvátal conjecture was that every sufficiently tough graph was pancyclic; this was proved false by Brandt, Faudree and Goddard [14].

Theorem 3.15 *For every t, there exists a t-tough graph that does not contain a cycle of length l, for some l between its girth and its circumference.*

4. Toughness and k-factors

A k-*factor* of a graph is a spanning subgraph that is regular of degree k; a 1-factor is also known as a *perfect matching*. Note also that a Hamiltonian cycle is simply a connected 2-factor. Chvátal observed the following immediate consequence of Tutte's matching theorem.

Theorem 4.1 *Every* 1*-tough graph of even order has a* 1*-factor.*

Chvátal then conjectured the following result, which was later proved by Enomoto *et al.* [27].

Theorem 4.2 *For each* k *and* $n \geq k+1$ *with* kn *even, every* k*-tough graph of order* n *has a* k*-factor.*

This result is sharp in the sense that, for every $\varepsilon > 0$, there is a $(k-\varepsilon)$-tough graph of order n with kn even that does not have a k-factor.

A result similar to Theorem 4.2 was obtained for $[r, s]$-factors by Katerinis [41]. An $[r, s]$*-factor* of a graph G is a spanning subgraph of G in which the degree of every vertex lies in the interval $[r, s]$.

Theorem 4.3 *Let* n, r *and* s *be such that* $r \leq s$, *with* $r < s$ *or with* sn *even. Then every* $(r + r/s - 1)$*-tough graph* G *has an* $[r, s]$*-factor.*

For bipartite graphs, for which being 1-tough is as tough as a graph can be, Katerinis [40] obtained the following result.

Theorem 4.4 *Every* 1*-tough bipartite graph of order* $n \geq 3$ *has a* 2*-factor.*

Other special families of graphs for which the bound for toughness in Theorem 4.2 can be reduced have also been considered. For example, Bauer *et al.* [8] proved the following result.

Theorem 4.5 *Every* $\frac{3}{2}$*-tough chordal graph has a* 2*-factor.*

Chvátal [21] had originally conjectured that every $\frac{3}{2}$-tough graph has a 2-factor, but this was disproved (see the remark after Theorem 4.2).

We now consider a concept called extendibility. A graph G is h*-extendible* if it contains a set of h disjoint edges, and every such set can be extended to a 1-factor. Plummer [51] proved the following result.

Theorem 4.6 *If* G *has even order* $n \geq 2h+2$ *and* $\tau(G) > h$, *then* G *is* h*-extendible.*

The condition $\tau(G) > h$ cannot be weakened to include h; for example, the graph $G = K_{2h} + 2K_{2h+1}$ has toughness h, but a 1-factor of K_{2h} cannot be extended to a perfect matching of G.

Plummer also pointed out that there exist h-extendible graphs with arbitrarily small toughness. The graphs he constructed have large order n, so he asked:
Is there a function $f(n)$ *such that if* G *is* $f(n)$*-extendible then* $\tau(G)$ *must be at least* 1?
The answer is in the affirmative [51].

Theorem 4.7 *If G has order n and is $f(n)$-extendible, where $f(n) = \lfloor \frac{1}{6}(n-2) \rfloor + 1$, then $\tau(G) \geq 1$.*

An interesting variation on Chvátal's observation on Tutte's theorem mentioned earlier applies to graphs of odd order, as noted by Enomoto [25].

Theorem 4.8 *If G is a 1-tough graph of odd order, then for every vertex v, $G - v$ has a 1-factor.*

This idea of removing vertices leads us to define a graph G to be *(k, s)-factor-critical* if, for every set S of s vertices, $G - S$ has a k-factor. (Thus, Theorem 4.8 deals with (1, 1)-factor-critical graphs.) Favaron [28] observed that every $(1, 2r)$-factor-critical graph is r-extendible, and proved the following result.

Theorem 4.9 *If G has even order $n > 2r + 2$, and if $\tau(G) > r$, then G is $(1, 2r)$-factor-critical.*

Liu and Yu [44] then showed that if $n \geq s + 3$ and $\tau(G) \geq s \geq 3$, then G is $(2, s)$-factor-critical; they also conjectured the following result, which was proved by Enomoto [25].

Theorem 4.10 *If G has order $n \geq 2r + 1$ and if $\tau(G) \geq r \geq 2$, then G is $(2, 2r - 2)$-factor-critical.*

Enomoto's proof introduced a variation on Chvátal's toughness τ, which we denote by τ': for a non-complete graph G,

$$\tau'(G) = \min\{|S|/(\omega(G - S) - 1)\},$$

where the minimum is taken over all cutsets S. This new measure has the property that, for every connected graph G,

$$\tau(G) < \tau'(G) \leq \kappa(G);$$

its relation with τ is that, for all sets S of k vertices,

$$\tau(G) \geq k \Leftrightarrow \tau'(G - S) \geq k.$$

Enomoto [25] and Enomoto and Hagita [26] proved that there are conditions on τ' that guarantee the existence of factors and are similar to the conditions on τ in Theorems 4.1 and 4.2.

Theorem 4.11

- *If $\tau'(G) \geq 1$ and n is even, then G has a 1-factor.*
- *If $\tau'(G) \geq 2$ and $n \geq 3$, then G has a 2-factor.*
- *If $\tau'(G) \geq k \geq 3$, kn is even and $n \geq k^2 - 1$, then G has a k-factor.*

In the last result, the exclusion of certain small graphs by some requirement such as $n \geq k^2 - 1$ is necessary since, for example, the wheel graph W_6 has no 3-factor, although its τ'-toughness is 3; however, it is likely that the bound $k^2 - 1$ can be reduced.

Another variation on toughness recently introduced relates to bipartite graphs. Recall that $\tau(G) \le 1$ for all bipartite graphs G, and that every 1-tough bipartite graph has a 2-factor (Theorem 4.4). For the study of the existence of k-factors (with $k > 2$) in bipartite graphs, toughness is not useful, and so Liu *et al.* [45] introduced a bipartite version of toughness and established some implications for k-factors. The *bipartite toughness* $\tau^*(G)$ of a bipartite graph G with partite sets X and Y is exactly as for toughness, except that the minimum is taken over all cutsets S that are proper subsets of either X or Y.

Theorem 4.12 *Let G be an r-by-r bipartite graph.*

- *If* $\tau^*(G) > s/(s+2)$ *where* $s = \lfloor \frac{1}{2}(r-1) \rfloor$, *then G has a* 1*-factor.*
- *If* $2 \le k \le \frac{1}{4}r + 1$ *and* $\tau^*(G) > (2k-1)(r-1)/(kr+1)$, *then G has a k-factor.*

These results are sharp.

Further variations on toughness have been introduced: among these are edge-toughness (see Katona [43]), topological toughness (see Göring and Katona [33]) and isolated toughness (see Bian [11]).

5. Binding number

Binding numbers were introduced in Section 1, where we gave some of their elementary properties. The following result, due to Kane, Mohanty and Straus [39], gives another basic property for graphs with binding number at most 1.

Theorem 5.1 *Let G be a graph.*

- *If* $\beta(G) < 1$, *then every binding set of G is an independent set.*
- *If* $\beta(G) = 1$ *and G is connected, then G has an independent binding set.*

Theorems 1.1 and 1.2 show similarities between toughness and binding number. The next result (see [32] and [61]) shows how the two parameters are related to one another.

Theorem 5.2

(a) $\beta(G) \le 1 + \tau(G)$.
(b) *If* $\beta(G) \le 1$, *then* $\tau(G) \le \beta(G)$.

Proof (a) The result is trivial if $\beta(G) \le 1$, so suppose that $\beta(G) = b > 1$. We must show that if S is a cutset with $\omega(G - S) = k$, then $|S| \ge (b-1)k$.

Let X be the set of vertices in all but one of the smallest components of $G - S$ if all components are non-trivial, and let X be $V(G - S)$ if at least one component is trivial. In either case, $|X| \geq k$, $N(X) \neq V(G)$ and $|S| + |X| \geq b|X|$, and so

$$|S| \geq (b-1)|X| \geq (b-1)k.$$

(b) The proof is trivial if G is not connected. If G is connected then, by Theorem 5.1, G has an independent binding set Y. Since Y is independent, $\omega(G - N(Y)) \geq |Y|$.

If $|Y| = 1$, then

$$\beta(G) - |N(Y)| \geq \delta(G) \geq 2\tau(G) > \tau(G).$$

If $|Y| \geq 2$, then $N(Y)$ is a cutset, and so

$$\tau(G) \leq |N(Y)|/\omega(G - N(Y)) \leq |N(Y)|/|Y| = \beta(G). \qquad \blacksquare$$

It follows from Theorem 5.2 that $\tau(G)$ must be large if $\beta(G)$ is. The converse is, however, not true. For example, take $G = K_n \times K_n$; then $\tau(G) = n - 1$, and

$$\beta(G) = (n^2 - 1)/(n^2 - 2n + 2) \to 1 \text{ as } n \to \infty.$$

By arguments similar to the above, we obtain another result of Woodall [61].

Theorem 5.3 *The binding number of any graph of order n and connectivity k is at most $(n + k)/(n - k)$.*

It follows that if $\beta(G) > 1$, then $k > 0$, and so G is connected.

A useful characterization of binding number was given by Woodall [60].

Theorem 5.4 *If G is a graph of order n, then*

$$\beta(G) = \max\{c : c|N(S)| \geq n(c-1) + |S|, S \text{ a non-empty subset of } V(G)\}.$$

It follows that, if $\beta(G) = b$, then $\delta(G) \geq (n(b-1) + 1)/b$. Use of this yields the bound in the next theorem.

Theorem 5.5 *The binding number of any graph of order n and minimum degree δ is at most $(n-1)/(n-\delta)$.*

Thus, the binding number of every tree is at most 1, a fact that also follows from Theorem 1.2.

We conclude this section by considering the chromatic number. Kane and Mohanty [38] showed that if $r \geq 3$ and $\beta(G) \geq r - \frac{4}{3}$, then G contains K_r as a subgraph. It follows immediately that $\chi(G) > \beta(G)$, and so $\chi(G)$ is large whenever $\beta(G)$ is. However, the converse is not true since, for example, there are triangle-free graphs with arbitrarily large chromatic number, and every triangle-free graph has binding number less than $\frac{5}{3}$. This result is very different from the corresponding result for toughness, described in Theorem 3.14.

Some improvements on the bounds of Kane and Mohanty have been obtained by Lyle and Plummer [46]. They showed that, if $r \geq 5$, then G must contain K_r as a subgraph whenever $\beta(G) \geq r - \frac{3}{2} - 1/(4r-6)$. For $r = 4$, the Kane–Mohanty bound was reduced to $\frac{1}{6}(5 + \sqrt{91}) \approx 2.423$, and an example was given showing that it cannot be reduced below $\frac{1}{4}(3 + \sqrt{33}) \approx 2.186$.

6. Binding number and *k*-factors

We now study the existence of k-factors under various binding number conditions. The first such result, due to Anderson [2], is a direct application of Tutte's matching theorem.

Theorem 6.1 *Every $\frac{4}{3}$-bound graph of even order has a 1-factor.*

Since $rK_1 + (r+2)K_2$ has no 1-factor but has binding number $(4r+3)/(3r+3)$, it follows that the bound of $\frac{4}{3}$ is sharp.

Woodall [61] then showed that applying the defect form of Tutte's theorem yields results in the case when $\beta(G) \leq \frac{4}{3}$.

Theorem 6.2 *Let $b \leq \frac{4}{3}$, and let G be a graph of order n and binding number at least b.*

- *If $0 \leq b \leq \frac{1}{2}$, then G has at least $nb/(b+1)$ independent edges.*
- *If $\frac{1}{2} \leq b \leq 1$, then G has at least $\frac{1}{3}n$ independent edges.*
- *If $1 \leq b \leq \frac{4}{3}$, then G has at least $(n-2) - \frac{2}{3}(n-3)/b$ independent edges.*

Furthermore, these results are sharp.

Another result, due to Chen [19], concerns the extension of a set of independent edges to a 1-factor and parallels Theorem 4.6.

Theorem 6.3 *If G has even order and $\beta(G) > \max\{h, \frac{1}{12}(7h+13)\}$, then G is h-extendible.*

It follows from this that G is 1-extendible if $\beta(G) > \frac{5}{3}$, is 2-extendible if $\beta(G) > \frac{9}{4}$, and is h-extendible if $\beta(G) > h \geq 3$.

Chen's bounds are the best possible among bounds that depend only upon h. Robertshaw and Woodall [54] later showed how to improve on them by giving bounds that depend on n as well as h.

Theorem 6.4 *Let G be a graph with even order n and let $h \geq 1$. Then G is h-extendible if*

- *$\beta(G) > (2n-2)/(n-2h+2)$, when $n \leq 2h+2+2\sqrt{(2h+1)}$;*
- *$\beta(G) > (n+2h)/(n-2h)$, when $2h+2+2\sqrt{(2h+1)} < n \leq 6h$;*
- *$\beta(G) > 4(n-3)/3(n-2h-2)$, when $n > 6h$.*

This result is sharp whenever $n - 2h \equiv 0$ or $2 \pmod 4$ and $n - 2h \geq 6$. It does not contain Chen's theorem, but Chen's theorem can be deduced from it. Note also that, for fixed h, as $n \to \infty$ the bound in the last result tends to the $\frac{4}{3}$ that appears in Theorem 6.1.

We now turn to the existence of k-factors where $k > 1$, beginning with a theorem of Katerinis and Woodall [42].

Theorem 6.5 *Let $k \geq 2$. Then every graph G of order $n \geq 4k - 6$ with kn even and*

$$\beta(G) > (2k-1)(n-1)/(kn-2k+3)$$

has a k-factor.

As one might expect, the proof of this result uses Tutte's k-factor theorem. It has several important consequences.

Corollary 6.6

- *If G is a graph of order $n \geq 4k - 6$, with kn even and $\beta(G) > 2$, then G has a k-factor.*
- *If $\beta(G) \geq \frac{3}{2}$, then G has a 2-factor.*
- *If G has even order and $\beta(G) > \frac{5}{3}$, then G has a 3-factor.*

The bounds in the last two parts are sharp; for example, $G = K_{4n-2} + (3n-1)K_2$ has binding number $\frac{5}{3}$, but has no 3-factor. However, the condition $n \geq 4k - 6$ in Theorem 6.5 can be weakened. It was shown by Egawa and Enomoto [23] that it can be replaced by the weaker condition $n \geq 4k - 1 - 4\sqrt{(k+2)}$, and that this bound is sharp. These and several other related results were later shown to be derivable from a general theorem of Woodall [62].

The binding number can also be combined with minimum-degree conditions to obtain other results on factors. The results in the next theorem are due respectively to Anderson [3] and Tokushige [57].

Theorem 6.7 *Let G be a graph of order n.*

- *If n is even, $\delta(G) \geq cn$, and $\beta(G) \geq (3-4c) / (2-2c)$, then G has a 1-factor.*
- *If $\delta(G) \geq cn$ and $\beta(G) \geq (2-3c) / (1-c)$, then G has a 2-factor.*

Toskushige also gave a more complicated condition for the existence of a k-factor for larger values of k.

Next recall the definition of an $[r, s]$-factor of a graph. The following result is due to Chen [17].

Theorem 6.8 *If $n \geq (r+s-1)(r+s-2)/s$, where $1 \leq r \leq s$, and if*

$$\beta(G) > (n-1)(r+s-1)/(ns-2s+3),$$

then G has an $[r, s]$-factor.

Note that if $r = s = k$, then this result reduces to Theorem 6.5. Note too that the bound for $\beta(G)$ is greater than $1+(r-1)/s$ when $r \leq s$ and $s \geq 3$. Chen [17], [18] has investigated under what conditions the above bound on $\beta(G)$ can in fact be replaced by $1 + (r-1)/s$.

Recently Zhou [64] has studied what are called (r, s, k)-factor critical graphs, those graphs with an $[r, s]$-factor after any k vertices are removed.

Theorem 6.9 *Let $s > 2$ and $k \geq 0$ be integers. Then any graph with $\delta(G) \geq k + 2$ and $\beta(G) \geq (ks + 3s + 4)/3s$ is $(2, s, k)$-factor critical.*

7. Binding numbers and cycles

We now turn to the existence of cycles in graphs with given binding number. The main result in Woodall's 1973 paper [61] is the following.

Theorem 7.1

- *If $\beta(G) \geq \frac{3}{2}$, then G has a Hamiltonian cycle.*
- *If $n \neq 8$ and $\beta(G) \geq c$, where $1 < c < \frac{3}{2}$, then G has a cycle whose length is at least* $3(n-1)(c-1)/c$.

These bounds are sharp: consider $sK_1 + rK_2$, where $s < r$ and $n = 2r + s$, as this graph has binding number $(2r - 1 + s)/(2r - 1)$.

Woodall [61] conjectured that every $\frac{3}{2}$-bound graph contains a triangle, and indeed is pancyclic (that is, it contains cycles of all lengths from 3 to n); these conjectures were eventually proved by Shi [55], [56].

Theorem 7.2 *Every $\frac{3}{2}$-bound graph is pancyclic.*

Shi proved the existence of a triangle by looking at the structure of a maximal possible counter-example; a shorter proof was later given by Goddard and Kleitman [30]. Shi's proof that G is pancyclic was much longer, but Robertshaw and Woodall [53] have since observed that the pancyclic property follows from the existence of a triangle and a Hamiltonian cycle, by a result of Brandt [13] that says that if a graph with minimum degree at least $\frac{1}{3}(n + 2)$ is not bipartite, then it has a cycle of each length between its girth and its circumference. Note that, by Theorem 5.4, a graph is $\frac{3}{2}$-bound if and only if $|N(S)| \geq \frac{1}{3}(n + 2|S|)$, for every non-empty subset S of $V(G)$. Thus, G is pancyclic if $|N(S)| \geq \frac{1}{3}(n + 2|S|)$ for all such S.

Robertshaw and Woodall [53] used Brandt's result to obtain the following theorem.

Theorem 7.3 *If G is 2-connected and $|N(S)| \geq \frac{1}{3}(n + |S|)$ for every independent set S of vertices, then G is pancyclic.*

8. Other measures of vulnerability

The connectivity $\kappa(G)$ of a graph G tells us how many vertices have to be removed to disconnect G. The toughness $\tau(G)$ tells us something about how many components there are in the resulting disconnected graph. But we may be interested in the *size* of the components. If the graph represents a communications network, and if S is a cutset, then the resulting components may all be fairly small, with the result that few vertices can then communicate with each other. For this reason, the concept of the *integrity* of a graph was introduced by Barefoot, Entringer and Swart [4]: it is defined as

$$I(G) = \min\{|X| + s(G - X) : X \subseteq V(G)\},$$

where $s(G - X)$ is the order of the largest component of $G - X$. Thus, the integrity measures how easy it is to create only small components by removing a small number of vertices.

However, it was shown by Chartrand *et al.* [16] that the integrity can be completely determined by a subgraph that forms a relatively small part of the graph, the remainder of the graph making no contribution. Consequently, they introduced the *mean integrity* $J(G)$ of G, defined as

$$J(G) = \min\{|X| + \text{average } p_v(G - X) : v \in V(G - X) \text{ and } X \subseteq V(G)\},$$

where $p_v(G - X)$ is the order of the component of $G - X$ containing v.

The two ideas of toughness and integrity were combined by Cozzens, Moazzami and Stueckle [22] to form the concept of *tenacity*, defined as

$$T(G) = \min\{(|X| + s(G - X))/\omega(G - X) : X \text{ is a cutset of } G\}.$$

Clearly $T(G) > \tau(G)$, so that $T(G) > \beta(G) - 1$. A discussion of the advantages of tenacity over the other parameters can be found in [48]. A similar parameter is the *edge-tenacity* (see [49]),

$$T'(G) = \min\{(|X| + r(G - X))/\omega(G - X) : X \text{ is an edge-cut of } G\},$$

where $r(G - X)$ is the number of edges in the largest component of $G - X$.

References

1. N. Alon, Tough Ramsey graphs without short cycles, *J. Alg. Combin.* **4** (1995), 189–195.
2. I. Anderson, Perfect matchings of a graph, *J. Combin. Theory (B)* **10** (1971), 183–186.
3. I. Anderson, Sufficient conditions for matchings, *Proc. Edinburgh Math. Soc.* **18** (1972), 131–136.
4. C. A. Barefoot, R. Entringer and H. Swart, Vulnerability in graphs – a survey, *J. Combin. Math. Combin. Comput.* **1** (1987), 13–22.
5. D. Bauer, H. J. Broersma, J. van den Heuvel and H. J. Veldman, Long cycles in graphs with prescribed toughness and minimum degree, *Discrete Math.* **141** (1995), 1–10.
6. D. Bauer, H. J. Broersma and H. J. Veldman, Not every 2-tough graph is Hamiltonian, *Discrete Applied Math.* **99** (2000), 317–321.
7. D. Bauer, J. van den Heuvel and E. Schmeichel, Toughness and triangle-free graphs, *J. Combin. Theory* (*B*) **65** (1995), 208–221.
8. D. Bauer, G. Y. Katona, D. Kratsch and H. J. Veldman, Chordality and 2-factors in tough graphs, *Discrete Appl. Math.* **99** (2000), 323–329.

9. D. Bauer, A. Morgana, E. Schmeichel and H. J. Veldman, Long cycles in graphs with large degree sums, *Discrete Math.* **79** (1989/90), 59–70.
10. D. Bauer and E. Schmeichel, On a theorem of Häggkvist and Nicoghossian, *Graph Theory, Combinatorics, Algorithms, and Applications* (eds. Y. Alavi, F. Chung, R. L. Graham and D. S. Hsu), SIAM (1991), 20–25.
11. Q. Bian, On toughness and (g, f)-factors in bipartite graphs, *J. Appl. Math. Comput.* **22** (2006), 299–304.
12. T. Böhme, J. Harant and M. Tkáč, More than one-tough chordal planar graphs are Hamiltonian, *J. Graph Theory* **32** (1999), 405–410.
13. S. Brandt, *Sufficient Conditions for Graphs to Contain all Subgraphs of a Given Type*, Ph.D. thesis, Freie Universität, Berlin, 1994.
14. S. Brandt, R. Faudree and W. Goddard, Weakly pancyclic graphs, *J. Graph Theory* **27** (1998), 141–176.
15. H. J. Broersma, L. Xiong and K. Yoshimoto, Toughness and hamiltonicity in k-trees, *Discrete Math.* **307** (2007), 832–838.
16. G. Chartrand, S. F. Kappor, T. A. McKee and O. R. Oellermann, The mean integrity of a graph, *Recent Studies in Graph Theory* (edn. V. R. Kulli), Vishwa International Publications (1989), 70–80.
17. C. Chen, Binding number and minimum degree for $[a, b]$-factors, *Syst. Sci. Math. Sci.* **6** (1993), 179–185.
18. C. Chen, Binding number of graphs and $[a, b]$-factors, *Ars Combin.* **41** (1995), 217–224.
19. C. Chen, Binding number and toughness for matching extensions, *Discrete Math.* **149** (1995), 303–306.
20. G. Chen, M. S. Jacobson, A. E. Kézdy and T. Lehel, Tough enough chordal graphs are Hamiltonian, *Networks* **31** (1998), 29–38.
21. V. Chvátal, Tough graphs and Hamiltonian circuits, *Discrete Math.* **5** (1973), 215–228.
22. M. Cozzens, D. Moazzami and S. Stueckle, The tenacity of a graph, *Graph Theory, Combinatorics and Algorithms* (eds. Y. Alavi and A. Schwenk), Wiley (1995), 1111–1122.
23. Y. Egawa and H. Enomoto, Sufficient conditions for the existence of k-factors, *Recent Studies in Graph Theory* (ed. V. R. Kulli), Vishwa International Publications (1989), 96–105.
24. M. N. Ellingham and X. Zha, Toughness, trees and walks, *J. Graph Theory* **33** (2000), 125–137.
25. H. Enomoto, Toughness and existence of k-factors III, *Discrete Math.* **189** (1998), 277–282.
26. H. Enomoto and M. Hagita, Toughness and existence of k-factors IV, *Discrete Math.* **216** (2000), 111–120.
27. H. Enomoto, B. Jackson, P. Katerinis and A. Saito, Toughness and existence of k-factors, *J. Graph Theory* **9** (1985), 87–95.
28. O. Favaron, On k-factor-critical graphs, *Discuss. Math. Graph Theory* **16** (1996), 41–51.
29. R. Faudree, R. Gould, M. Jacobson, L. Lesniak and A. Saito, Toughness, degrees and 2-factors, *Discrete Math.* **286** (2004), 245–249.
30. W. Goddard and D. J. Kleitman, A note on maximal triangle-free graphs, *J. Graph Theory* **17** (1993), 629–631.
31. W. Goddard, M. D. Plummer and H. Swart, Maximum and minimum toughness of graphs of small genus, *Discrete Math.* **167/8** (1997), 329–339.
32. W. Goddard and H. Swart, On the toughness of a graph, *Quaestiones Math.* **13** (1990), 217–232.

33. F. Göring and G. Y. Katona, Local topological toughness and local factors, *Graphs Combin.* **23** (2007), 387–399.
34. C. T. Hoàng, Hamiltonian degree conditions for tough graphs, *Discrete Math.* **142** (1995), 121–139.
35. B. Jackson and N. C. Wormald, k-walks of graphs, *Austral. J. Combin.* **2** (1990), 135–146.
36. H. A. Jung, On maximal circuits in finite graphs, *Ann. Discrete Math.* **3** (1978), 129–144.
37. H. A. Jung and P. Wittmann, Longest cycles in tough graphs, *J. Graph Theory* **31** (1999), 107–127.
38. V. G. Kane and S. P. Mohanty, Binding numbers, cycles and complete graphs, *Combinatorics and Graph Theory*, Lecture Notes in Mathematics **885**, Springer (1981), 290–295.
39. V. G. Kane, S. P. Mohanty and E. G. Straus, Which rational numbers are binding numbers?, *J. Graph Theory* **5** (1981), 379–384.
40. P. Katerinis, Two sufficient conditions for a 2-factor in a bipartite graph, *J. Graph Theory* **11** (1987), 1–6.
41. P. Katerinis, Toughness of graphs and the existence of factors, *Discrete Math.* **80** (1990), 81–92.
42. P. Katerinis and D. R. Woodall, Binding numbers of graphs and the existence of k-factors, *Quart. J. Math.* **38** (1987), 221–228.
43. G. Y. Katona, Toughness and edge-toughness, *Discrete Math.* **164** (1997), 187–196.
44. G. Liu and Q. Yu, k-factors and extendability with prescribed components, *Congr. Numer.* **139** (1997), 77–88.
45. J. Liu, J. Qian, J. Z. Sun and R. Xu, Bipartite toughness and k-factors in bipartite graphs, *Int. J. Math., Math. Sci.*, Art. ID 597408, 8pp., 2008.
46. J. Lyle and M. D. Plummer, The binding number of a graph and its cliques, *Discrete Appl. Math.* **157** (2009), 3336–3340.
47. M. Matthews and D. P. Sumner, Hamiltonian results in $K_{1,3}$-free graphs, *J. Graph Theory* **8** (1984), 139–146.
48. D. Moazzami, Vulnerability in graphs – a comparative survey, *J. Combin. Math. Combin. Comput.* **30** (1999), 23–31.
49. D. Moazzami, On the edge-tenacity of graphs, *Intern. Math. Forum* **3** (2008), 929–936.
50. R. E. Pippert, On the toughness of a graph, *Graph Theory and Applications,* Lecture Notes in Mathematics **303**, Springer (1972), 225–233.
51. M. D. Plummer, Toughness and matching extensions in graphs, *Discrete Math.* **72** (1988), 311–320.
52. M. D. Plummer, A note on toughness and tough components, *Congr. Numer.* **125** (1977), 179–192.
53. A. M. Robertshaw and D. R. Woodall, Triangles and neighbourhoods of independent sets in graphs, *J. Combin. Theory (B)* **80** (2000), 122–129.
54. A. M. Robertshaw and D. R. Woodall, Binding number conditions for matching extension, *Discrete Math.* **248** (2002), 169–179.
55. R. Shi, The binding number of a graph and its triangle, *Acta Math. Applic. Sinica* **2** (1985), 80–86.
56. R. Shi, The binding number of a graph and its pancyclism, *Acta Math. Applic. Sinica* **3** (1987), 257–267.
57. N. Tokushige, Binding number and minimum degree for k-factors, *J. Graph Theory* **13** (1989), 607–617.
58. W. T. Tutte, A theorem on planar graphs, *Trans. Amer. Math. Soc.* **82** (1956), 99–116.

59. L. A. Vinh and D. P. Dung, Explicit tough Ramsey graphs, *Proc. Intern. Conf. on Relations, Orders and Graphs,* Nouha Editions (2008), 139–146.
60. S. Win, On a connection between the existence of k-trees and the toughness of a graph, *Graphs Combin.* **5** (1989), 201–205.
61. D. R. Woodall, The binding number of a graph and its Anderson number, *J. Combin. Theory* (B) **15** (1973), 225–255.
62. D. R. Woodall, k-factors and neighbourhoods of independent sets in graphs, *J. London Math. Soc.* **41** (1990), 385–392.
63. R. Xu and Q. Yu, Note on toughness and (k, r)-factor-critical graphs, *Bull. Inst. Combin. Applic.* **29** (2000), 85–87.
64. S. Zhou, Binding number conditions for (a, b, k)-critical graphs, *Bull. Korean Math. Soc.* **45** (2008), 53–57.

9

Graph fragmentability

KEITH EDWARDS and GRAHAM FARR

Fragmentability concerns the extent to which a graph can be broken up into small (bounded-sized) components by removing as few vertices as possible. To make this concept precise, we define the coefficient of fragmentability of a class of graphs, and summarize the known upper and lower bounds for this parameter, for various graph classes. We discuss the relationship of fragmentability to the planarization of graphs (by removing vertices), and to colourings with small monochromatic components. We also mention a number of open problems and potential applications of fragmentability.

1. Introduction

How vulnerable is a graph to being *fragmented* – that is, broken into small pieces by removing some of its vertices?

To make this question precise, we use two parameters: the proportion ε of vertices to be removed, and an upper bound C on the number of vertices in any connected component that is left behind by that removal. We say that a graph $G = (V, E)$ is (C, ε)*-fragmentable* if there exists $X \subseteq V$ with $|X| \leq \varepsilon|V|$ such that every connected component of $G - X$ has at most C vertices; the set X is called the *fragmenting set*. For example, the complete graph K_n is (C, ε)-fragmentable for all $\varepsilon \geq (n - C)/n$, and the complete bipartite graph $K_{n,n}$ is $(1, \varepsilon)$-fragmentable for all $\varepsilon \geq \frac{1}{2}$. The cycle C_n is easily shown to be (C, ε)-fragmentable for any $\varepsilon \geq \lceil n/(C+1) \rceil / n$.

Our focus will be on *classes* of graphs. A class Γ is *(C, ε)-fragmentable* if every graph in the class is (C, ε)-fragmentable, and *ε-fragmentable* if it is (C, ε)-fragmentable for some C.

In general, there is a trade-off between C and ε: if you want to reduce the proportion of vertices removed, then you may need to relax the bound on the sizes of the components left behind. But if the proportion of vertices you remove is too low, then the components left behind may be unbounded in size. For a given class of graphs, we wish to know how small the proportion of vertices removed can be, while still keeping the remaining components bounded in size.

The *coefficient of fragmentability* $c_f(\Gamma)$ of a class Γ is defined as

$$c_f(\Gamma) = \inf\{\varepsilon : \Gamma \text{ is } \varepsilon\text{-fragmentable}\}.$$

This infimum may or may not be attained, depending on the class. We return to this point in Section 4.

These definitions were proposed by us in [15], and extend the definition of a *fragmentable* class of graphs by Edwards and McDiarmid [20], which, in our terminology, is a class with coefficient of fragmentability 0 (one that is ε-fragmentable for all $\varepsilon > 0$). Results for a number of classes of graphs were given in [18] and [19]. We summarize these results in the next section.

One useful strategy for fragmenting a graph is *reduction*, where we first remove vertices so as to leave a graph that belongs to some class whose coefficient of fragmentability is known, and then fragment that graph. In Section 3 we show how this strategy gives an upper bound for the coefficient of fragmentability of one class, using an upper bound for that of another. We make most use of this technique when the second class has coefficient of fragmentability 0.

Our work has emphasized finding bounds for the coefficient of fragmentability of classes of graphs of bounded maximum degree, or bounded average degree. Our upper bounds use reduction to various fragmentable classes, especially planar graphs and series-parallel graphs. We discuss this in Section 4.

Our extensive use of reduction to the class of planar graphs, or some of its subclasses, motivates us to consider this problem in its own right. In Section 5 we discuss *planarization*, the removal of vertices to leave a planar graph. Our results give approximation algorithms with the best known performance for the classical Maximum Induced Planar Subgraph (MIPS) problem.

In Section 6 we discuss some applications of fragmentability and planarization, including improved bounds on the size of a regular expression representing a finite automaton.

In Section 7 we look at colouring graphs so that the colour classes induce subgraphs with small components. We describe a parameter that measures how few colours are required, for a class of graphs, if the sizes of components induced by colour classes are to be bounded. This parameter is related to the chromatic number of a graph, in the same way that the coefficient of fragmentability is related to the independence number of a graph.

This chapter is about the *theory* of fragmentability, as a topic in pure graph theory. But it is worth briefly considering potential applications.

There is a huge body of work on network vulnerability, where graphs are used to represent communication networks of various kinds (electronic, social, and so forth). Vertices may be removed from such a graph, either through random faults or through the action of an adversary. There are various 'levels' of survival of a damaged network; for example, if the surviving vertices induce a connected graph, then it is still possible for any surviving vertex to communicate, along some path, with any other vertex, although the efficiency of these communications may be reduced.

Fragmentability concerns a weaker level of survival. Suppose that there is some 'critical mass' of vertices (say, $C+1$ vertices) that, if interconnected, can still operate at some useful level, but where groups of fewer interconnected vertices are not useful. The network fails completely if no surviving group of interconnected vertices reaches the 'critical mass' of $C+1$ vertices – or, in graph-theoretic terms, if every connected component of the subgraph left behind has at most C vertices. If the network is not (C, ε)-fragmentable, then it can survive any removal of a fraction ε of its vertices. Beygelzimer *et al.* [9] used the proportion of vertices in the largest connected component of the surviving network as a measure of the availability of the network after the removal of vertices. From this viewpoint, a network with n vertices is (C, ε)-fragmentable if any removal of at most εn vertices leaves a largest connected component having at most a proportion C/n of the network's vertices.

Fragmentability may be useful in the modelling of vaccination against the spread of infectious diseases; see, for example, the model described in Britten, Janson and Martin-Löf [10]. Here, the vertices represent people and edges represent interactions that may allow the disease to pass from one to the other. Vaccination effectively removes vertices from the graph, and limiting the spread of the epidemic means limiting the sizes of the remaining components.

In view of the ubiquity of graphs as models of real-world networks, and of the importance of network robustness and vulnerability, it is not surprising that many graph parameters have been devised to measure robustness and vulnerability. Some of these parameters have some of the flavour of fragmentability. We mention two of these briefly.

Caro and Yuster [11] defined a graph G to be *k-slim* if, for every subgraph H of G with at least k vertices, there exists $X \subseteq V(G)$ with $|X| \leq k$ for which $H - X$ has at least two components, each with at most $\frac{2}{3}\,|V(H)|$ vertices. By a result of Edwards and McDiarmid [20], if Γ is a class in which every graph is k-slim, for some number k independent of the graph, then $c_f(\Gamma) = 0$.

Barefoot, Entringer and Swart [6] defined the *integrity* of a graph G as

$$I(G) = \min\{|X| + |V(\text{largest component of } G - X)| \; : \; X \subseteq V(G)\};$$

see the survey by Bagga *et al.* [5]. Integrity is related to the coefficient of fragmentability by an easy, but sometimes very loose, inequality:

$$c_f(\Gamma) \geq \inf_n \max \left\{ \frac{I(G)}{n} : G \in \Gamma, |V(G)| = n \right\}.$$

The coefficient of fragmentability has at least two distinctive features. One is its focus on classes of graphs, rather than on individual graphs. Another is that it does not depend on the exact value of the maximum component size C, requiring only that it be bounded. This means that, as a measure of network robustness, it is itself 'robust' with respect to the 'critical mass' of interconnected vertices needed for network survival. So the measure should still be useful, even if the exact value of the critical mass in a given situation is not known.

2. Values and bounds for fragmentability

We begin with two simple examples to illustrate the definitions.

The family of cycles has coefficient of fragmentability 0, since (roughly speaking), for each $\varepsilon > 0$, removing every $(1/\varepsilon)$-th vertex around a cycle leaves components with at most about $(1/\varepsilon) - 1$ vertices, so their size is bounded by a constant (independent of n). (Recall our remark about the fragmentation of C_n in Section 1.)

At the other extreme, the family of complete graphs has coefficient of fragmentability 1. This family is not ε-fragmentable for any $\varepsilon < 1$ since, for any positive integer constant C, there is a graph in the family, the complete graph $K_{\lceil (C+1)/(1-\varepsilon) \rceil}$, that cannot be reduced to components of size at most C by removing at most a fraction ε of its vertices.

In fact, anything between these extremes is possible: any $\varepsilon \in [0, 1]$ is the coefficient of fragmentability of some class. (This is a routine exercise, see [15, p. 31].)

Another elementary observation is that, for any classes Γ and Γ', if $\Gamma \subseteq \Gamma'$ then $c_f(\Gamma) \leq c_f(\Gamma')$. From this, and our above remark about complete graphs, it follows that $c_f(\{\text{all graphs}\}) = 1$.

A number of classes of graphs with $c_f = 0$ were given in [20]. These include trees, and indeed graphs of bounded tree-width, planar graphs, graphs of bounded genus, graphs with any fixed excluded minor, and subgraphs of grid graphs of bounded dimension.

The identification of interesting classes of graphs with $c_f > 0$ began with [15]. The focus there was on families of graphs with bounded maximum degree; we consider such families further in Section 4. Now we consider some other interesting classes.

The *thickness* $\theta(G)$ of a graph $G = (V, E)$ is the minimum number t for which there exists a partition $E = E_1 \cup \cdots \cup E_t$ for which each $G_i = (V, E_i)$ is planar $(1 \leq i \leq t)$; for further information, see the surveys [7] and [38]. Let Θ_t be the class of graphs of thickness t or less. Note that Θ_1 is the class of planar graphs, so $c_f(\Theta_1) = 0$. The following result appears in [18].

Theorem 2.1

$$\tfrac{1}{3} \le c_f(\Theta_2) \le \tfrac{3}{4} \quad \textit{and} \quad \frac{t-1}{2t-1} \le c_f(\Theta_t) \le \frac{6t-2}{6t+1} \quad \textit{for } t \ge 3.$$

We now turn to classes based on colourability. Let $\mathrm{Col}(k)$ be the class of k-colourable graphs. The following results appear in [18].

Theorem 2.2 *For* $k \ge 1$,

$$c_f(\mathrm{Col}(k)) = \frac{k-1}{k}.$$

Theorem 2.3 *If* Γ *is a class of bipartite graphs containing regular graphs of arbitrarily high degree, then* $c_f(\Gamma) = \frac{1}{2}$.

Corollary 2.4 *If* Ξ *is the set of k-dimensional cubes* (*one for each* $k \in \mathbb{N}$), $\mathbb{N}$, *then* $c_f(\Xi) = \frac{1}{2}$.

3. Reduction and separation

Planar graphs are intuitively easy to fragment: we can imagine repeatedly slicing the graph (or pieces of it) 'in half', until all the pieces are small. This idea can be made precise using the separator theorems of Lipton and Tarjan [33] and later authors. Although we cannot quite cut a planar graph 'in half', Lipton and Tarjan proved that by removing at most $\sqrt{8n}$ vertices from a planar graph with n vertices, we can split it into two pieces each containing at most $\frac{2}{3}n$ vertices. This was generalized to graphs of bounded genus by Gilbert, Hutchinson and Tarjan [21], and to any excluded-minor class by Alon, Seymour and Thomas [4]. Since the pieces are subgraphs of the original graph, and are therefore in the same class, we can repeatedly separate the pieces until all are small enough. Adding up the number of vertices removed shows that this is small. Thus we can state the following result.

Theorem 3.1 *Let* Γ *be a class of graphs. Suppose that there exist real numbers* $A > 0$, $0 \le \lambda < 1$, *and* $0 < \alpha < 1$, *such that every graph* $G \in \Gamma$ *with n vertices has a set of at most* An^{λ} *vertices whose removal leaves every component with at most* αn *vertices, and each component a member of* Γ. *Then* Γ *is* ε*-fragmentable for any* $\varepsilon > 0$, *so that* $c_f(\Gamma) = 0$.

Using the separator results mentioned above, we can show that $c_f(\Gamma) = 0$ for any excluded minor class Γ, such as planar graphs. The same is true, for example, when Γ is the class of subgraphs of a cubic grid (a class with no excluded minor).

In addition to these direct results, we can also use these classes to obtain upper bounds on the coefficient of fragmentability for other classes. We do this by reducing one class to another: if we can show that removing a certain proportion of vertices from a graph in Γ gives a graph in Γ', then we can obtain a bound for $c_f(\Gamma)$ in terms of $c_f(\Gamma')$. We use the following theorem.

Theorem 3.2 *Let Γ and Γ' be classes of graphs, let $\varepsilon \geq 0$, and let A be a non-negative integer. If, for every $G \in \Gamma$, a set of at most $\varepsilon|V(G)| + A$ vertices can be removed from G to leave a graph in Γ', then $c_f(\Gamma) \leq c_f(\Gamma') + \varepsilon - \varepsilon c_f(\Gamma')$.*

In particular, if Γ' is one of the classes mentioned above, with $c_f(\Gamma') = 0$, then we get $c_f(\Gamma) \leq \varepsilon$. This technique has been used to obtain upper bounds on c_f, using the class of planar graphs or of series-parallel graphs as the intermediate class Γ'.

4. Bounded degree classes

In many real-world applications, the graphs used as models tend to be sparse, and will often have bounded maximum degree, or at least bounded average degree. Thus it is of interest to consider the coefficient of fragmentability of these classes.

For any bounded-degree class (except one in which all components of the graphs already have bounded size), the coefficient of fragmentability is not attained – that is, it is genuinely an infimum, rather than a minimum. This is shown by the following result from [15].

Theorem 4.1 *Let d be a positive integer and let Γ be an ε-fragmentable class of graphs with maximum degree at most d. If $\varepsilon > 0$, then there is a constant $\varepsilon' < \varepsilon$ such that Γ is ε'-fragmentable.*

We consider three classes. For any positive integer d, we denote by Γ_d the class of all graphs with maximum degree at most d. For any positive real number d, we denote by $\overline{\Gamma}_d$ the class of all graphs with average degree at most d, and by $\overline{\Gamma}_d^c$ the class of all *connected* graphs with average degree at most d.

Upper bounds for bounded degree classes

In [15] it is shown that, for $d \geq 2$,

$$c_f(\Gamma_d) \leq \frac{d-2}{d+1}. \tag{1}$$

This was extended to average degree classes in [18], where it is shown that, for real $d \geq 4$,

$$c_f(\overline{\Gamma}_d) \leq \frac{d-2}{d+1}, \tag{2}$$

and for real $d \geq 2$,

$$c_f(\overline{\Gamma}_d^c) \leq \frac{d-2}{d+1}. \tag{3}$$

In [19] these bounds are improved by replacing the fraction $(d-2)/(d+1)$ by a function $g(d)$. This function is defined as follows.

Set $g(2)=0$, $g(3)=\frac{1}{4}$, and, for any integer $d \geq 4$,

$$dg(d)=(d-2)g(d-1)+g(d-2)+1.$$

Then extend the definition to real numbers by linear interpolation – that is, for any integer $d \geq 2$ and real number r satisfying $0<r<1$, set $g(d+r)=(1-r)g(d)+rg(d+1)$. It can then be shown that

- $g(d)<(d-2)/(d+1)$, for all $d>3$;
- $g(d)=(d-\frac{9}{4})/(d+1)+O(1/d^3)$;
- $g(d)=\frac{13}{4}A(d)/d!+\frac{5}{4}(-1)^d/d!-\frac{9}{4}$ when d is an integer, where $A(d)$ is the alternating factorial function given by $A(d)=d!-(d-1)!+\ldots-(-1)^d 1!$.

Thus we obtain the following results.

Theorem 4.2

- $c_f(\Gamma_d) \leq g(d)$, *for any integer* $d \geq 2$;
- $c_f(\overline{\Gamma}_d) \leq g(d)$, *for any real* $d \geq 5$;
- $c_f(\overline{\Gamma}^c_d) \leq g(d)$, *for any real* $d \geq 2$.

Lower bounds for bounded degree classes

In [15] we showed that $c_f(\Gamma_d) \geq (d-2)/(2d-2)$ for $d \geq 2$, with the extremal graphs being d-regular graphs of high girth. In particular, this shows (together with the upper bound above) that $c_f(\Gamma_3)=\frac{1}{4}$. We also conjectured that $c_f(\Gamma_d) \to 1$ as $d \to \infty$, and this was proved by Haxell, Pikhurko and Thomason in [25]. They also proved, this time using line graphs of graphs of high girth, that for $d \geq 4$,

$$c_f(\Gamma_d) \geq \begin{cases} 1-\dfrac{4}{d+2} & \text{if } d \text{ is even,} \\[2ex] 1-\dfrac{4(d+2)}{(d+1)(d+3)} & \text{if } d \text{ is odd.} \end{cases}$$

For integer values of d, these lower bounds also hold for the classes $\overline{\Gamma}_d$ and $\overline{\Gamma}^c_d$. Thus, for small values of d, we have the following bounds for $c_f(\Gamma_d)$:

d	2	3	4	5	6	7	8
upper bound	0	$\frac{1}{4}$	$\frac{3}{8}$	$\frac{19}{40}$	$\frac{131}{240}$	$\frac{1009}{1680}$	$\frac{8651}{13440}$
lower bound	0	$\frac{1}{4}$	$\frac{1}{3}$	$\frac{5}{12}$	$\frac{1}{2}$	$\frac{11}{20}$	$\frac{3}{5}$
gap	0	0	$\frac{1}{24}$	$\frac{7}{120}$	$\frac{11}{240}$	$\frac{17}{336}$	$\frac{587}{13440}$

Random regular and sparse graphs

It is natural to ask about the fragmentability of randomly generated graphs. Janson and Thomason [27] considered both (C, ε)-fragmentability and a relaxation of it, where $C = \delta n$ for constant δ (rather than the usual $C = O(1)$), to sparse random graphs (the classical Erdős–Rényi model with edge probability c/n, with c constant) and to random d-regular graphs. The relaxation was motivated by the work of Britton, Janson and Martin-Löf [10] on modelling vaccination strategies. It allows fewer vertices to be removed, in general, since the size bound for components is greater (for large enough n). However, for small δ it makes little difference, in the following sense. The main result of [27] implies that, for a random graph, with probability approaching 1 as $n \to \infty$, the proportion of vertices that must be removed to leave components of size at most δn is 'almost' at least as great as the proportion to be removed if the components remaining have to have size at most $1/\delta$. By making δ small enough, we can make the 'slippage' in the inequality arbitrarily small.

5. Planarization

One useful strategy for finding an upper bound for $c_f(\Gamma)$ is to show that any graph in the class Γ can be made planar by removing at most a fraction ε of its vertices. Such a result, together with Theorem 3.2 (where Γ' is the class of planar graphs, for which $c_f(\Gamma') = 0$), establishes that $c_f(\Gamma) \leq \varepsilon$.

This motivates us to consider the computational problem of finding the size of the smallest set of vertices of a graph whose removal leaves a planar subgraph. The complement of this set is the largest set of vertices which induces a planar subgraph of the original graph. The latter gives us the following well-known problem.

Maximum Induced Planar Subgraph (MIPS)
INPUT: Graph G, on n vertices.
OUTPUT: A largest set P of vertices for which the induced subgraph $\langle P \rangle$ is planar.

Finding a large planar subgraph of a graph is an important first step in many graph drawing algorithms (see [14]). There, the Maximum Planar Subgraph (MPS) problem (where the planar subgraph sought does not have to be induced) has attracted more attention than MIPS; see, for example, the survey by Liebers [31]. The process of removing vertices (for MIPS) or edges (for MPS) in order to obtain a planar subgraph is known as *planarization*.

MIPS was shown to be NP-hard in [29] and [30]. The number $|P|$ of vertices in a largest induced planar subgraph is also hard to approximate: Lund and Yannakakis [35] showed that there exists $\varepsilon > 0$ such that $|P|$ cannot be approximated in polynomial time with performace ratio at most $n^{-\varepsilon}$ unless P = NP. On the positive side, Halldórsson [23] gave a polynomial-time approximation algorithm that finds an induced planar subgraph of size $\Omega(n^{-1}(\log n/\log\log n)^2)$ times the optimum. This lower bound grows more slowly than $n^{-\varepsilon}$, for any $\varepsilon < 1$, but is an improvement on trivial algorithms with performance ratio $\Omega(n^{-1})$, such as simply taking any four vertices of the graph.

As we remarked in Section 4, practical applications of graphs tend to feature sparse graphs, with bounded maximum or average degree. We therefore consider approximation algorithms for restrictions of MIPS to such classes, and we begin by looking at graphs of maximum degree at most d.

Halldórsson and Lau [24] gave a linear-time algorithm that finds an induced planar subgraph of at least $1/\lceil \frac{1}{3}(d+1) \rceil$ vertices; for example, for graphs of maximum degree 3, it finds an induced planar subgraph with at least half the vertices of the original graph. The induced planar subgraphs that it finds are of a very special form: they have maximum degree at most 2, and so every component is a path or a cycle. The strength of Halldórsson and Lau's approach is its generality, since it gives a suite of algorithms for any problem of finding the maximum induced subgraph with some hereditary property (in the sense of being preserved by vertex-removal) – and all the algorithms, for all such properties, have the same performance ratio. The algorithm uses a lemma of Lovász [34] to partition the vertex-set of a graph so that each part induces a subgraph of maximum degree 2. Then a part with the most vertices is chosen. For MIPS, this is the output. For other induced subgraph problems, we just solve the problem on the chosen part, which has maximum degree at most 2 and admits an easy solution.

The proof in [15, Theorem 3.2] of our first upper bound (1) for $c_f(\Gamma_d)$ yielded an approximation algorithm for the restriction of MIPS to Γ_d; that algorithm is presented and analysed in [16]. The induced planar subgraph that it finds has at least $3n/(d+1)$ vertices. For example, for graphs of maximum degree 3, it finds an induced planar subgraph with $\frac{3}{4}$ of the vertices of the original graph, a substantial improvement on the ratio $\frac{1}{2}$ for the Halldórsson–Lau algorithm. Its performance ratio 'removes the ceiling' from the Halldórsson–Lau performance ratio, and so is an improvement when $d \not\equiv 2 \pmod 3$. The subgraphs that it finds are of more general form, although they still do not have the full generality of planar graphs since they are always series-parallel. Its time-complexity is $O(nm)$, where m is the number of edges of the input graph.

This algorithm belongs to a family of algorithms where we form a large induced subgraph with some property by starting with an empty set P and repeatedly adding vertices to P (provided that some simple condition involving vertex-degrees is satisfied) and sometimes swapping a vertex in the set with one outside it (according to some other condition). Simpler members of this family include a classical heuristic for finding a large independent set of size at least $n/(d+1)$ (see, for example, [28]), and another for finding a large induced forest of size at least $2n/(d+1)$. The algorithm and analysis of [16] involve some features and complications not seen in its simpler predecessors. Morgan and Farr [37] added an algorithm to this family, one that finds a large induced outerplanar subgraph with $3n/(d+\frac{5}{3})$ vertices.

There is another family of large-induced-subgraph algorithms whose members start with $P = V$ and repeatedly remove vertices of high degree in some sense. An early 'vertex-removal' algorithm of this kind, due to Alon, Kahn and Seymour [3], finds large induced k-degenerate subgraphs. A graph is *k-degenerate* if it can be reduced to an empty graph by repeated removal of vertices of degree less than k; equivalently, every subgraph has a vertex of degree less than k. So the 1-degenerate graphs are those with no edges, and the 2-degenerate graphs are precisely the forests. The approach of Alon, Kahn and Seymour repeatedly removes a vertex of degree less than k (if one exists) or of maximum degree, and finds an induced k-degenerate subgraph with at least $kn/(d+1)$ vertices in any graph whose average degree is at most d, provided that $d \geq 2k-2$. In particular, for $d \geq 2$, it finds an induced forest with at least $2n/(d+1)$ vertices in any graph in $\overline{\Gamma}_d$. For the simpler case $k=1$, it reduces to the lower bound $n/(d+1)$ for the size of the maximum independent set in a graph of average degree not exceeding d.

A different vertex-removal algorithm, which we will call the *reduction-removal algorithm*, for finding large induced series-parallel subgraphs, was introduced in [18]. Starting with $P = V$, each iteration removes from P a vertex of maximum degree, not in $\langle P \rangle$ itself but in the *reduced graph* of $\langle P \rangle$. This reduced graph is formed by repeatedly removing an isolated vertex, or removing a leaf, or replacing a vertex of degree 2 by an edge between its neighbours (if none exists already), until none of these operations is possible any more. (Note that, although this vertex is chosen using the reduced graph, it is removed from $\langle P \rangle$. The reductions are used only to help to choose this vertex; the reduced graph never replaces $\langle P \rangle$.) The algorithm stops when the reduction process produces an empty graph, which happens if and only if $\langle P \rangle$ is series-parallel. This use of reductions is a key difference from [3]. Note also that series-parallel graphs are 3-degenerate, but that the converse is not necessarily true. Our result is stronger than the $k=3$ case of [3], even though the lower bound of $3n/(d+1)$ on induced subgraph size is the same in each case, since we find graphs from a more restricted class.

This algorithm performs better than the earlier one of [16]. The performance ratio proved for it has similar form when the degree bound is an integer, but is expressed in terms of the average degree rather than the maximum degree. The algorithm has time-complexity $O(nm)$ for a graph with n vertices and m edges.

Theorem 5.1 *Let G be a graph with n vertices and average degree at most d. Then the reduction-removal algorithm finds an induced series-parallel subgraph with at least $3n/(d+1)$ vertices if either $d \geq 4$ or G is connected and $d \geq 2$.*

In fact, [18, Theorem 10] is slightly stronger than this when d is not an integer, which is worth considering now that d bounds the average degree rather than the maximum degree. The proportion of vertices in the subgraph found actually interpolates linearly between the values of $3/(d+1)$ for integer d.

Theorem 5.1 is the basis of the upper bounds for $c_f(\overline{\Gamma}_d)$ and $c_f(\overline{\Gamma}_d^c)$ given respectively in (2) and (3) of Section 4, again using Theorem 3.2.

The improved upper bounds for these coefficients of fragmentability given in Theorem 4.2 come from a new approximation algorithm for MIPS (and another application of Theorem 3.2). This algorithm improves on the reduction-removal algorithm described above in its choice of vertex to be removed, by taking into account the degrees of its neighbours.

The link between planarization and fragmentability also works in the other direction: a lower bound on $c_f(\Gamma)$ gives an upper bound on the proportion of vertices that must belong to some induced planar subgraph. Define

$$c_p(\Gamma) = \sup\{\varepsilon \ : \ \text{for all } G \in \Gamma \text{ there exists } P \subseteq V(G) \text{ for which } |P| \geq \varepsilon |V(G)| \text{ and } \langle P \rangle \text{ is planar}\}.$$

So, if $c_f(\Gamma) \geq \varepsilon$, then $c_p(\Gamma) \leq (1 - \varepsilon)$; for example, the fact that $c_f(\Gamma_3) \geq \frac{1}{4}$ (in fact we have equality in this case) implies that $c_p(\Gamma_3) \leq \frac{3}{4}$, so that a graph in Γ_3 cannot be guaranteed to have an induced planar subgraph with more than three-quarters of the vertices of the original graph. Each of the lower bounds for $c_f(\Gamma_d)$ described in Section 4 gives an upper bound of this nature for planarization. These bounds are stronger than the trivial upper bound $c_p(\Gamma_d) \leq 4/(d+1)$, obtained from the observation that the maximum induced planar subgraph of K_{d+1} is K_4.

Morgan and Farr [37] reported on an experimental study of approximation algorithms for MIPS. Their study compared the performance (in terms of the size of the induced subgraph found) of implementations of several algorithms, including the Halldórsson–Lau algorithm, the vertex-addition algorithm of [16], the reduction-removal algorithm of [18] (referred to there as the vertex-removal algorithm), and the induced outerplanar subgraph algorithm of [37]. The study also included a *vertex-subset removal* algorithm which modifies the reduction-removal algorithm so that, when removing a vertex of maximum degree in the reduced graph, it chooses the one with the most neighbours of lower degree. This algorithm is based on an earlier version of the work reported in [19]. As simple comparators, the sequential algorithm for finding a maximal independent set and a simple heuristic for finding a large induced forest were included as well. (The latter works as follows: take vertices in order of increasing degree, and add a vertex if and only if it has at most one neighbour in each component of the induced forest found so far.) The algorithms were compared on classical random graphs of expected average degree d, and on random d-regular graphs, for $3 \leq d \leq 9$ and various numbers of vertices up to 10,000 (although the behaviour had always settled down by about 1000 vertices).

It was found that, apart from the sequential independent set algorithm and the Halldórsson–Lau algorithm, the numbers of vertices in the induced subgraph found do not vary greatly from one algorithm to the next. For a given order n, considering the output number of vertices as a function of d, these algorithms give very similar-looking curves, all close together. The algorithms seem to be running up against a barrier that prevents them from finding a large induced planar subgraph beyond a certain size. This is perhaps surprising, considering the variety of the algorithms: from a simple heuristic for finding large induced forests, through to the vertex-subset removal algorithm (which was the most successful). The numbers of vertices of the induced subgraphs found were also very concentrated, in that the standard deviation of the number of vertices in the subgraph found was less than 1% of the mean in almost all cases (that is, for almost all choices of d, n, the random generation method and the planarization algorithm). It would be interesting to find mathematical explanations for these observed phenomena.

6. Applications

Although it is not our primary motivation, it is worth noting a couple of ways in which the idea of fragmentability has been used in other branches of mathematics and computer science. In each case, the technique is to exploit the existence of a large subgraph with a particularly simple structure – namely, having bounded component size. This bounded component size implies that the components belong to one of a finite number of isomorphism classes; this is so in many cases, such as bounded-degree graphs, even when adjacencies (in the original graph) outside the components are taken into account.

The original motivation for the idea of fragmentability (in [20]) came from the theorem of Wilson [40] concerning the decomposition of a complete graph into edge-disjoint copies of a fixed graph. Graphs in a class with coefficient of fragmentability 0 can be reduced to a disjoint union of a number of copies of a fixed graph by removing an arbitrarily small proportion of the vertices. This may allow us to exploit Wilson's theorem or other properties of such a special graph.

Another idea uses an alternative formulation of the coefficient of fragmentability in terms of tree-width. The coefficient of fragmentability is, roughly, the smallest proportion of the vertices whose removal can guarantee a graph of bounded tree-width. Formally, $c_f(\Gamma)$ is the infimum of the set of numbers λ for which there exists an integer k such that, for every $G \in \Gamma$, a graph of tree-width at most k can be obtained by deleting at most $\lambda|V(G)|$ vertices from G.

The idea is to improve the constant in an exponential upper bound for a graph property or algorithm by using the special properties of the large induced subgraph of bounded tree-width, as follows. Suppose we have a bound of the form c^n, which is derived from a multiplicative contribution from each vertex of a graph G. If G belongs to a class with coefficient of fragmentability α, then we can find an induced subgraph G' of size $(1 - \alpha - o(1))n$ with bounded tree-width. For this subgraph, we may well be able to improve the upper bound to a polynomial, while still having a contribution of c for each vertex of $G - G'$. Thus the overall upper bound is improved to $\text{poly}(n)c^{(1-\alpha)n}$ – that is, the base is improved from c to $c^{1-\alpha}$.

This idea has been exploited to improve the upper bound on the length of the shortest regular expression representing a deterministic finite automaton (see [22]), using a large induced series-parallel subgraph – that is, one of tree-width 2. A similar idea was used by McDermid and Irving [36] to improve the exponential time bound of some matching algorithms on bounded-degree graphs. It seems likely that this idea may be useful in other situations.

7. Monochromatic components

In the study of fragmentability, we are interested in subgraphs that have only small components. This motivates the study of partitions of the vertex-set of a graph into subsets, each of which induces a subgraph that has only small components. In taking this step, we follow a long tradition in graph theory of moving between the study of subgraphs with some property and the study of partitions of the graph into subgraphs with that property.

Suppose we have a set Q of q available labels. An *assignment*, or *q-assignment*, of a graph G is a map from $V(G)$ to Q. We call the labels *colours*, although the assignment does not have to be a proper colouring in the usual sense. A *colour class* is a preimage, under the q-assignment, of a single colour.

A *chromon* of G under an assignment c is a component of a subgraph of G induced by a colour class – that is, a maximal connected monochromatic subgraph (sometimes called a *monochromatic component*). A *k-chromon* is a chromon with k vertices.

A graph G is *$[q, C]$-colourable* if it has a q-assignment in which every chromon has at most C vertices. An ordinary (proper) colouring is thus a $[q, 1]$-colouring, and the chromons under such a colouring are just the individual vertices.

A class of graphs Γ is *$[q, C]$-colourable* if every graph in Γ is $[q, C]$-colourable, and is *q-metacolourable* if there is some C for which Γ is $[q, C]$-colourable. The *metachromatic number* $\chi(\Gamma)$ is the smallest value of q for which Γ is q-metacolourable.

These definitions have appeared in various forms, implicitly or explicitly, in a number of papers over the years. We mention some of these now, emphasizing results about the classes of graphs that we have considered in this chapter.

The above generalization of graph colouring falls within the class of generalized colourings introduced by Weaver and West [39], and studied further by Deuber and Zhu [13]. They investigated q-assignments in which each chromon is a member of some fixed hereditary class of graphs. If we use the class of all graphs with at most C vertices, then we are looking at $[q, C]$-colourability.

Akiyama *et al.* [1] considered q-assignments in which each chromon is a path with at most C vertices. The existence of such a colouring makes the graph $[q, C]$-colourable. It follows from one of their results that $\chi(\Gamma_d) \leq \lceil \frac{1}{2}(d+1) \rceil$. Alon *et al.* [2] showed that $\chi(\{\text{planar graphs}\}) = 4$.

We next consider graphs of bounded degree. It is immediate that $\chi(\Gamma_0) = \chi(\Gamma_1) = 1$, and it is clear that $\chi(\Gamma_2) = 2$, since paths and even cycles are bipartite, while odd cycles are $[2, 2]$-colourable. For $G \in \Gamma_3$, a lemma of Lovász [34] (in fact, the one used by the Halldórsson–Lau algorithm, mentioned in Section 5) can be used to obtain a $[2, 2]$-colouring, so that $\chi(\Gamma_3) = 2$.

This brings us to the first non-trivial case. Alon *et al.* [2] showed that Γ_4 is $[2, 57]$-colourable, and hence that $\chi(\Gamma_4) = 2$. The question of how low the chromon size can be here is intriguing. Haxell, Szabó and Tardos [26, Section 2.2] lowered the upper bound on the chromon size to 6, so that Γ_4 is [2, 6]-colourable. It is natural to ask for the minimum number C for which every graph in Γ_4 is $[2, C]$-colourable. Haxell, Szabó and Tardos showed that $C \geq 4$, as attained by the complement of the 7-cycle.

An efficient algorithm for finding [2, 6]-colourings for graphs in Γ_4 can be extracted from [26]; see also an algorithm, due to the authors [17], which appears to be the simplest known $[2, o(n)]$-colouring algorithm for Γ_4.

The next bounded-degree case is Γ_5. This was also resolved by Haxell, Szabó and Tardos [26], answering a question in [2] by showing that $\chi(\Gamma_5) = 2$. They also showed that $\chi(\Gamma_8) \leq 3$.

For a general $d \geq 3$, the state of the art is $\frac{1}{4}(d+3) \leq \chi(\Gamma_d) \leq \lceil \frac{1}{3}(d+1) \rceil$. The lower bound can be found in Alon *et al.* [2], while the upper bound is due to Haxell *et al.* [26] and uses the same upper bound on chromon size for all d. This upper bound improves slightly on $\lceil \frac{1}{3}(d+2) \rceil$, due to Alon *et al.* [2] who showed that Γ_d is $[\lceil \frac{1}{3}(d+2) \rceil, 12d^2 - 36d + 9]$-colourable (which is where the 57 comes from in their above-mentioned result on Γ_4), again using the lemma of Lovász [34].

For results on chromon size for minor-closed classes of graphs, and further results on chromon size for graphs of bounded degree (emphasizing Γ_7), see [32]. For results on the complexity of determining whether a graph in Γ_d has a $[2, C]$-colouring, see Berke and Szabó [8] who, among other results, showed that the problem is NP-complete for $d \geq 4$ and $C = 2$ or 3, and also for $d = 6$ and all $C \geq 2$.

8. Open problems

We list here some of the open problems in the area.

- What are the true values of $c_f(\Gamma_d)$ and of $c_f(\overline{\Gamma}_d)$?
  In particular, is the lower bound, of $(d-2)/(d+2)$, given by Haxell *et al.*, sharp?
- Does the lower bound of $(d-2)/(d+2)$ hold for odd d as for well as for even d?
- What is the optimum constant for planarization of (average) degree bounded graphs? – that is, what is p_d, where

$$p_d = \min\{\mu : p(G) \leq (\mu + o(1))|V(G)| \text{ for all connected } G \in \Gamma_d\}?$$

(It is easy to see that the minimum is attained here.) We know that $p_3 = \frac{1}{4}$, but what about p_d for $d \geq 4$? In particular, what is p_4? We know that $\frac{1}{3} \leq p_4 \leq \frac{3}{8}$, but it seems doubtful whether either of these bounds is the correct value.

We can also generalize the notion of planarization considerably. If W is a set of graphs, we define $\mu(W, \Gamma)$ to be the minimum number μ such that, for all $G \in \Gamma$, there is a set of at most $(\mu + o(1))|V(G)|$ vertices of G whose deletion leaves a graph which is W-minor-free. Thus, $p_d = \mu(\{K_5, K_{3,3}\}, \Gamma_d^c)$.

It would be of interest to determine values of $\mu(W, \Gamma)$. We know that for $\Gamma = \Gamma_3$, these values are almost independent of W, since provided that each member of W

has a K_4-minor, then $\mu(W, \Gamma_3) = \frac{1}{4}$; this follows from the chain of inequalities

$$\tfrac{1}{4} = c_f(\Gamma_3) \le \mu(W, \Gamma_3) \le \mu(\{K_4\}, \Gamma_3) \le \tfrac{1}{4}.$$

Of particular interest are values of $\mu(\{K_r\}, \Gamma_d)$. It is clear that $\mu(\{K_{r+1}\}, \Gamma_d) \le \mu(\{K_r\}, \Gamma_d)$, and it is easy to see that, the sequence converges to the coefficient of fragmentability – that is, $\lim_{r\to\infty} \mu(\{K_r\}, \Gamma_d) = c_f(\Gamma_d)$. Moreover, from above we see that, in the case $d = 3$, $\mu(\{K_r\}, \Gamma_d)$ is equal to $c_f(\Gamma_d)$ for all $r \ge 4$. Is something similar true for every bounded degree class Γ_d?

References

1. J. Akiyama, H. Era, S. V. Gervacio and M. Watanabe, Path chromatic numbers of graphs, *J. Graph Theory* **13** (1989), 569–575.
2. N. Alon, G. Ding, B. Oporowski and D. Vertigan, Partitioning into graphs with only small components, *J. Combin. Theory (B)* **87** (2003), 231–243.
3. N. Alon, J. Kahn and P. D. Seymour, Large induced degenerate graphs, *Graphs Combin.* **3** (1987), 203–211.
4. N. Alon, P. Seymour and R. Thomas, A separator theorem for graphs with an excluded minor and its applications, *Proc. 22nd Annual ACM Symposium on Theory of Computing*, Association for Computing Machinery (1990), 293–299.
5. K. S. Bagga, L. W. Beineke, W. D. Goddard, M. J. Lipman and R. E. Pippert, A survey of integrity, *Discrete Appl. Math.* **37/38** (1992), 13–28.
6. C. A. Barefoot, R. Entringer and H. Swart, Vulnerability in graphs – a comparative survey, *J. Combin. Math. Combin. Comput.* **1** (1987), 13–22.
7. L. W. Beineke, Biplanar graphs: a survey, *Comput. Math. Appl.* **34** (1997), 1–8.
8. R. Berke and T. Szabó, Deciding relaxed two-colourability: a hardness jump, *Combin. Probab. Comput.* **18** (2009), 53–81.
9. A. Beygelzimer, G. Grinstein, R. Linsker and I. Rish, Improving network robustness, *Physica (A)* **357** (2005), 593–612.
10. T. Britton, S. Janson and A. Martin-Löf, Graphs with specified degree distributions, simple epidemics and local vaccination strategies, *Adv. Appl. Probab.* **39** (2007), 922–948.
11. Y. Caro and R. Yuster, Graph decomposition of slim graphs, *Graphs Combin.* **15** (1999), 5–19.
12. H. de Fraysseix and P. Ossona de Mendez, A characterization of DFS cotree critical graphs, *Graph Drawing 2001* (Vienna) (eds. P. Mutzel, M. Jünger and S. Leipert), Lecture Notes in Computer Science **2265**, Springer (2002), 84–95.
13. W. Deuber and X. Zhu, Relaxed coloring of a graph, *Graphs Combin.* **14** (1998), 121–130.
14. G. Di Battista, P. Eades, R. Tamassia and I. Tollis, *Graph Drawing: Algorithms for the Visualization of Graphs*, Prentice-Hall, 1999.
15. K. J. Edwards and G. E. Farr, Fragmentability of graphs, *J. Combin. Theory (B)* **82** (2001), 30–37.
16. K. J. Edwards and G. E. Farr, An algorithm for finding large induced planar subgraphs, *Graph Drawing 2001* (Vienna) (eds. P. Mutzel, M. Jünger and S. Leipert), Lecture Notes in Computer Science **2265**, Springer (2002), 75–83.
17. K. J. Edwards and G. E. Farr, On monochromatic component size for improper colourings, *Discrete Appl. Math.* **148** (2005), 89–105.
18. K. J. Edwards and G. E. Farr, Planarization and fragmentability of some classes of graphs, *Discrete Math.* **308** (2008), 2396–2406.

19. K. J. Edwards and G. E. Farr, Improved upper bounds for planarization and series-parallelization of degree-bounded graphs, *Electron. J. Combin.* **19**(2) (2012), P 25.
20. K. J. Edwards and C. J. H. McDiarmid, New upper bounds on harmonious colorings, *J. Graph Theory* **18** (1994), 257–267.
21. J. R. Gilbert, J. P. Hutchinson and R. E. Tarjan, A separator theorem for graphs of bounded genus, *J. Algorithms* **5** (1984), 391–407.
22. H. Gruber and M. Holzer, Provably shorter regular expressions from deterministic finite automata (extended abstract), *Proc. 12th Intern. Conf. Developments in Language Theory* (Kyoto, 2008) (eds. M. Ito and M. Toyama), Lecture Notes in Computer Science **5257**, Springer (2008), 383–395.
23. M. M. Halldórsson, Approximations of weighted independent set and hereditary subset problems, *J. Graph Algorithms Appl.* **4** (2000), 1–16.
24. M. M. Halldórsson and H. C. Lau, Low-degree graph partitioning via local search with applications to constraint satisfaction, max cut, and coloring, *J. Graph Algorithms Appl.* **1** (1997), 1–13.
25. P. Haxell, O. Pikhurko and A. Thomason, Maximum acyclic and fragmented sets in regular graphs, *J. Graph Theory* **57** (2008), 149–156.
26. P. Haxell, T. Szabó and G. Tardos, Bounded size components – partitions and transversals, *J. Combin. Theory (B)* **88** (2003), 281–297.
27. S. Janson and A. Thomason, Dismantling sparse random graphs, *Combin. Probab. Comput.* **17** (2008), 259–264.
28. David S. Johnson, Approximation algorithms for combinatorial problems, *J. Comput. System Sci.* **9** (1974), 256–278.
29. M. S. Krishnamoorthy and N. Deo, Node-deletion NP-complete problems, *SIAM J. Comput.* **8** (1979), 619–625.
30. J. M. Lewis and M. Yannakakis, The node-deletion problem for hereditary properties is NP-complete, *J. Comput. System Sci.* **20** (1980), 219–230.
31. A. Liebers, Planarizing graphs – a survey and annotated bibliography, *J. Graph Algorithms Appl.* **5** (2001), 1–74.
32. N. Linial, J. Matoušek, O. Sheffet and G. Tardos, Graph colouring with no large monochromatic components, *Combin. Probab. Comput.* **17** (2008), 577–589.
33. R. J. Lipton and R. E. Tarjan, A separator theorem for planar graphs, *SIAM J. Appl. Math.* **36** (1979), 177–189.
34. L. Lovász, On decomposition of graphs, *Stud. Sci. Math. Hungar.* **1** (1966), 237–238.
35. C. Lund and M. Yannakakis, The approximation of maximum subgraph problems, *Proc. 20th Intern. Colloq. on Automata, Languages and Programming* (ICALP), Lecture Notes in Computer Science **700**, Springer-Verlag (1993), 40–51.
36. E. McDermid and R. Irving, Sex-equal stable matchings: complexity and algorithms, *Algorithmica*, (to appear).
37. K. Morgan and G. Farr, Approximation algorithms for the maximum induced planar and outerplanar subgraph problems, *J. Graph Algorithms Appl.* **11** (2007), 165–193.
38. P. Mutzel, T. Odenthal and M. Scharbrodt, The thickness of graphs: a survey, *Graphs Combin.* **14** (1998), 59–73.
39. M. L. Weaver and D. B. West, Relaxed chromatic numbers of graphs, *Graphs Combin.* **10** (1994), 75–93.
40. R. M. Wilson, Decomposition of complete graphs into subgraphs isomorphic to a given graph, *Proc. Fifth British Combinatorial Conference* (Aberdeen, 1975) (eds. C. St.J. A. Nash-Williams and J. Sheehan), *Congr. Numer.* **15** (1976), 647–659.

10

The phase transition in random graphs

BÉLA BOLLOBÁS and OLIVER RIORDAN

Fifty years ago Erdős and Rényi proved a surprising result about the sudden change in the component structure of the 'typical' random graph $G_{n,m}$ with n vertices and m edges, as $m = m(n)$ grows from less than $\frac{1}{2}n$ to greater than $\frac{1}{2}n$. After a quiet start, this result has exerted a tremendous influence on the theory of random graphs: a host of deep results have been proved about the nature of this phase transition. In this chapter we give a brief account of these results, from the original theorem to some of the most recent developments.

1. Introduction

The theory of random graphs was founded by Erdős and Rényi in a series of papers around 1960 (see [23]–[31]). By now thousands of papers have been written about random graphs, with many difficult and deep results. In this chapter we concentrate on a celebrated topic in this theory, the phase transition in the component structure.

The basic theorem was proved by Erdős and Rényi in 1960, and was soon hailed as the most striking result in this new field. For about a quarter of a century it was thought that, except for refinements of the approximations, this result was the final word on the topic. Only in the mid-1980s was it recognized that, as stated and proved, the deepest part of this theorem was actually incorrect, and interesting phenomena happen close to the critical probability. Since then, a host of difficult and exciting results have been proved in related areas, with research continuing to this day. Our aim in this chapter is to describe some of the basic results and to draw attention to some of the latest developments.

The basic models

There are three basic models of random graphs: the size model, the binomial model and the space of graph processes. All of these models concern labelled graphs on n vertices: as is customary, we consider graphs with vertex-set $V = [n] = \{1, 2, \ldots, n\}$. We write $N = \binom{n}{2}$ for the total number of possible edges – that is, the number of edges of the complete graph on V. Given a function $m = m(n)$ with $0 \le m \le N$, let $\mathcal{G}(n, m)$ be the set of all $\binom{N}{m}$ graphs with vertex-set $[n]$ and m edges. We turn $\mathcal{G}(n, m)$ into a probability space, the *size model*, by defining its graphs to be equiprobable. Also, we write $G_{n,m}$ for a random member of $\mathcal{G}(n, m)$. Thus, if H is a graph on $[n]$ with m edges, then the probability that $G_{n,m}$ is equal to H is $\mathbf{P}(G_{n,m} = H) = 1/\binom{N}{m}$.

To define the next model, for $0 < p = p(n) < 1$, let $G_{n,p}$ be the random graph with vertex-set $[n]$ in which any two vertices are joined by an edge with probability p, with all choices independent. The space $\mathcal{G}(n, p)$ of these random graphs $G_{n,p}$ is the *binomial model*. Note that we commit a standard abuse of notation in using $\mathcal{G}(n, x)$ and $G_{n,x}$ for both the size model and the binomial model. However, there is no danger of confusion since when $x < 1$ we mean the binomial model, and when $x \ge 1$ we mean the size model.

Note that the probability space $\mathcal{G}(n, p)$ has 2^N points, independently of p; different values of p give different probability measures on the set $\mathcal{G}_n$ of all 2^N graphs on $[n]$. More explicitly,

$$\mathbf{P}(G_{n,p} = H) = p^{e(H)}(1 - p)^{N-e(H)}$$

for every graph H on $[n]$, where $e(H)$ is the number of edges of H. The space $\mathcal{G}(n, \frac{1}{2})$ occupies a central position: it is simply the set $\mathcal{G}_n$ with all graphs taken to be equiprobable.

The concentration of the binomial distribution tells us that, for 'most' questions, the models $\mathcal{G}(n, p)$ and $\mathcal{G}(n, m)$ are indistinguishable, provided that $p = m/N$, or even just that m is close enough to pN – that is, $m = pN + o(\sqrt{pN})$.

Finally, a *graph process* on $V = [n]$ is a nested sequence

$$G_{n,0} \subset G_{n,1} \subset \cdots \subset G_{n,N}$$

of graphs on $[n]$, such that $G_{n,t}$ has precisely t edges. We write $\widetilde{\mathcal{G}}_n$ for the set of all $N!$ graph processes on $[n]$, and turn it into a probability space by taking all processes to be equiprobable. Clearly, a random graph process $\widetilde{G} = (G_{n,t})_0^N$ is a Markov chain: given the state $G_{n,t}$ of $\widetilde{G}$ at time t, the next state $G_{n,t+1}$ is obtained by adding to $G_{n,t}$ one of the $N - t$ possible edges, with all choices equiprobable. Stopping a random graph process $\widetilde{G} = (G_{n,t})_0^N$ at time m, we obtain the random graph $G_{n,m}$ defined above, justifying the further re-use of the notation $G_{n,x}$.

Questions concerning the random graphs $G_{n,m}$ are *counting questions*: What is the number of connected graphs on $[n]$ with m edges? What is the number of Hamiltonian graphs on $[n]$ with m edges?, and so on. For example, if $C(n,m)$ denotes the number of connected graphs on $[n]$ with m edges, then

$$\mathbf{P}(G_{n,m} \text{ is connected}) = C(n,m)/\binom{N}{m}.$$

Before Erdős and Rényi came along, this is precisely how these problems had been viewed and, to solve them, generating functions were used. The revolutionary idea of Erdős and Rényi was that, rather than count the number of certain graphs precisely, one can adopt a probabilistic point of view. They considered every graph invariant as a random variable and used the methods of probability theory to study them. Thus they turned counting questions into questions concerning the distributions of random variables. Over the years, more and more powerful probabilistic techniques have been used to tackle questions concerning random graphs.

It should be noted that the binomial model was introduced by Gilbert [36], rather than by Erdős and Rényi, at about the same time that Erdős and Rényi introduced their size model. However, Gilbert studied the binomial model the 'old-fashioned' way, setting himself the task of determining *exactly* the probabilities of certain sets of graphs, rather than adopting the probabilistic point of view by considering $\mathcal{G}(n,p)$ as a probability space with a host of variables on it. For this reason, both kinds of classical random graphs, $G_{n,m}$ and $G_{n,p}$, are now called *Erdős–Rényi random graphs*.

Threshold phenomena

An event is said to happen *with high probability*, abbreviated to *whp*, if the probability that it happens tends to 1 as the main parameter, in our case n, tends to ∞. One of the major discoveries of Erdős and Rényi was that, belying their name, random graphs are not very random, in the sense that in many respects they 'look the same' whp. Putting it differently, there is a 'typical' random graph, a graph to which most of the graphs in that class are rather similar. For example, when $m = \lfloor n \log n \rfloor$, a typical random graph $G_{n,m}$ contains a Hamiltonian cycle, contains triangles but no complete graph of order 4 and has maximum degree about $c \log n$, where $2 < c < 2e$ and $c \log(2e/c) = 1$.

Another aspect of the existence of 'typical' random graphs is that monotone increasing properties have so-called 'threshold functions'. A *monotone increasing* (or simply *monotone*) property Q_n of graphs on $[n]$ is a collection of graphs on $[n]$ (that is, $Q_n \subseteq \mathcal{G}_n$) which is closed under isomorphism and is invariant under the addition of edges: if $G \subseteq H$ and $G \in Q_n$, then $H \in Q_n$. Erdős and Rényi showed that many monotone increasing properties arise suddenly: there is a *threshold function* $t(n)$ such that, whenever $\omega(n) \to \infty$, if $t_0 \le t/\omega$, then whp G_{n,t_0} fails to have Q_n, while if $t_1 \geq \omega t$, then whp G_{n,t_1} has Q_n. The point is that the convergence $\omega(n) \to \infty$ can be as slow as we like, so that the lower threshold t_0 and the upper threshold t_1 are about the same size. In many cases, there is a much sharper threshold. For example, if Q_n is the property that a random graph $G_{n,m}$ is connected, then

$$t_0 = \tfrac{1}{2}n(\log n - \omega(n)) \quad \text{and} \quad t_1 = \tfrac{1}{2}n(\log n + \omega(n))$$

will do, provided that $\omega(n) \to \infty$ (no matter how slowly).

Similar threshold phenomena hold for the binomial model $\mathcal{G}(n, p)$. Indeed, using the random graph process $\widetilde{G}$, we can easily transfer results about monotone properties between $G_{n,p}$ and $G_{n,m}$. To be explicit, let $\widetilde{G} = (G_{n,t})_0^N$ be the random graph process defined above, and let T have the binomial distribution with parameters N and p, with $\widetilde{G}$ and T independent. Note that T has the same distribution as the number of edges of $G_{n,p}$; we shall take T to be the number of edges of $G_{n,p}$. Then, given T, the graph $G_{n,p}$ is equally likely to be any graph on $[n]$ with T edges. It follows that the graph $G_{n,T}$ selected from $\widetilde{G}$ has the distribution of $G_{n,p}$. Suppose that Q_n is a monotone property, and that $p = p(n)$ and $m = m(n)$ are chosen so that $\mathbf{P}(T \geq m) \to 1$, and that $G_{n,m}$ has property Q_n whp. From the definition of the process, $G_{n,m} \subseteq G_{n,T}$ whenever $T \geq m$, and it follows that $G_{n,p}$ has Q_n whp. This, and a similar argument for the reverse comparison, show that the connectedness result for $G_{n,m}$ above is equivalent to the following result for $G_{n,p(n)}$: whenever $\omega(n) \to \infty$, then

$$\mathbf{P}(G_{n,p} \text{ is connected}) \to 1 \quad \text{if} \quad p \geq \frac{\log n + \omega(n)}{n}, \tag{1}$$

and

$$\mathbf{P}(G_{n,p} \text{ is connected}) \to 0 \quad \text{if} \quad p \leq \frac{\log n - \omega(n)}{n}.$$

For another easy example, let us consider the property of containing the complete graph K_r. If $pn^{2/(r-1)} \to 0$, then whp $G_{n,p}$ does not contain K_r, and if $pn^{2/(r-1)} \to \infty$, then whp $G_{n,p}$ contains K_r. In fact, more precise results hold: if $pn^{2/(r-1)} \to c > 0$, then

$$\mathbf{P}(G_{n,p} \text{ has no } K_r) \to \exp\left(-c^{\binom{r}{2}}/r!\right). \tag{2}$$

Although to the uninitiated the above examples may seem impressive, they are in fact very simple. In particular, (2) is a consequence of the stronger assertion that the distribution of the number μ_r of K_r-subgraphs of $G_{n,p}$ converges to a Poisson distribution with mean $c^{\binom{r}{2}}/r!$; in turn, this can be shown by estimating the moments of μ_r and applying the inclusion–exclusion formula.

Bollobás and Thomason [14] noticed that not only are there threshold functions for many increasing properties of graphs, as shown by Erdős and Rényi, but *every* increasing property of graphs has a threshold function. Considerably sharper and deeper results were later proved by Friedgut and Kalai [35], based on results of Kahn, Kalai and Linial [40] and Bourgain *et al.* [15], and by Friedgut [34].

The phase transition

The other major achievement of Erdős and Rényi in the theory of random graphs was their theorem (really, a group of theorems) describing the emergence of the 'giant component' – the 'phase transition' that the random graph process $(G_{n,t})_0^N$ goes through as t passes through $\frac{1}{2}n$. The main aim of this chapter is to describe the beginning of the substantial theory that Erdős and Rényi launched with this milestone result, and to sketch some of the recent deep results. For space reasons we cannot go into much detail, but in addition to describing several major results, we emphasize the rich variety of methods and tools used to prove them. In fifty years the theory has come a long way: a wide variety of methods have been used to study the phase transition – not only probabilistic ones, but also generating functions to count numbers of graphs exactly before deducing their asymptotics.

In the next section we present the original Erdős–Rényi theorem. In Section 3 we describe the rebirth of the theory over two decades later. In Section 4 we show that the most basic probabilistic methods can be used to prove weak forms of the main phase-transition results – this is the only section that contains full proofs. In Section 5 we briefly describe some of the more modern proof techniques used in this area. Next, in Section 6, we state some of the very detailed results concerning the exact 'speed' of the phase transition. Finally, in Section 7, we give some substantial recent results about the structure of the random graph $G_{n,m}$ close to its phase transition.

To conclude this introduction, let us draw attention to the remarkable fact that, shortly before Erdős and Rényi founded the theory of random graphs, Broadbent and Hammersley [16] had founded the theory of percolation, the study of the component structure of random subgraphs of lattices and lattice-like graphs. Although in spirit the two theories are very close, for several decades they did not influence each other, and only in the last decade has there been a fair amount of cross-fertilization. In particular, the model $\mathcal{G}(n, p)$ is called the *mean-field model* of percolation, and the emergence of the giant component in $G_{n,p}$ is the *phase transition* in this mean-field model.

2. The Erdős–Rényi theorem: the double jump

What can one say about the component structure of a typical random graph? We have already remarked that, for $m = \frac{1}{2}n(\log n + \omega(n))$, a typical $G_{n,m}$ is disconnected if $\omega(n) \to -\infty$, and connected if $\omega(n) \to \infty$. Also, if $m = \varepsilon(n)n$ with $\varepsilon(n) \to 0$, then a typical $G_{n,m}$ is a forest, and the largest component has about $(\log n)/\log(1/\varepsilon)$ vertices. What happens when m is about cn for a positive constant c? As shown by Erdős and Rényi [24] in 1960, it is in this range that the component structure is most interesting.

To state their fundamental theorem, we recall some notation. Given functions $f(n)$ and $g(n)$, the notation $f(n) \sim g(n)$ means that $f(n)/g(n) \to 1$ as $n \to \infty$, and $f(n) = \Theta(g(n))$ means that

$$0 < \liminf f(n)/g(n) \leq \limsup f(n)/g(n) < \infty.$$

We write $L_i(G)$ for the number of vertices in the ith largest component of a graph G. Note that in the statement below, i is not allowed to depend on n.

Theorem 2.1 *Let $m(n) \sim \frac{1}{2}cn$, where c is a positive constant.*

(a) *If $0 < c < 1$, then for all $i \geq 1$, $L_i(G_{n,m(n)}) = \Theta(\log n)$ whp.*
(b) *If $c = 1$, then for all $i \geq 1$, $L_i(G_{n,m(n)}) = \Theta(n^{2/3})$ whp.*
(c) *If $c > 1$, then $L_1(G_{n,m(n)}) \sim \alpha_c n$ whp, where $\alpha_c > 0$, and for all $i \geq 2$, $L_i(G_{n,m(n)}) = \Theta(\log n)$ whp.*

The most striking aspect of this theorem was summarized by Erdős and Rényi [24] as follows. (We have changed their $N(n)$ to $m(n)$ to conform to our notation.)
The largest component of $G_{n,m(n)}$ is of order $\log n$ for $m(n)/n \sim c < \frac{1}{2}$, of order $n^{2/3}$ for $m(n)/n \sim \frac{1}{2}$, and of order n for $m(n)/n \sim c > \frac{1}{2}$. This 'double jump' of the size of the largest component when $m(n)/n$ passes the value $\frac{1}{2}$ is one of the most striking facts concerning random graphs.

This is a remarkable result indeed: for $m(n) \sim \frac{1}{2}cn$, the component structure undergoes a dramatic phase transition as c increases from less than 1 to greater than 1. (The reason for the artificial-looking usage of c, so that $m(n)$ is asymptotic to $\frac{1}{2}cn$, rather than cn, will be clear later.) Following Erdős and Rényi, the unique largest component in (c), guaranteed to exist whp, is called the *giant component*, so the phase transition concerns the emergence of the giant component.

Unfortunately, Theorem 2.1 is incorrect as stated above. The problem is with the case $c = 1$: as we shall see, for $m(n) = \frac{1}{2}n + n^{6/7} \sim \frac{1}{2}n$, say, the giant component has about $4n^{6/7}$ vertices, whereas for $m(n) = \frac{1}{2}n + n/\log n \sim \frac{1}{2}n$, it has about $4n/\log n$ vertices.

Erdős and Rényi [24] proved a number of other results about $G_{n,m}$, including an expression for the constant α_c in Theorem 2.1:

$$\alpha_c = 1 - \frac{1}{c} \sum_{k=1}^{\infty} \frac{k^{k-1}}{k!} \left(ce^{-c}\right)^k. \tag{3}$$

This function is shown in Fig. 1. Note that $\alpha_c = 0$ if $0 \leq c \leq 1$, and $\alpha_c > 0$ if $c > 1$. The right-hand derivative of α_c at $c = 1$ is 2, and $\alpha_c \to 1$ as $c \to \infty$.

At first sight, the expression in (3) is rather surprising; however, as we shall see later, it has several natural explanations. In particular, for $c > 1$, the constant α_c is also the unique positive root of the equation $x = 1 - e^{-cx}$ (see Section 5).

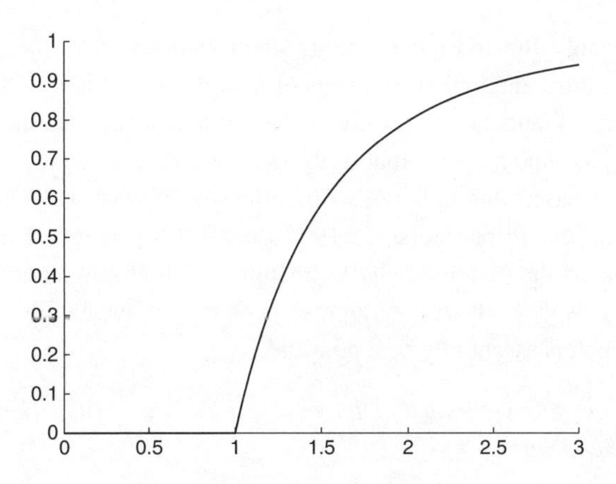

Fig. 1. The graph of α_c against c.

3. Correction: no double jump

Strangely, it went unnoticed for almost a quarter of a century that Theorem 2.1 is incorrect, and that what Erdős and Rényi actually proved is considerably less: if $m(n) = \lfloor \frac{1}{2}cn \rfloor$ (rather than $m(n) \sim \frac{1}{2}cn$), then the assertions in Theorem 2.1 hold. Actually, for $c \neq 1$ it makes no difference whether we take $m(n) = \lfloor \frac{1}{2}cn \rfloor$ or $m(n) \sim \frac{1}{2}cn$, since for $0 < c < 1$ whp $L_1(G_{n,\lfloor cn/2 \rfloor}) = O(\log n)$, and for $c > 1$ whp $L_1(G_{n,\lfloor cn/2 \rfloor}) = (\alpha_c + o(1))n$, where $\alpha_c > 0$. However, the situation is completely different for $c = 1$: for a 'typical' $G_{n,m(n)}$, the maximum order of a component depends greatly on $m(n) - \frac{1}{2}n$. In 1984, Bollobás [8] (see also [9]) realized that there is no double jump close to $m(n) = \frac{1}{2}n$ when $m(n)$ is taken to be a finer function of n. He proved that if $m(n) = \frac{1}{2}(1 + \varepsilon)n$ with $4(\log n)^{1/2} n^{-1/3} \leq \varepsilon = \varepsilon(n) = o(1)$, then whp $G_{n,m(n)}$ has a unique giant component with about $2\varepsilon n$ vertices, and all other components have $O((\log n)/\varepsilon^2)$ vertices; this result implies the corresponding statement about $G_{n,p}$. As the random graph $G_{n,p}$ is the mean-field model of percolation, this result was the first theorem about the exact nature of the phase transition in a non-trivial percolation model.

To prove this, Bollobás studied the random graph process $(G_{n,t})_0^N$ by considering each $G_{n,t}$ separately and concentrating on the distribution of not-too-large components. This distribution depends on the number of connected graphs with k vertices and l edges. Indeed, writing $C(k,l)$ for the number of connected graphs with k labelled vertices and l edges, as before, the expected number of components of $G_{n,t}$ with k vertices and l edges is

$$\binom{n}{k} C(k,l) \binom{N - \binom{k}{2} - k(n-k)}{t-l} \binom{N}{t}^{-1}.$$

An important ingredient of the proof in [8] was a reasonable (but not very sharp) upper bound on $C(k,l)$ in the entire range $k - 1 \leq l \leq \binom{k}{2}$. Like Erdős and Rényi,

Bollobás used only the most basic tools of probability theory – bounds on the expectation, variance and higher moments of a random variable.

This result in [8] launched the study of the *phase transition* in the random graph process $(G_{n,t})_0^N$ around $t=\frac{1}{2}n$ – that is, the dramatic change in the component structure of $G_{n,t}$ as t passes through $\frac{1}{2}n$, with emphasis on what happens very close to the critical value $\frac{1}{2}n$. In particular, in 1990 Łuczak [44] tightened the calculations in [8] and replaced the factor $(\log n)^{1/2}$ in $(\log n)^{1/2}n^{-1/3}$ by an arbitrary function tending to ∞. This gives the following result concerning the emergence of the giant component, which is essentially best possible.

Theorem 3.1 *Let $\omega(n)\to\infty$, and let $m=m(n)=\frac{1}{2}n+s(n)$, where $\omega(n)n^{2/3}\le s(n)=o(n)$. Then whp*

$$L_1(G_{n,m(n)}) = (4+o(1))s \text{ and } L_2(G_{n,m(n)}) \le (\log n)n^2/s^2.$$

As shown in [8], if $m(n)$ is substantially larger than $\frac{1}{2}n$, then whp the component structure of $G_{n,m}$ is surprisingly simple.

Theorem 3.2 *Let $c_0>1$ be constant, and let $\omega(n)\to\infty$. If $c=2m(n)/n\ge c_0$, then whp the random graph $G_{n,m(n)}$ has a giant component with $\alpha_c n+O(\omega(n)n^{1/2})$ vertices, where α_c is the constant given by* (3), *and every other component has $O(\log n)$ vertices and contains at most one cycle. Furthermore, for $m(n)\ge\omega(n)n$, whp every component of $G_{n,m(n)}$ other than the giant component is a tree.*

As we have already remarked, these results, and many others in a similar vein, were proved by following in the footsteps of Erdős and Rényi. In this approach, the random graphs $G_{n,m}$ and $G_{n,p}$ are considered as 'static' random variables, and their structures are studied by estimating the moments of certain functions defined on them. Later, very different methods were used to study the components. In particular, Martin-Löf [50] and Karp [41] made use of a natural process 'exploring' the component containing a given vertex to study the component sizes; in Section 5 we say a few words about the advantages of this point of view, as well as of a related point of view based on branching processes.

There are now much sharper results concerning the phase transition, including the limiting distribution of the order of the giant component, rescaled by subtracting the mean and dividing by the standard deviation. Some of these results appear in Section 6.

As with many a groundbreaking paper, the Erdős–Rényi giant component paper [24] has several blemishes; in particular, Theorem 2.1 is not its only false theorem. Over the years, these 'results' have been corrected. For example, in 1989, Łuczak and Wierman [49] corrected a false claim in [24] concerning the number of cycles with a certain number of diagonals. They proved that at the critical probability $p=1/n$, whp every cycle in $G_{n,p}$ is a 'hole' – that is, a cycle without a diagonal. A consequence of this is that whp $G_{n,1/n}$ has chromatic number 3.

In [24], the claim was used as a stepping stone towards establishing that the threshold for $G_{n,p}$ to become non-planar is at $p = 1/n$. Although the proof given by Erdős and Rényi was incorrect, the result is true. Indeed, as noted by Łuczak and Wierman [49], this follows from a theorem of Ajtai, Komlós and Szemerédi [2] stating that if $c > 1$ and k are constant, then whp $G_{n,c/n}$ contains a subdivision of K_k.

4. The phase transition – simple results

In this section we prove weak forms of the phase transition results described in the previous section, in order to illustrate the surprising power that even the most basic tools from probability have in this context. Indeed, essentially the only probabilistic techniques that we use are elementary properties of expectation and variance. Concerning $C(k, l)$, the number of connected graphs with k labelled vertices and l edges, we use only Cayley's classic formula that the number of trees with k labelled vertices is k^{k-2}.

For more precise calculations, unicyclic components are also easily dealt with using the formula

$$C(k,k) = \frac{(k-1)!}{2} \sum_{j=0}^{k-3} \frac{k^j}{j!} \sim \left(\tfrac{1}{8}\pi\right)^{1/2} k^{k-1/2}.$$

In fact, there are good approximations for $C(k, k+l)$ with l fixed and $k \to \infty$, of the form $C(k, k+l) \sim c_l k^{k+(3l-1)/2}$ with c_l constant (see Wright [61], who proved considerably more). However, here we use only Cayley's formula.

As noted in the Introduction, for $p = m/N$, where $N = \binom{n}{2}$, the models $G_{n,m}$ and $G_{n,p}$ are very similar, in the sense that for 'most' questions the answers for the two models are essentially the same. In fact, we need not take p to be exactly m/N – that is, m to be exactly pN: being close enough is good enough. It often suffices to use this similarity in a very crude way. For example, if Q is any property of graphs, then

$$\mathbf{P}(G_{n,m} \in Q) = O(\sqrt{pqN})\mathbf{P}(G_{n,p} \in Q), \tag{4}$$

whenever $q = 1 - p$, and p and m satisfy $m - pN = O(\sqrt{pqN})$. Indeed, this is an immediate consequence of the asymptotic normality of the binomial random variable $\mathrm{Bi}(N, p)$: the probability that it takes the value m is $\Theta(1/\sqrt{pqN})$, for any m satisfying $m - pN = O(\sqrt{pqN})$.

In what follows, we write $\mathbf{P}_{n,p}$ for the probability measure associated with the random graph $G_{n,p}$, and $\mathbf{P}_{n,m}$ for that associated with $G_{n,m}$. Thus (4) can be written as $\mathbf{P}_{n,m}(Q) = O(\sqrt{pqN})\mathbf{P}_{n,p}(Q)$. Similarly, we write $\mathbf{E}_{n,p}$ for the expectation corresponding to $\mathbf{P}_{n,p}$, so if X is some function of a graph G, then $\mathbf{E}_{n,p}(X)$ denotes the expected value of the random variable $X(G_{n,p})$.

We know from Theorems 2.1 and 3.1 that the component structure of $G_{n,m}$ changes dramatically as m passes through $\frac{1}{2}n$. Accordingly, we call $G_{n,m}$ *subcritical* if m is appreciably less than $\frac{1}{2}n$, *critical* if m is very close to $\frac{1}{2}n$, and *supercritical* if m is appreciably larger than $\frac{1}{2}n$. The same terminology is used for $G_{n,p}$, with $p=1/n$ in place of $m=\frac{1}{2}n$. Putting it slightly differently, if $p=\lambda(n)/n$, then $G_{n,p}$ is subcritical if $\lambda(n)$ is appreciably less than 1, critical if $\lambda(n)$ is very close to 1, and supercritical if $\lambda(n)$ is appreciably larger than 1.

Needless to say, the first important question is what we mean by 'appreciably' and 'very'. The best results show that $G_{n,m}$ is subcritical if $(\frac{1}{2}n-m)/n^{2/3} \to \infty$, critical if $m-\frac{1}{2}n=O(n^{2/3})$ – that is, m belongs to the 'critical window' or 'scaling window' – and supercritical if $(m - \frac{1}{2}n)/n^{2/3} \to \infty$. More precisely, these results show that, within each of these regimes, the behaviour of $G_{n,m}$ is qualitatively similar (in a sense that we do not make precise here), and that the behaviour in different regimes is qualitatively different. For $p=\lambda(n)/n=\left(1+\theta(n)n^{-1/3}\right)/n$ the three regimes are given by $\theta(n) \to -\infty$, $\theta(n)=O(1)$, and $\theta(n) \to \infty$.

In this section we prove some results that are far from best possible, but are more than enough to illustrate that the rather crude division of ranges in Theorem 2.1 is not adequate, and that the study of $G_{n,m}$ is especially challenging when m is close to $\frac{1}{2}n$. When studying $G_{n,p}$, we have switched to writing p as λ/n, rather than c/n, to emphasize that λ is (usually) a function of n rather than a constant.

Throughout this section, our arguments are based primarily on estimates for the expectations of certain simple quantities. An advantage of working with $G_{n,p}$ (rather than $G_{n,m}$) is that such expectations are very simple to write down. For example, the expected number of k-vertex subgraphs of $G_{n,p}$ that are trees (not necessarily tree components or induced trees) is exactly

$$\binom{n}{k} k^{k-2} p^{k-1}, \tag{5}$$

since there are $\binom{n}{k}$ possible choices for the vertex-set of the tree, then k^{k-2} trees with this vertex-set, and any given tree with k vertices is present with probability p^{k-1}. It is only slightly harder to write down the expectation of the number X_k of k-vertex *tree components* – that is, components of $G_{n,p}$ that are trees of order k:

$$\mathbf{E}_{n,p}(X_k) = \binom{n}{k} k^{k-2} p^{k-1} (1-p)^{k(n-k)+\binom{k}{2}-k+1}. \tag{6}$$

The first three factors are as in (5); the final factor gives the probability that $G_{n,p}$ contains no edges joining vertices of our tree to other vertices, and no edges among the tree vertices other than the $k-1$ edges forming the tree.

Throughout we use the following standard bounds. A very crude form of Stirling's formula (or a simple inductive argument) gives $k! \geq k^k e^{-k}$, so

$$\binom{n}{k} \leq \left(\frac{en}{k}\right)^k. \tag{7}$$

When we wish to be a little more precise, we use the following formula, valid for any $k \le \frac{1}{2}n$, say:

$$\binom{n}{k} = \frac{n^k}{k!}\prod_{i=1}^{k-1}\left(1-\frac{i}{n}\right) = \frac{n^k}{k!}\exp(O(k^2/n)). \tag{8}$$

Finally, one of the most frequently used identities in probabilistic combinatorics is the expansion of $\log(1-x)$, for $0<x<1$:

$$\log(1-x) = -x - \tfrac{1}{2}x^2 - \tfrac{1}{3}x^3 - \cdots.$$

This implies that, for $0<x<\frac{1}{2}$, say,

$$\exp(-x-x^2) \le 1-x \le \exp(-x-\tfrac{1}{2}x^2).$$

Combining the above formulas, we obtain a simple estimate for the number of tree components of moderate size in $G_{n,p}$.

Theorem 4.1 *If* $\frac{1}{10} \le \lambda = \lambda(n) \le 10$ *and* $1 \le k = k(n) \le n^{2/5}$, *then*

$$\mathbf{E}_{n,\lambda/n}(X_k) = (1+o(1))\frac{n}{\lambda}\frac{k^{k-2}}{k!}(\lambda e^{-\lambda})^k \tag{9}$$

and

$$\mathbf{E}_{n,\lambda/n}(X_k) \le (1+o(1))\frac{n}{\lambda}k^{-5/2}(\lambda e^{1-\lambda})^k. \tag{10}$$

Proof Setting $p=\lambda/n$, combining (6) and (8), and then using $(1-\lambda/n) = \exp\big(-\lambda/n + O(1/n^2)\big)$, we have

$$\begin{aligned}\mathbf{E}_{n,p}(X_k) &= \frac{n^k}{k!}k^{k-2}\left(\frac{\lambda}{n}\right)^{k-1}\left(1-\frac{\lambda}{n}\right)^{kn-k^2/2-3k/2+1}\exp(O(k^2/n))\\ &= \frac{n}{\lambda}\frac{k^{k-2}}{k!}\lambda^k e^{-\lambda k}\exp(O(k^2/n))\\ &= (1+o(1))\frac{n}{\lambda}\frac{k^{k-2}}{k!}(\lambda e^{-\lambda})^k,\end{aligned}$$

since $k^2/n = o(1)$. This proves (9).

To deduce (10), simply note that $k! \ge k^{k+1/2}e^{-k}$, by a crude form of Stirling's formula. ∎

As noted above, we can bound the numbers of non-tree components using estimates for $C(k,l)$, but for the results in this section, the following much simpler approach suffices. Let Y_k denote the number of k-vertex components in the graph under consideration.

Theorem 4.2 *Let* $k=k(n)\ge 1$ *and* $p=p(n)$ *satisfy* $k^2p \le \frac{1}{2}$. *Then*

$$\mathbf{E}_{n,p}(X_k) \le \mathbf{E}_{n,p}(Y_k) \le \mathbf{E}_{n,p}(X_k)(1+2k^2p). \tag{11}$$

Proof The first inequality is trivial, since $X_k \le Y_k$.

A component of order k with $k - 1 + l$ edges may be obtained from a tree component by adding l edges to it – that is, changing l non-edges between vertices of the component to edges. Since we have fewer than $\binom{k}{2}^l < k^{2l}2^{-l}$ choices for the non-edges, the sum of the probabilities of the resulting components is at most $k^{2l}2^{-l}(p/(1-p))^l \le (k^2p)^l$ times the probability of the original tree component. Since $k^2p \le \frac{1}{2}$, we have $\sum_{l\ge 1}(k^2p)^l \le 2k^2p$, giving the upper bound in (11). ■

We now examine the three regimes separately.

The subcritical regime

Our first aim is to show that whp $G_{n,n/20}$ has no component with at least $6\log n$ vertices; to take advantage of the simplicity of (5), we first calculate in $G_{n,p}$, and then transfer the result to $G_{n,m}$. For $p = 1/(10n)$ it follows from (5) and (7) that the expected number of trees with $k = \lceil 6\log n \rceil$ vertices present as subgraphs of $G_{n,p}$ is exactly

$$\binom{n}{k}k^{k-2}p^{k-1} \le \left(\frac{en}{k}\right)^k k^{k-2}\left(\frac{1}{10n}\right)^{k-1} \le n2^{-(k-1)} \le n^{-2}.$$

Thus, the probability that $G_{n,1/(10n)}$ has a component of order at least $6\log n$ is at most n^{-2} and, from (4), the probability that $G_{n,n/20}$ has such a component is at most n^{-1}.

For $p = (1+\varepsilon)/n$, $\varepsilon \ne 0$, call a component C of $G_{n,p}$ *small* if $|C| < 20(\log n)/\varepsilon^2$, *large* if $|C| > 100(\log n)/\varepsilon^2$, and *mid-size* otherwise. For $G_{n,m}$, $m \ne \frac{1}{2}n$, we use the same terminology as for $G_{n,p}$, where $p = 2m/n^2$. Note that in this case $p = (1+\varepsilon)/n$, where $\varepsilon = \varepsilon(m) = (2m/n) - 1$.

Theorem 4.3 *Whp the random graph process $\widetilde{G} = (G_{n,t})_0^N$ is such that, for all t in the range $\frac{1}{20}n \le t \le t_0 = \lfloor \frac{1}{2}n(1 - n^{-1/6}) \rfloor$, every component of $G_{n,t}$ is small. Also, for any given $t = t(n)$ in this range, whp all but $o(n)$ vertices of $G_{n,t}$ are on small tree components.*

Proof Let $[k_0, k_1]$ be the mid-size range for $G_{n,t}$, so $k_0 = k_0(\varepsilon) = \lceil 20(\log n)/\varepsilon^2 \rceil$ and $k_1 = k_1(\varepsilon) = \lfloor 100(\log n)/\varepsilon^2 \rfloor$, where $\varepsilon = \varepsilon(t) = 1 - 2t/n$. We claim that whp every $G_{n,t}$ in the appropriate range has no mid-size components. Given $\widetilde{G} = (G_{n,t})_0^N$, suppose that, for some $t < \frac{1}{2}n$, the graph $G_{n,t}$ has only small components and $G_{n,t+1}$ has no mid-size components, then $G_{n,t+1}$ has no large components: then the maximum possible order of a component is $2k_0(t) < k_1(t) \le k_1(t+1)$. Since whp $G_{n,n/20}$ has only small components, if we prove the claim, then the first assertion of the theorem follows.

Since the claim holds with plenty of room to spare, we give a very crude argument. As before, let Y_k denote the number of k-vertex components, and X_k be the number of tree components with k vertices, so $X_k \leq Y_k$. We will show that

$$\sum_{t=n/20}^{t_0} \mathbf{P}_{n,t}\left(\sum_{k=k_0(t)}^{k_1(t)} Y_k \geq 1\right) = o(1).$$

Switching from $G_{n,t}$ to $G_{n,p}$ with $p = (1-\varepsilon)/n$, it suffices to show that

$$\mathbf{P}_{n,p}\left(\sum_{k=k_0(\varepsilon)}^{k_1(\varepsilon)} Y_k \geq 1\right) \leq n^{-8/5},$$

say, whenever $p = (1-\varepsilon)/n$ with $n^{-1/6} \leq \varepsilon \leq \frac{9}{10}$. In turn, this follows (again very crudely) if we show that

$$\mathbf{E}_{n,p}(Y_k) \leq n^{-2} \tag{12}$$

whenever $n^{-1/6} \leq \varepsilon \leq \frac{9}{10}$ and $k_0(\varepsilon) \leq k \leq k_1(\varepsilon)$.

We first bound $\mathbf{E}_{n,p}(X_k)$ using Theorem 4.1, noting that $k_1 \leq n^{2/5}$. Let $\lambda = 1-\varepsilon$, so $p = \lambda/n$. Since $1-\varepsilon \leq e^{-\varepsilon-\varepsilon^2/2}$, we have

$$\lambda e^{-\lambda} \leq e^{-1-\varepsilon^2/2} \leq e^{-1-\varepsilon^2/6}. \tag{13}$$

Since $\lambda \geq \frac{1}{10}$, it follows from Theorem 4.1 that, for $k_0 \leq k \leq k_1$,

$$\mathbf{E}_{n,p}(X_k) \leq (10+o(1))nk^{-5/2}e^{-\varepsilon^2 k/6} \leq 11nk^{-5/2}e^{-\varepsilon^2 k_0/6} \leq 11k^{-5/2}n^{-7/3}.$$

Noting that $k_1^2 p = o(1)$, and applying Theorem 4.2, we see that

$$\mathbf{E}_{n,p}(Y_k) \leq 2\mathbf{E}_{n,p}(X_k) \leq n^{-7/3},$$

which implies (12).

For the second assertion, Theorem 4.2 shows that the expected number of vertices in small components that are not trees is $o(n)$, since it is small compared to the expected number of vertices in small tree components, which is at most n. Hence whp only $o(n)$ vertices are in small non-tree components, and none are in mid-size or large components. ■

An immediate consequence of Theorem 4.3 is the following purely analytic identity.

Corollary 4.4 *For* $0 \leq \lambda \leq 1$,

$$\sum_{k=1}^{\infty} \frac{k^{k-1}}{k!}(\lambda e^{-\lambda})^k = \lambda. \tag{14}$$

Proof In proving this, we may assume that $0 < \lambda < 1$. Set $p = \lambda/n$, with λ constant as $n \to \infty$. By Theorem 4.3, whp all but $o(n)$ vertices of $G_{n,p}$ are in small tree components of order at most $k_0 = \lfloor 20(\log n)/(1-\lambda)^2 \rfloor$. In particular, the expected number of vertices in small tree components is $(1+o(1))n$ – that is,

$$\sum_{k=1}^{k_0} k\mathbf{E}_{n,p}(X_k) = (1+o(1))n.$$

It follows from this and (9) that

$$\frac{n}{\lambda}\sum_{k=1}^{k_0} \frac{k^{k-1}}{k!}\left(\lambda e^{-\lambda}\right)^k = (1+o(1))n,$$

which implies (14). ∎

It is tempting to assume that an identity like (14), which holds for $0 \le \lambda \le 1$, must therefore hold for all $\lambda \ge 0$. However, a moment's thought tells us that this is not the case. Indeed, on $[0, \infty)$, the function $\lambda e^{-\lambda}$ first strictly increases from 0, reaching $1/e$ at $\lambda = 1$, and then strictly decreases to 0. Thus, for $0 < x < 1/e$, the equation $\lambda e^{-\lambda} = x$ has precisely two solutions, which we call duals. In other words, if $0 < \mu < 1 < \lambda$, the numbers λ and μ are *dual* to each other if $\mu e^{-\mu} = \lambda e^{-\lambda}$; we define $\lambda^\star = \mu$ and $\mu^\star = \lambda$. For $\lambda > 1$, applying (14) to $\lambda^\star < 1$ shows that

$$\sum_{k=1}^{\infty} \frac{k^{k-1}}{k!}(\lambda e^{-\lambda})^k = \lambda^\star. \tag{15}$$

More to the point, Theorem 4.3 enables us to pin down the maximum order of a component of $G_{n,m}$ or $G_{n,p}$ in the subcritical regime.

Theorem 4.5 *Let $\lambda = 1-\varepsilon$, where $n^{-1/6} \le \varepsilon = \varepsilon(n) \le \frac{9}{10}$, and let $\delta = -\log\left(\lambda e^{1-\lambda}\right)$. Then whp the maximum order L_1 of a component of $G_{n,\lambda/n}$ satisfies*

$$L_1 = (1+o(1))(\log(\varepsilon^3 n))/\delta. \tag{16}$$

If, in addition, $\varepsilon \to 0$, then whp $L_1 = (2+o(1))(\log(\varepsilon^3 n))/\varepsilon^2$, and if $\varepsilon \to 0$ and $1/\varepsilon = n^{o(1)}$, then $L_1 = (2+o(1))(\log n)/\varepsilon^2$.

Proof Since

$$\delta = -\log\lambda + \lambda - 1 = -\log(1-\varepsilon) - \varepsilon = \tfrac{1}{2}\varepsilon^2 + \tfrac{1}{3}\varepsilon^3 + \cdots,$$

it suffices to prove (16).

Set $l_0 = \log(\varepsilon^3 n)/\delta$. Our first aim is to show that $G_{n,p}$ is unlikely to have a component with substantially more than l_0 vertices. In the light of Theorem 4.3, we need consider only components of order up to k_0.

For $\eta > 0$ constant and n large we have, by Theorems 4.1 and 4.2,

$$\mathbf{E}_{n,p}\left(\sum_{k=(1+\eta)l_0}^{k_0} Y_k\right) \le 2\mathbf{E}_{n,p}\left(\sum_{k=(1+\eta)l_0}^{k_0} X_k\right)$$

$$\le \frac{3n}{\lambda} \sum_{k=(1+\eta)l_0}^{k_0} k^{-5/2}\left(\lambda e^{1-\lambda}\right)^k$$

$$= \frac{3n}{\lambda} \sum_{k=(1+\eta)l_0}^{k_0} k^{-5/2} e^{-\delta k}.$$

The terms in the final sum decrease by at least a factor of $e^{-\delta}$ at each stage, so the sum is at most $O(\delta^{-1})$ times the first term, giving the bound

$$Cn\delta^{-1} l_0^{-5/2} e^{-\delta(1+\eta)l_0} = Cn\delta^{-1} l_0^{-5/2} (\varepsilon^3 n)^{-(1+\eta)}.$$

Since $\delta^{-1} = O(\varepsilon^{-2})$ and $l_0^{-5/2} = O(\varepsilon^5)$, this is $O((\varepsilon^3 n)^{-\eta})$ and so tends to 0.

We claim that whp $G_{n,p}$ contains a tree component whose order is not much smaller than l_0. To justify this claim, let l_1 be the maximal integer for which $\mu = \mathbf{E}_{n,p}(X) \ge \log\log\log n$, where $X = \sum_{k=l_1}^{k_0} X_k$. It is easily checked that $l_1 \ge (1 - o(1))l_0$, and that $\mathbf{E}_{n,p}(X^2) = (1 + o(1))\mu^2$, so by Chebyshev's inequality, $\mathbf{P}_{n,p}(X \ge 1) = 1 + o(1)$. ■

For $G_{n,m}$, one can either follow the same argument with somewhat less pleasant calculations for the expectations, or transfer the result from $G_{n,p}$.

With a little work, we can sharpen considerably the second assertion in Theorem 4.3. For this we need to estimate the binomial coefficient a little more carefully: for $k \le n^{2/3}$,

$$\log\left(\binom{n}{k}\frac{k!}{n^k}\right) = \sum_{i=0}^{k-1} \log\left(1 - \frac{i}{n}\right) = -\sum_{i=0}^{k-1}\left(\frac{i}{n} + \frac{i^2}{2n^2} + O(i^3/n^3)\right)$$

$$= -\frac{k^2}{2n} - \frac{k^3}{6n^2} + O(k/n). \tag{17}$$

We state the next result for $G_{n,p}$ rather than $G_{n,m}$.

Corollary 4.6 *Let $p = \lambda/n = (1 - \varepsilon)/n$ where $n^{-1/6} \le \varepsilon = \varepsilon(n) \le \frac{9}{10}$, and let $\omega(n) \to \infty$. Then whp all but $\omega(n)/\varepsilon^2$ vertices of $G_{n,p}$ are in small tree components.*

Proof We bound from below the expected number of vertices in small tree components, using (17) and $1-p=e^{-p-O(p^2)}$, and omitting some steps in the calculations:

$$\begin{aligned}\mathbf{E}_{n,p}\left(\sum_{k=1}^{k_0} kX_k\right) &= \sum_{k=1}^{k_0}\binom{n}{k}k^{k-1}p^{k-1}(1-p)^{k(n-k)+\binom{k}{2}-k+1}\\ &= \frac{n}{\lambda}\sum_{k=1}^{k_0}\frac{k^{k-1}}{k!}(\lambda e^{-\lambda})^k \exp\Big(-\tfrac{1}{2}\varepsilon k^2/n - O(k/n)\Big)\\ &\geq \frac{n}{\lambda}\sum_{k=1}^{k_0}\frac{k^{k-1}}{k!}(\lambda e^{-\lambda})^k\Big(1-\tfrac{1}{2}\varepsilon k^2/n - O(k/n)\Big).\end{aligned}$$

Recalling Corollary 4.4, after some calculation, we see that

$$\mathbf{E}_{n,p}\left(\sum_{k=1}^{k_0} kX_k\right) \geq n - O(1/\varepsilon^2),$$

proving the assertion. ■

The critical regime

As noted earlier, for $G_{n,p}$ the critical window is $p-1/n=O(n^{-4/3})$; to make the calculations less cumbersome, we confine our attention to the centre of the window, $p=1/n$.

Theorem 4.7 *If $\omega(n)\to\infty$, then whp $n^{2/3}/\omega \leq L_1(G_{n,1/n}) \leq \omega n^{2/3}$.*

Proof We may assume that $\omega(n)$ is large, in particular, at least 1. First we prove the upper bound on $L_1(G_{n,1/n})$, and then the lower bound. All our calculations concern tree components.

Write $Z=Z(G_{n,1/n})$ for the number of vertices in tree components of order at most $n^{2/3}/\omega$. We shall show that

$$\mathbf{E}_{n,1/n}(Z) \geq n - 4\sqrt{\omega}n^{2/3}; \tag{18}$$

this clearly implies that whp $L_1 \leq \omega n^{2/3}$. To prove (18), note that

$$\mathbf{E}_{n,1/n}(Z) = \sum_{k=1}^{n^{2/3}/\omega}\binom{n}{k}k^{k-1}\left(\frac{1}{n}\right)^{k-1}\left(1-\frac{1}{n}\right)^{t_{n,k}}, \tag{19}$$

where $t_{n,k}=k(n-k)+\binom{k}{2}-(k-1)=kn-\frac{1}{2}k^2-\frac{3}{2}k+1$. Since

$$\left(1-\frac{1}{n}\right)^{t_{n,k}} = \exp\Big(-t_{n,k}/n + O(t_{n,k}/n^2)\Big) = \exp\Big(-k+\tfrac{1}{2}k^2/n + O(k/n)\Big),$$

it follows from (17) that

$$\binom{n}{k}\left(1-\frac{1}{n}\right)^{t_{k,n}} = \frac{n^k}{k!}\exp\left(-k-\tfrac{1}{6}k^3/n^2+O(k/n)\right). \tag{20}$$

Hence, from (19),

$$\begin{aligned}\mathbf{E}_{n,1/n}(Z) &= n\sum_{k=1}^{n^{2/3}/\omega}\frac{k^{k-1}}{k!}\exp\left(-k-\tfrac{1}{6}k^3/n^2+O(k/n)\right)\\ &\geq (1-n^{-1/3})n\sum_{k=1}^{n^{2/3}/\omega}\frac{k^{k-1}}{k!}e^{-k}\left(1-\frac{k^3}{n^2}\right),\end{aligned}$$

say. Now

$$\sum_{n^{2/3}/\omega}^{\infty}\frac{k^{k-1}}{k!}e^{-k} \leq \sum_{n^{2/3}/\omega}^{\infty} k^{-3/2} \leq 3\sqrt{\omega}/n^{1/3},$$

and

$$\frac{1}{n^2}\sum_{k=1}^{n^{2/3}/\omega}\frac{k^{k+2}}{k!}e^{-k} \leq \frac{1}{n^2}\sum_{k=1}^{n^{2/3}/\omega}k^{3/2} \leq \frac{1}{n^2}(n^{2/3}/\omega)^{5/2} = n^{-1/3}\omega^{-5/2}.$$

Consequently, since $\sum_{k=1}^{\infty}k^{k-1}e^{-k}/k! = 1$,

$$\mathbf{E}_{n,1/n}(Z) \geq n(1-n^{-1/3})(1-3\sqrt{\omega}n^{-1/3}-n^{-1/3}\omega^{-5/2}) \geq n-4\sqrt{\omega}n^{2/3},$$

as claimed.

Turning to the lower bound on L_1, let W be the number of tree components of order k with $n_0 \leq k \leq 2n_0$, where $n_0 = \lfloor n^{2/3}/\omega\rfloor$. We claim that $\mu = \mathbf{E}_{n,1/n}(W) \to \infty$, and that the variance of W is $o(\mu^2)$. Indeed, using (20), we have

$$\begin{aligned}\mu &= \sum_{k=n_0}^{2n_0}\binom{n}{k}k^{k-2}n^{-(k-1)}\left(1-\frac{1}{n}\right)^{t_{n,k}}\\ &\geq \tfrac{1}{2}n\sum_{k=n_0}^{2n_0}\frac{k^{k-2}}{k!}e^{-k} \geq \tfrac{1}{6}n\sum_{k=n_0}^{2n_0}k^{-5/2}\\ &\geq \tfrac{1}{36}nn_0^{-3/2} \geq \tfrac{1}{36}\omega^{3/2}.\end{aligned}$$

To bound the variance of W, we consider the expectation of the second factorial moment of W – that is, $(W)_2 = W(W-1)$:

$$\begin{aligned}&\mathbf{E}_{n,1/n}((W)_2)\\ &= \sum_{l_1=n_0}^{2n_0}\sum_{l_2=n_0}^{2n_0}\binom{n}{l_1}\binom{n-l_1}{l_2}l_1^{l_1-2}l_2^{l_2-2}n^{-(l_1-1+l_2-1)}\left(1-\frac{1}{n}\right)^{t_{n,l_1}+t_{n,l_2}-l_1l_2}.\end{aligned}$$

This formula is almost the same as the obvious expansion of μ^2: the only difference is that each term is multiplied by

$$\binom{n-l_1}{l_2}\binom{n}{l_2}^{-1}\left(1-\frac{1}{n}\right)^{-l_1 l_2}.$$

A little calculation shows that these factors are $1+O(n_0^3/n^2)=1+o(1)$. Hence,

$$\mathrm{Var}(W)=\mathbf{E}((W)_2)+\mu-\mu^2=o(\mu^2),$$

and Chebyshev's inequality implies that $W>1$ whp. ∎

The above bounds change very little if $p=1/n$ is replaced by $p=(1+o(n^{-1/3}))/n$, or indeed by $p=(1+\theta n^{-1/3})/n$, with θ constant.

The supercritical regime

As in the subcritical case, our main weapon is the fact that whp the random graph process $\widetilde{G}$ is such that, appreciably after the critical time $\frac{1}{2}n$, there is a substantial gap in the sequence of component sizes.

The proof of the next theorem is almost identical to that of Theorem 4.3, using Theorems 4.1 and 4.2, once it is noted that, for $\varepsilon\le 1$, we have

$$1+\varepsilon\le e^{\varepsilon-\varepsilon^2/2+\varepsilon^3/3}\le e^{\varepsilon-\varepsilon^2/6},$$

so the bound (13) remains valid.

Theorem 4.8 *Whp the random graph process $\widetilde{G}=(G_{n,t})_0^N$ is such that, for all t in the range $t_1=\lceil\frac{1}{2}n(1+n^{-1/6})\rceil\le t\le n$, the graph $G_{n,t}$ has no mid-size components.*

Corollary 4.9 *Whp $\widetilde{G}=(G_{n,t})_0^N$ is such that, for all $t=\frac{1}{2}(1+\varepsilon)n$ in the range $t_1=\lceil\frac{1}{2}n(1+n^{-1/6})\rceil<t\le n$, each large component of $G_{n,t}$ contains a large component of $G_{n,t-1}$, and $G_{n,t}$ has no component of order k with $20(\log n)/\varepsilon^2<k<100(\log n)n^{1/3}$.*

Proof We know that whp $\widetilde{G}=(G_{n,t})_0^N$ is such that none of the graphs G_{n,t_1}, $G_{n,t_1+1},\ldots,G_{n,n}$ has a mid-size component. The maximum order of a component of $G_{n,t}$ not containing a large component of $G_{n,t-1}$ is thus $2k_0(t-1)<k_1(t)$. Hence every large component of $G_{n,t}$ contains a large component of $G_{n,t-1}$ and so, by induction, of G_{n,t_1}. In particular, every component of $G_{n,t}$ is either small, or at least as large as a large component of G_{n,t_1}. ∎

The number of vertices in small tree components can be estimated as in Corollary 4.6. This time, using (17), we have

$$\mathbf{E}_{n,p}\left(\sum_{k=1}^{k_0} kX_k\right) = \sum_{k=1}^{k_0} \binom{n}{k} k^{k-1} p^{k-1} (1-p)^{k(n-k)+\binom{k}{2}-k+1}$$
$$= \frac{n}{\lambda} \sum_{k=1}^{k_0} \frac{k^{k-1}}{k!} (\lambda e^{-\lambda})^k \left(1 + O(\varepsilon k^2/n + k/n)\right).$$

Easy calculations, using (15) in place of (14), thus give

$$\mathbf{E}_{n,p}\left(\sum_{k=1}^{k_0} kX_k\right) = \frac{\lambda^\star n}{\lambda} - O(1/\varepsilon^2) = (1-\alpha_\lambda)n - O(1/\varepsilon^2).$$

Considering the expected number of pairs of tree components of various sizes, it is not hard to show that the variance of $\sum_{k=1}^{k_0} kX_k$ is $O(n/\varepsilon)$, and certainly that it is $o(\varepsilon^2 n^2)$. Since $1/\varepsilon^2$ is small compared to εn, it follows that whp $(1-\alpha_\lambda + o(\varepsilon))n$ vertices are in small tree components.

Using Theorem 4.2, we may crudely bound the expected number of vertices in small components that are not trees by $2k_0^2 pn \le 4k_0^2 = o(\varepsilon n)$. Combined with Corollary 4.9, since α_λ is of order ε, this gives the following.

Corollary 4.10 *If $t = \frac{1}{2}\lambda n$ with $t_1 < t \le n$, then whp $G_{n,t}$ has $(1+o(1))\alpha_\lambda n$ vertices in components of order at least $100(\log n)n^{1/3}$, and the remaining vertices in small components.*

Finally, it is now not hard to show that, if t is only a little larger than t_1, then there is only one large component. The technique we use is called 'sprinkling', and was first applied by Erdős and Rényi [24].

Theorem 4.11 *Whp the random graph process $\widetilde{G} = (G_{n,t})_0^N$ is such that, for all t in the range $t_2 = \lceil \frac{1}{2}n(1 + 2n^{-1/6}) \rceil \le t \le n$, there is a unique large component. Furthermore, for a given $t = \frac{1}{2}\lambda n$ in this range, whp $L_1 \sim \alpha_\lambda n$.*

Proof It suffices to prove the first statement; the second then follows by Corollary 4.10.

In the light of Corollary 4.9, it suffices to prove that whp G_{n,t_2} has only one large component. Furthermore, if $C_1, C_2, \ldots, C_r$ are the large components of G_{n,t_1}, it suffices to show that whp there is a component C of G_{n,t_2} that contains all of the C_i. From now on, we condition on G_{n,t_1}. By Corollary 4.10, we may assume that

$$|V(C_1)| + |V(C_2)| + \cdots + |V(C_r)| \ge \tfrac{3}{4}\alpha_{1+n^{-1/6}} n \ge n^{5/6}, \tag{21}$$

where we use $\alpha_{1+\varepsilon} \sim 2\varepsilon$ for the second inequality.

Set $q = \frac{1}{2}n^{-7/6}$. To obtain independence, we consider the graph G obtained from G_{n,t_1} by adding each non-edge with probability q, independently of the others (and of the graph G_{n,t_1}). Let M be the random number of edges added. Note that, conditional on M, each subset of M non-edges of G_{n,t_1} is equally likely to be added, so G has the distribution of G_{n,t_1+M}. Now M has a binomial distribution with mean $q(\binom{n}{2} - t_1) \le \frac{1}{2}qn^2 \le \frac{1}{4}n^{5/6}$. Since a binomial distribution is concentrated around its mean, we see that $M \le \frac{1}{2}n^{5/6}$ holds whp, so whp $t_1 + M \le t_2$. Thus it suffices to show that whp the components $C_1, C_2, \ldots, C_r$ are contained in a single component of G.

For a reason that will become clear in a moment, it is convenient to split the vertex-set of each C_i into sets of size between $s = 100(\log n)n^{1/3}$ and $2s$; this is clearly possible, since each C_i contains at least s vertices. Let $V_1, V_2, \ldots, V_a$ denote the resulting sets. Since $|V_i| \le 2s$ for each i, we see from (21) that

$$a \ge \frac{n^{5/6}}{2s} \ge \frac{n^{5/6}}{200(\log n)n^{1/3}} = \frac{n^{1/2}}{200 \log n}.$$

We now construct an auxiliary graph H with vertex-set $\{1, 2, \ldots, a\}$, as follows. Each possible edge ij is included in H if and only if at least one of our added edges joins V_i to V_j. Since $|V_i|, |V_j| \ge s$, the expected number of added edges joining them is at least $s^2 q \ge 1000(\log n)^2 n^{-1/2}$. It follows easily that each edge-probability in H is at least $p = 900(\log n)^2 n^{-1/2}$. Since different edges are present in H independently, we may form a subgraph of H whose distribution is exactly that of $G_{a,p}$. Now $ap \ge 3 \log n \ge 3 \log a$, so from a crude form of the connectedness result (1) mentioned in the Introduction, H is connected whp.

It remains only to show that, whenever H is connected, the graph G includes a component C' containing $C_1, C_2, \ldots, C_r$. But this is immediate from the fact that, for each edge $ij \in H$, the corresponding V_i and V_j are joined by an edge in G, so the components $C_{i'}$ and $C_{j'}$ containing V_i and V_j must be part of the same component of G. ■

5. Exploring components

Suppose that we are given a graph G and a chosen vertex v of G, and wish to find the set of vertices in the component C_v containing v. There is a natural family of search algorithms for accomplishing this, which may be described as follows. The algorithm maintains three sets (or lists) of vertices that are updated as it proceeds: the lists represent *explored* vertices, *active* vertices and *unseen* vertices. We write $\mathcal{E}_t$, $\mathcal{A}_t$ and $\mathcal{U}_t$ for the corresponding lists after t steps of the algorithm.

At the beginning, $\mathcal{A}_0 = \{v\}$, $\mathcal{E}_0 = \varnothing$ and $\mathcal{U}_0 = V(G) \setminus \{v\}$. The algorithm terminates when $\mathcal{A}_t$ is first empty. At each step, if $\mathcal{A}_t$ is not empty, then a vertex w_t is chosen from $\mathcal{A}_t$, and all neighbours $z_{t,1}, z_{t,2}, \ldots, z_{t,r_t}$ of w_t among the vertices in $\mathcal{U}_t$ are found. Then w_t is moved from the active list to the explored list, and the $z_{t,i}$ (if there are any) are moved from the unseen list to the active list. More precisely, we set $\mathcal{E}_{t+1} = \mathcal{E}_t \cup \{w_t\}$, $\mathcal{A}_{t+1} = \mathcal{A}_t \cup \{z_{t,1}, z_{t,2}, \ldots, z_{t,r_t}\} \setminus \{w_t\}$, and

$\mathcal{U}_{t+1} = \mathcal{U}_t \setminus \{z_{t,1}, z_{t,2}, \ldots, z_{t,r_t}\}$. Often one says that the edges from w_t to the vertices in $\mathcal{U}_t$ are *tested* at step t of the algorithm.

Note that we have described a family of algorithms: there are many possible rules for picking which active vertex to explore at a given step. If we pick the 'oldest' (the one that was added earliest to the active list), then we obtain the standard breadth-first search. If we pick the 'newest', we obtain depth-first search.

Karp [41] was the first to use the above algorithm to prove results about random graphs; he applied it to a random *directed* graph, rather than to $G_{n,p}$, but this makes essentially no difference. His key observation was that each possible edge of G is 'tested' at most once. Hence, if we apply the algorithm to $G = G_{n,p}$, conditional on the history, each test succeeds with probability p. Thus, the number r_t of neighbours of w_t that we find is binomially distributed with parameters $|\mathcal{U}_t|$ and p. This means that we can study the behaviour of the algorithm by keeping track only of the sizes of the lists. Indeed, let A_t denote the number of vertices in $\mathcal{A}_t$, and so on for the other sets. Then the sequence $(E_t, A_t, U_t)_{t=0}^{l}$ is a Markov chain that may be described as follows.

Start with $E_0 = 0$, $A_0 = 1$, and $U_0 = n-1$. Given $(E_i, A_i, U_i)_{i=0}^{t}$, if $A_t = 0$, we set $l = t$ and stop the process. Otherwise, let r_t be binomially distributed with parameters U_t and p, and set $E_{t+1} = E_t + 1$, $A_{t+1} = A_t - 1 + r_t$ and $U_{t+1} = U_t - r_t$.

Since $E_t = t$, and $E_t + A_t + U_t = n$ for all t, it suffices to keep track of the sequence (A_t), which we may think of as a random walk that terminates when it first hits 0. The time l at which it does so corresponds to the size of C_v, so results about the distribution of l imply results about the distribution of the component sizes in $G_{n,p}$. In 1990, Karp used this point of view to prove results about the component structure of random directed graphs; a little earlier, a closely related approach had been used by Martin-Löf [50] in the context of epidemics, where it arises even more naturally. Recently, a slightly modified form of this exploration has been used by Nachmias and Peres [53] and by the present authors [13] to give simple new proofs of strong results about the giant component of $G_{n,p}$.

There is another way of viewing the above exploration that is perhaps even more informative. Suppose that we explore the component in a breadth-first manner. Then the vertices of C_v are reached in *rounds* or *generations*: generation 0 consists of v itself, generation 1 consists of the vertices reached directly from v, generation 2 those reached from vertices in generation 1, and so on. Moreover, the exploration reveals a spanning tree of C_v, given by the edges that we used to reach the various vertices in C_v.

Let $\mathrm{Po}(c)$ denote the Poisson distribution with mean c. Then a $\mathrm{Po}(c)$-*Galton–Watson tree* is a random rooted tree constructed recursively as follows. Start with a single vertex as the root. Each vertex has a Poisson number of 'children' in the next level with mean c, independently of all other vertices. This random tree is the canonical example of a *branching process*.

If c is constant, $n \to \infty$ and $p \to 0$ with $np \to c$, then the binomial distribution $\mathrm{Bi}(n, p)$ converges (in distribution) to the Poisson distribution $\mathrm{Po}(c)$. Because of this, it is essentially trivial to see that if $p = c/n$ with c constant, then, to start with, the random tree given by the exploration above is essentially a Po(c)-Galton–Watson tree. This gives a simple intuitive understanding of Theorem 2.1: if $c < 1$, then the branching process always dies out (that is, it gives rise to a finite tree), and indeed the chance of its reaching size k decreases exponentially with k. This suggests that the probability that a given vertex will be in a component of size at least k decays as $e^{-\beta k}$, for some $\beta > 0$. Since there are n possible starting vertices, this indicates that we should expect $L_1(G_{n,c/n})$ to be of order $\log n$, as it is.

If $c > 1$, then the branching process may *survive* forever – that is, it may give rise to an infinite tree; let ρ_c denote the probability of this event. Since the component of a vertex v of $G_{n,p}$ is always finite, it is clear that the branching-process approximation described above breaks down for very large components. However, it is valid up to at least some size $\omega(n)$ that tends to infinity. Very roughly, it follows that a fraction ρ_c of the vertices of $G_{n,c/n}$ lie in 'large' components (of size at least $\omega(n)$), and the rest in 'small' components. One can show, for example using a 'sprinkling' argument such as that illustrated in the final subsection of Section 4, that whp almost all vertices in large components are in a single giant component.

In this way, one can obtain a different proof of parts (a) and (c) of Theorem 2.1. A great advantage of the branching-process approach is that it tells us what the constant α_c 'really is': it is simply ρ_c, the survival probability of the branching process. From basic results about branching processes, one can easily check that $\alpha_c = \rho_c$ is the maximal solution to $\alpha_c = 1 - e^{-c\alpha_c}$, which is often a more convenient formula for α_c than (3).

It is well known that the branching-process approach that we have just outlined (omitting the non-trivial technical details) easily gives weak results about $G_{n,p}$, and that, unlike tree-counting methods, it lends itself well to generalizations (see, for example, Bollobás, Janson and Riordan [11]). Perhaps less well known is that it can be combined with tree counting to prove strong results about the component distribution in $G_{n,p}$ (see [12]).

Exploration and branching-process methods were not the first more powerful alternatives to the Erdős–Rényi type arguments used by Bollobás and Łuczak: in 1993, Janson *et al.* [37] wrote a monumental paper in which they combined algebraic, analytic and combinatorial techniques to study $G_{n,p}$ around its phase transition. This paper marked the re-entry of generating functions into mainstream random graph theory. Among the striking results they obtained is the formula $\frac{5}{18}\pi + o(1)$ for the probability that the process $\widetilde{G}$ has the property that no $G_{n,t}$ has more than one component with more than one cycle.

6. The phase transition – finer results

In Section 4 we considered $p = (1 \pm \varepsilon)/n$, where we assumed throughout that $\varepsilon = \varepsilon(n) \geq n^{-1/6}$. This latter assumption simplifies the calculations somewhat, but is not essential. In the 1984 paper [8], where the case $\varepsilon \to 0$ was first considered, the results corresponding to those in the previous section were proved under the weaker assumptions $\varepsilon \geq 4(\log n)^{1/2} n^{-1/3}$ and $\varepsilon = o(1/\log n)$. To cover this range of ε, one must consider not only tree components, but also components containing cycles. Later, Łuczak [44] extended the range of ε to 'what it should be' (see later) – namely, $\varepsilon = o(1)$ with $\varepsilon^3 n \to \infty$. Unfortunately, his main results as he stated them were not quite correct, although it is easy to correct them to obtain the next two theorems.

A random variable X_n is of *probabilistic order* $f(n)$, written $X_n = O_{\mathrm{p}}(f(n))$, if $X_n/f(n)$ is bounded in probability – that is, for every constant $\delta > 0$ there exists a constant C such that $\mathbf{P}\big(|X_n| \leq C f(n)\big) \geq 1 - \delta$ for all n. This is equivalent to the assertion that, whenever $\omega(n) \to \infty$, we have $|X_n| \leq \omega(n) f(n)$ whp. As with deterministic O-notation, one writes (for example) $X_n = \log n + O_{\mathrm{p}}(1)$ to mean that $X_n - \log n = O_{\mathrm{p}}(1)$ – that is, for any $\omega(n) \to \infty$ we have $\log n - \omega(n) \leq X_n \leq \log n + \omega(n)$ whp.

Theorem 6.1 *Let $\lambda = 1 - \varepsilon$, where $\varepsilon = \varepsilon(n) > 0$ satisfies $\varepsilon \to 0$ and $\Lambda = \varepsilon^3 n \to \infty$. Then the order L_1 of the largest component of $G_{n,\lambda/n}$ satisfies*

$$L_1 = \delta^{-1} \left(\log \Lambda - \tfrac{5}{2} \log \log \Lambda + O_{\mathrm{p}}(1) \right),$$

where $\delta = \lambda - 1 - \log \lambda = -\varepsilon - \log(1 - \varepsilon) = \frac{1}{2}\varepsilon^2 + \frac{1}{3}\varepsilon^3 + O(\varepsilon^4)$.

Recall that, for $\lambda > 1$, the constant α_λ appearing in Theorem 2.1 may be defined by (3). Alternatively, as noted in Section 5, it is the positive solution to the equation $x = 1 - e^{-\lambda x}$, and is the survival probability of a certain branching process.

Theorem 6.2 *Let $\lambda = 1 + \varepsilon$, where $\varepsilon = \varepsilon(n) > 0$ satisfies $\varepsilon \to 0$ and $\Lambda = \varepsilon^3 n \to \infty$. Then the orders L_1, L_2 of the two largest components of $G_{n,\lambda/n}$ satisfy*

$$L_1 = \alpha_\lambda n + O_{\mathrm{p}}(\varepsilon n/\sqrt{\Lambda})$$

and

$$L_2 = \delta^{-1} \left(\log \Lambda - \tfrac{5}{2} \log \log \Lambda + O_{\mathrm{p}}(1) \right),$$

where $\delta = \lambda - 1 - \log \lambda = \frac{1}{2}\varepsilon^2 - \frac{1}{3}\varepsilon^3 + O(\varepsilon^4)$.

For relatively simple proofs of Theorems 6.1 and 6.2 that combine tree counting arguments and branching-process methods, see [12].

Theorems 6.1 and 6.2 are tight, in the sense that the gap between the upper and lower bounds is of the same order as the typical variation (standard deviation) of the value of L_1 or L_2, respectively. Nevertheless, one can ask even more: what is the limiting distribution of this error term?

In the subcritical case, Łuczak [44] answered this question: the distribution turns out to be an 'extreme-value' distribution associated to the Poisson distribution. The supercritical case is more complicated, but has a more pleasant answer; the distribution is asymptotically normal. When the average degree is constant, this was proved by Stepanov [60]. In this, and many results concerning $G_{n,\lambda/n}$, the dual parameter $\lambda^\star$ defined in Section 4 plays a key role. Recall that this is the solution $\lambda^\star < 1$ to $\lambda^\star e^{-\lambda^\star} = \lambda e^{-\lambda}$; it is sometimes known as the *dual branching process parameter* to λ, and is equal to $\lambda(1-\alpha_\lambda)$. The following result (originally due to Stepanov) was reproved by Pittel [55] by careful arguments based on tree counting.

Theorem 6.3 *Let $\lambda > 1$ be constant, let α_λ be the positive solution to $\alpha_\lambda = 1 - e^{-\lambda\alpha_\lambda}$ and set*

$$\sigma_\lambda^2 = \frac{\alpha_\lambda(1-\alpha_\lambda)}{(1-\lambda^\star)^2}.$$

If L_1 is the maximal order of a component of $G_{n,\lambda/n}$, then

$$\frac{L_1 - \alpha_\lambda n}{\sigma_\lambda \sqrt{n}} \xrightarrow{\mathrm{d}} \mathrm{No}(0,1),$$

where $\xrightarrow{\mathrm{d}}$ denotes convergence in distribution, and $\mathrm{No}(0,1)$ *is the standard normal distribution.*

As the average degree tends to 1 from above, the situation becomes more delicate: the size of the giant component decreases, but the fluctuations increase. Indeed, with $p = (1+\varepsilon)/n$ and $\alpha = \alpha_{1+\varepsilon}$ (so α depends on n), we have $\alpha \sim 2\varepsilon$ as $\varepsilon \to 0$, so we expect $L_1 \sim 2\varepsilon n$. On the other hand, $\sigma_{1+\varepsilon}^2 \sim 2\varepsilon^{-1}$, so we expect the standard deviation to be around $\varepsilon^{-1/2}\sqrt{2n}$. Pittel and Wormald [57] proved that these statements do indeed hold, and much more.

Theorem 6.4 *Let $\varepsilon = \varepsilon(n)$ satisfy $\varepsilon \to 0$ but $\varepsilon^3 n \to \infty$. Then the order L_1 of the largest component of $G_{n,(1+\varepsilon)/n}$ satisfies*

$$\frac{L_1 - \alpha_{1+\varepsilon} n}{\varepsilon^{-1/2}\sqrt{2n}} \xrightarrow{\mathrm{d}} \mathrm{No}(0,1).$$

Recently, Nachmias and Peres [53] combined martingale techniques with the exploration process described in Section 5 to give a much simpler proof of a weak form of Theorem 6.4. In [13] we show that the same strategy easily recovers the full strength of Theorem 6.4, as well as Stepanov's result (Theorem 6.3).

There is another way in which the results described in this section are tight: the assumption $\varepsilon^3 n \to \infty$ cannot be weakened. It was taken for granted from the beginning, and is clear from calculations in [8] that, when $p = (1+\theta n^{-1/3})/n$ with θ bounded, $G_{n,p}$ behaves very much like $G_{n,1/n}$. Furthermore, the behaviour in the scaling window $p = (1 + O(n^{-1/3}))/n$ is very different from either the subcritical or supercritical regime: for example, the size of the largest component is not concentrated. More precisely, when θ is constant and $p = (1+\theta n^{-1/3})/n$, the random variable $n^{-2/3}L_1(G_{n,p})$ converges in distribution to a certain non-trivial distribution S_θ. This was first stated explicitly by Aldous [3] in 1997, who gave a beautiful description of S_θ (and of the joint limit of $n^{-2/3}L_1, n^{-2/3}L_2, \ldots$) in terms of the excursions of a Brownian motion with a certain drift.

Further detailed results about the limiting distribution of the sequence of rescaled component sizes inside the scaling window were proved recently by Janson and Spencer [39].

In 2001, in a difficult and very technical paper, Pittel [56] gave a much more explicit form for the limiting distribution of $n^{-2/3}L_1$ inside the scaling window. In other words, for $a>0$ and θ constant, he gave a very complicated formula for the limiting value of the probability that $G_{n,p}$ contains no component with more than $an^{2/3}$ vertices, when $p=(1+\theta n^{-1/3})/n$. A special case implies that, if a is constant and $\theta \to \infty$, then this probability is

$$\exp\left(-\tfrac{1}{6}\theta^3 - (1+o(1))\frac{\theta \log(\theta \log \theta)}{a}\right).$$

7. The young giant

In this section we describe some recent results concerning the structure of the 'young' giant component – that is, the largest component of $G_{n,p}$ in the range where $p=(1+\varepsilon)/n$ with $\varepsilon=\varepsilon(n)>0$ satisfying $\varepsilon \to 0$ but $\varepsilon^3 n \to \infty$. The condition $\varepsilon^3 n \to \infty$ ensures that the giant component has already emerged as the whp unique largest component; the condition $\varepsilon \to 0$ ensures that it is 'young'.

The title of the Pittel–Wormald paper [57] from which Theorem 6.4 is taken is 'Counting connected graphs inside-out'. This refers to a counting strategy going back to Bollobás [7], who used it to establish the gap in the sequence of components near the phase transition; this is a key lemma in [8]. (In Section 4 we made use of a weaker form of this 'gap argument'.) Of course, in [57], this strategy is just the starting point, and the calculations are much more precise than the crude bounds that sufficed in [7] and [8].

First, studying the distribution of L_1 is closely connected to counting connected graphs with a given number of vertices and edges. Indeed, the expected number of components of $G_{n,p}$ with k vertices and l edges is exactly

$$C(k,l)\binom{n}{k}p^l(1-p)^{\binom{k}{2}-l+k(n-k)},$$

where, as before, $C(k,l)$ denotes the number of connected graphs on k labelled vertices with l edges. Indeed, there are $\binom{n}{k}$ choices for the vertex-set W of the component, and then $C(k,l)$ choices for exactly which graph H on W the component is. One particular H is present *as a component of* $G_{n,p}$ (with the given vertex-set) if and only if all l edges of H are present in $G_{n,p}$, none of the $\binom{k}{2}-l$ edges of H^c is present, and no edges are present from $W=V(H)$ to the rest of G, giving the formula above. Of course, the expectation and the probability of existence need not be similar in general, but in the range of values typical of the maximum component it turns out to be unlikely that there is more than one component of comparable size, and the formula above is a good estimate for the probability that the largest component has k vertices and l edges.

Given a connected graph G, its *2-core* (or simply *core*) is the subgraph $G^{(2)}$ defined as follows. Starting from the graph G, repeatedly delete vertices of degree 0 or 1 until no such vertices remain. In other words, $G^{(2)}$ is the maximal subgraph of G with minimum degree at least 2. (This concept and its natural generalization to minimum degree at least k were introduced in [7].) Note that $G^{(2)}$ may be the empty graph with no vertices. By definition, $G^{(2)}$ has minimum degree at least 2, and any graph with this property may arise as a 2-core (for example, of itself). Reversing the process, we see that G may be obtained from its 2-core by adding trees. In particular, if G is connected and not a tree (so it has a non-trivial 2-core), then G may be formed from $G^{(2)}$ by taking one rooted tree T_v for each vertex v of $G^{(2)}$, and then identifying the root of T_v with v. Note also that the 2-core of any connected graph that is not a tree is connected.

A key observation behind the arguments in [7] and [57] is that trees and forests are relatively easy to count, so the main difficulty in evaluating $C(k, l)$ concerns counting 2-cores. In fact, one can go a step further.

Let $G^{(2)}$ be any graph with minimum degree 2. Then the vertices of degree 2 in $G^{(2)}$ can be seen as playing a special role. More precisely, if $G^{(2)}$ is connected and not a cycle, then we can view $G^{(2)}$ as constructed from its *kernel* K as follows: the vertices of K are all vertices of $G^{(2)}$ with degree at least 3. With the possible exception of components consisting of only vertices of degree 2, each vertex of degree 2 sits on a path joining two vertices of K; for the edge-set of K we simply take one edge for each such path. Note that K may be a multigraph: two vertices v and w of $G^{(2)}$ with degree 3 may be joined by two or more paths, leading to parallel edges in K. Also, there may be a path joining v to itself, leading to a loop.

Starting from a kernel K, to form a corresponding core $G^{(2)}$ with a given number of vertices and edges, we simply have to subdivide the edges of K appropriately, making sure that all loops are subdivided, as well as all but one of any set of parallel edges. The number of such subdivisions is relatively easy to count. Hence, if one can obtain good bounds on the number of kernels K with certain properties, then one can transfer these to cores and thence to connected graphs. This approach was introduced in [7], and has been used many times since then. In [57], where Pittel and Wormald proved strong results on the joint asymptotic normality of the sizes of the giant component of $G_{n,p}$ and of its 2-core, this strategy was just the starting point, and the details are considerably more complicated than this description might suggest!

The strategy described above has a natural probabilistic interpretation. Let $\mathcal{L}_1(G)$ denote the largest component of a graph G – that is, the one with the most vertices, which will be whp unique in all cases that we consider. Given the numbers of vertices and edges, the 2-core of $\mathcal{L}_1(G_{n,p})$ is easily seen to be uniformly distributed on all connected 2-cores with these parameters, and a similar observation applies to the kernel. Łuczak [45] studied the degree distribution of the kernel of the young giant component, and used the fact that almost all vertices have degree 3 to show that whp the young giant component $\mathcal{L}_1$ contains a cycle of length at least $(\frac{4}{3}+o(1))\varepsilon^2 n$. Actually, this bound appears in [45] only as a conditional result, under the assumption that a random 3-regular graph is whp Hamiltonian. The latter result was proved just afterwards by Robinson and Wormald [59].

More recently, Ding *et al.* [22] have given a detailed description of the distribution of the young giant component of $G_{n,p}$, built from the inside out as above; this description is both more precise, and easier to use, than Łuczak's.

Let $\mathrm{No}(\mu, \sigma^2)$ denote the normal distribution with mean μ and variance σ^2, and let $\mathrm{Geom}(\varepsilon)$ denote the geometric distribution with mean $1/\varepsilon$. Let $\mathrm{Po}(\lambda)$ denote the Poisson distribution with mean λ. Recall that a $\mathrm{Po}(\lambda)$-*Galton Watson tree* is a random rooted tree constructed recursively as follows. Start with a single vertex as the root. Each vertex has a Poisson number of 'children' in the next level with mean λ, independently of all other vertices.

To state a special case of result of [22], consider the *model giant component* $\widetilde{\mathcal{L}}_1$ constructed in the following three steps. First, let $Z \sim \mathrm{No}(\frac{2}{3}\varepsilon^3 n, \varepsilon^3 n)$, and let K be a random 3-regular multigraph on $N = 2\lfloor Z \rfloor$ vertices. Next, replace each edge of K by a path, where the path lengths are independent and have the distribution $\mathrm{Geom}(\varepsilon)$. Finally, attach an independent $\mathrm{Po}(1-\varepsilon)$-Galton–Watson tree to each vertex. In the special case where $\varepsilon = o(n^{-1/4})$, the result of [22] may be stated as follows.

Theorem 7.1 *Let $\mathcal{L}_1 = \mathcal{L}_1(G_{n,p})$ be the largest component of the random graph $G_{n,p}$ for $p = (1+\varepsilon)/n$, where $\varepsilon^3 n \to \infty$ and $\varepsilon = o(n^{-1/4})$. Then $\mathcal{L}_1$ is contiguous to the model $\widetilde{\mathcal{L}}_1$, in the sense that, for any set of graphs Q, $\lim_{n\to\infty} \mathbf{P}(\widetilde{\mathcal{L}}_1 \in Q) = 0$ implies that $\lim_{n\to\infty} \mathbf{P}(\mathcal{L}_1 \in Q) = 0$.*

The general result of Ding *et al.* [22] covers the entire range where $\varepsilon^3 n \to \infty$ and $\varepsilon \to 0$; the statement is a little more involved, since in this range there are a significant number of vertices of degree greater than 3 in the kernel.

Theorem 7.1 and its more general companion make it easy to 'read off' many properties of the young giant. Here is one simple example from [22].

Corollary 7.2 *Let $\varepsilon^3 n \to \infty$ and $\varepsilon = o(n^{-1/4})$, and let $\mathcal{L}_1^{(2)}$ be the 2-core of the largest component of the random graph $G_{n,p}$ where $p = (1+\varepsilon)/n$. Then whp the maximal length of a 2-path in $\mathcal{L}_1^{(2)}$ is $(1/\varepsilon)\log(\varepsilon^3 n) + O_{\mathrm{p}}(1/\varepsilon)$.*

Here a 2-path is simply a path in the 2-core, each of whose internal vertices has degree 2 – that is, one of the paths corresponding to an edge of the kernel.

Recall that by the result of Robinson and Wormald [59], a random 3-regular graph is Hamiltonian whp. Since a Hamiltonian cycle in the cubic graph appearing in the above construction has around $\frac{4}{3}\varepsilon^3 n$ edges, and on average each corresponds to a path of length $1/\varepsilon$, it follows immediately that when $\varepsilon(n)$ satisfies the given conditions, then $\mathcal{L}_1$ contains a cycle of length at least $(\frac{4}{3} + o(1))\varepsilon^2 n$, as shown by Łuczak [45]. Recently, by analyzing the length (total weight) of the longest cycle in a random cubic graph with random weights on the edges, Kemkes and Wormald [42] have improved this bound to approximately $1.739\varepsilon^2 n$.

When $\varepsilon \to 0$, a geometric random variable $\mathrm{Geom}(\varepsilon)$ is well approximated by an exponential distribution; indeed, after multiplication by ε, the geometric distribution converges to the standard exponential distribution. The study of shortest paths (measured by sums of weights) in graphs with independent exponential weights on their edges is a well-developed topic known as 'first-passage percolation' (see, for example, Kesten [43]). The results of Ding *et al.* [22] allow one to use such results to study the young giant and its 2-core. In particular, using earlier results on first-passage percolation due to Bhamidi, Hooghiemstra and van der Hofstad [4], Ding *et al.* [22] obtained the following result.

Corollary 7.3 *Let $\varepsilon^3 n \to \infty$ and $\varepsilon \to 0$, and let $\mathcal{L}_1^{(2)}$ be the 2-core of the largest component of the random graph $G_{n,p}$ where $p = (1+\varepsilon)/n$. If v and w are two vertices of degree at least 3 in $\mathcal{L}_1^{(2)}$ chosen uniformly at random among all such vertices, then the distance between v and w is whp $(1/\varepsilon + O(1))\log(\varepsilon^3 n)$.*

What if one wishes to study the maximum distance between vertices, rather than the typical distance? In other words, what if we wish to study the *diameter* $\mathrm{diam}(G)$ of a connected graph G such as $\mathcal{L}_1(G_{n,p})$? Here the situation is much more complicated: taking the first-passage percolation viewpoint above, the diameter comes from 'exceptional' vertices, rather than typical ones, so strong extreme-value-type estimates for the first-passage percolation are needed. Using such methods, Ding *et al.* [21] obtained the following result.

Theorem 7.4 *Consider the random graph $G_{n,p}$ for $p = (1+\varepsilon)/n$, where $\varepsilon^3 n \to \infty$ and $\varepsilon \to 0$. Let $\mathcal{L}_1$ be the largest component of $G_{n,p}$, with 2-core $\mathcal{L}_1^{(2)}$ and kernel K. Then, whp,*

$$\begin{aligned}\mathrm{diam}(\mathcal{L}_1) &= (3+o(1))\,(1/\varepsilon)\log(\varepsilon^3 n),\\ \mathrm{diam}(\mathcal{L}_1^{(2)}) &= (2+o(1))\,(1/\varepsilon)\log(\varepsilon^3 n), \qquad (22)\\ \max_{v,w\in K} d_{\mathcal{L}_1^{(2)}}(v,w) &= (\tfrac{5}{3}+o(1))\,(1/\varepsilon)\log(\varepsilon^3 n).\end{aligned}$$

In the last statement, the distance is measured in the 2-core $\mathcal{L}_1^{(2)}$.

The diameter of a connected graph is defined as the maximum distance between any pair of its vertices, but what is the diameter of a disconnected graph G? There are two possible conventions. One is to say that if G is disconnected, then $\mathrm{diam}(G) = \infty$; this has the advantage that $\mathrm{diam}(G)$ can only decrease as edges are added. Alternatively, one can consider the maximum of the diameters of the components of G. In the past, the former definition was the more common; Theorem 7.4 implicitly uses the latter, as the diameter of $G_{n,p}$ is whp equal to that of its largest component $\mathcal{L}_1$ in this range.

Historically, the question of determining the diameter of $G_{n,p}$ was at first considered only when p is large enough that this graph has a reasonable chance of being connected. The diameter of $G_{n,p}$ was perhaps first studied by Burtin [17], [18]; Bollobás [7] obtained a very precise formula for the diameter at exactly the point where the graph first becomes connected, and (slightly earlier) when the average degree is at least $(\log n)^3$, say (see [6]).

Turning to the case of bounded average degree considered throughout this chapter, Łuczak [46] proved very precise results for the subcritical case. For the 'mature' giant component – that is, for $p = c/n$ with $c > 1$ constant – this question was neglected for many years. Chung and Lu [19] proved a partial result, but the correct asymptotic formula was obtained only much later, independently by Fernholz and Ramachandran [32] and by Bollobás, Janson and Riordan [11], in both papers as a special case of a result for a much more general model.

More recently, Riordan and Wormald [58] obtained essentially best-possible estimates for the diameter of $G_{n,p}$ throughout the supercritical regime. For the young giant component, they proved the following result.

Theorem 7.5 *Let $\varepsilon = \varepsilon(n)$ satisfy $0 < \varepsilon < \frac{1}{10}$ and $\varepsilon^3 n \to \infty$. Set $\lambda = \lambda(n) = 1 + \varepsilon$, and let $\lambda^\star < 1$ be the dual of λ. Then*

$$\operatorname{diam}(G_{n,\lambda/n}) = \frac{\log(\varepsilon^3 n)}{\log \lambda} + 2\frac{\log(\varepsilon^3 n)}{\log(1/\lambda^\star)} + O_{\mathrm{p}}(1/\varepsilon). \tag{23}$$

This result is sharp, in that the standard deviation of the diameter is of order $1/\varepsilon$. Of course one could ask for even more: what is the limiting distribution of the error term? This was also answered in [58], but the statement is a little involved.

When $\lambda = 1+\varepsilon$ with $\varepsilon \to 0$, it turns out that $\lambda^\star = 1-\varepsilon+O(\varepsilon^2)$, so the formula (23) does indeed refine (22). While the kernel–core–giant component approach, pioneered by Bollobás [7] and used by Ding *et al.* [21], is very powerful, it does seem to have its limits; the contiguity established in Theorem 7.1 and its companion is tailored to proving whp results, but cannot be directly applied to study limiting distributions.

Taking the graph-process view, Łuczak and Seierstad [48] gave estimates for the diameter that whp apply throughout the entire supercritical regime, but these are much weaker: their upper and lower bounds differ by a constant factor.

Let us note that, inside the window of the phase transition, the behaviour of the diameter of $G_{n,p}$ is even more complicated. Nachmias and Peres [54] showed that, throughout this window, the diameter is $O_{\mathrm{p}}(n^{1/3})$. More recently, for p inside the scaling window, Addario-Berry, Broutin and Goldschmidt [1] established convergence of $\mathcal{L}_1(G_{n,p})$ to a certain (complicated) continuum limit object in a certain sense, and thus gave a (very involved) description of the distribution of its diameter scaled by dividing by $n^{1/3}$.

8. Final words

The Erdős–Rényi theorem [24] about the sudden change in the component structure of a random graph, resulting in the emergence of the giant component, is a milestone not only in the theory of random graphs, but also in probabilistic combinatorics. Following its proof in 1960, for a quarter of a century this result about the so-called 'phase transition' was considered to be the complete description of a fascinating and curious phenomenon. In 1984 it was discovered [8] that this description is far from complete, and that the Erdős–Rényi theorem is only the tip of the iceberg: rather than the final word on the phase transition, it should be the starting point of much further work. Indeed, since then ever more precise results have been proved about the nature of the phase transition.

The phase transition of random graphs is the quintessential example of the phase transitions of random combinatorial structures, including those in percolation theory, so the results proved about it indicate what kind of results we can hope for in other structures. As we have seen, in recent years numerous deep results have been proved about the emergence of the giant component: the hope is that these results will help us to prove similar results about other structures.

References

1. L. Addario-Berry, N. Broutin and C. Goldschmidt, The continuum limit of critical random graphs, *Probab. Th. Related Fields* **152** (2012), 367–406.
2. M. Ajtai, J. Komlós and E. Szemerédi, Topological complete subgraphs in random graphs, *Studia Sci. Math. Hungar.* **14** (1979), 293–297 (1982).
3. D. Aldous, Brownian excursions, critical random graphs and the multiplicative coalescent, *Ann. Probab.* **25** (1997), 812–854.
4. S. Bhamidi, R. van der Hofstad and G. Hooghiemstra, First passage percolation on random graphs with finite mean degrees, *Ann. Appl. Probab.* **20** (2010), 1907–1965.
5. B. Bollobás, A probabilistic proof of an asymptotic formula for the number of labelled regular graphs, *Europ. J. Combin.* **1** (1980), 311–316.
6. B. Bollobás, The diameter of random graphs, *Trans. Amer. Math. Soc.* **267** (1981), 41–52.
7. B. Bollobás, The evolution of sparse graphs, *Graph Theory and Combinatorics (Cambridge, 1983)*, Academic Press (1984), 35–57.
8. B. Bollobás, The evolution of random graphs, *Trans. Amer. Math. Soc.* **286** (1984), 257–274.
9. B. Bollobás, *Random Graphs*, Academic Press (1985). Second edn: Cambridge Studies in Advanced Mathematics **73**, Cambridge University Press, 2001.
10. B. Bollobás, *Modern Graph Theory*, Graduate Texts in Mathematics **184**, Springer-Verlag, 1998.
11. B. Bollobás, S. Janson and O. Riordan, The phase transition in inhomogeneous random graphs, *Random Struct. Alg.* **31** (2007), 3–122.
12. B. Bollobás and O. Riordan, Random graphs and branching processes, *Handbook of Large-scale Random Networks* (eds. B. Bollobás, R. Kozma and D. Miklós), Bolyai Soc. Math. Stud. **18** (2009), 15–115.
13. B. Bollobás and O. Riordan, Asymptotic normality of the size of the giant component via a random walk, *J. Combin. Theory (B)* **102** (2012), 53–61.
14. B. Bollobás and A. Thomason, Threshold functions, *Combinatorica* **7** (1987), 35–38.
15. J. Bourgain, J. Kahn, G. Kalai, Y. Katznelson and N. Linial, The influence of variables in product spaces, *Israel J. Math.* **77** (1992), 55–64.
16. S. R. Broadbent and J. M. Hammersley, Percolation processes. I. Crystals and mazes, *Proc. Cambridge Philos. Soc.* **53** (1957), 629–641.
17. Ju. D. Burtin, Asymptotic estimates of the diameter and the independence and domination numbers of a random graph, *Dokl. Akad. Nauk SSSR* **209** (1973), 765–768; transl. in *Soviet Math. Dokl.* **14** (1973), 497–501.

18. Ju. D. Burtin, Extremal metric characteristics of a random graph. I, *Teor. Verojatnost. i Primenen.* **19** (1974), 740–754.
19. F. Chung and L. Lu, The diameter of sparse random graphs, *Adv. Appl. Math.* **26** (2001), 257–279.
20. C. Cooper and A. Frieze, The size of the largest strongly connected component of a random digraph with a given degree sequence, *Combin. Probab. Comput.* **13** (2004), 319–337.
21. J. Ding, J. H. Kim, E. Lubetzky and Y. Peres, Diameters in supercritical random graphs via first-passage percolation, *Combin. Probab. Comput.* **19** (2010), 729–751.
22. J. Ding, J. H. Kim, E. Lubetzky and Y. Peres, Anatomy of a young giant component in the random graph, *Random Struct. Alg.* **39** (2011), 139–178.
23. P. Erdős and A. Rényi, On random graphs I, *Publ. Math. Debrecen* **5** (1959), 290–297.
24. P. Erdős and A. Rényi, On the evolution of random graphs, *Magyar Tud. Akad. Mat. Kutató Int. Közl.* **5** (1960), 17–61.
25. P. Erdős and A. Rényi, On a classical problem of probability theory, *Magyar Tud. Akad. Mat. Kutató Int. Közl.* **6** (1961), 215–220.
26. P. Erdős and A. Rényi, On the evolution of random graphs, *Bull. Inst. Internat. Statist.* **38** (1961), 343–347.
27. P. Erdős and A. Rényi, On the strength of connectedness of a random graph, *Acta Math. Acad. Sci. Hungar.* **12** (1961), 261–267.
28. P. Erdős and A. Rényi, Asymmetric graphs, *Acta Math. Acad. Sci. Hungar.* **14** (1963), 295–315.
29. P. Erdős and A. Rényi, On random matrices, *Magyar Tud. Akad. Mat. Kutató Int. Kőzl.* **8** (1964), 455–461.
30. P. Erdős and A. Rényi, On the existence of a factor of degree one of a connected random graph, *Acta Math. Acad. Sci. Hungar.* **17** (1966), 359–368.
31. P. Erdős and A. Rényi, On random matrices II, *Studia Sci. Math. Hungar.* **3** (1968), 459–464.
32. D. Fernholz and V. Ramachandran, The diameter of sparse random graphs, *Random Struct. Alg.* **31** (2007), 482–516.
33. N. Fountoulakis and B. A. Reed, The evolution of the mixing rate of a simple random walk on the giant component of a random graph, *Random Struct. Alg.* **33** (2008), 68–86.
34. E. Friedgut, Sharp thresholds of graph properties, and the k-sat problem, with an appendix by Jean Bourgain, *J. Amer. Math. Soc.* **12** (1999), 1017–1054.
35. E. Friedgut and G. Kalai, Every monotone graph property has a sharp threshold, *Proc. Amer. Math. Soc.* **124** (1996), 2993–3002.
36. E. N. Gilbert, Random graphs, *Ann. Math. Statist.* **30** (1959), 1141–1144.
37. S. Janson, D. E. Knuth, T. Łuczak and B. G. Pittel, The birth of the giant component, *Random Struct. Alg.* **4** (1993), 233–358.
38. S. Janson, T. Łuczak and A. Ruciński, *Random Graphs*, Wiley, 2000.
39. S. Janson and J. Spencer, A point process describing the component sizes in the critical window of the random graph evolution, *Combin. Probab. Comput.* **16** (2007), 631–658.
40. J. Kahn, G. Kalai and N. Linial, The influence of random variables on Boolean functions, *Proc. 29th Ann. Symp. on Foundations of Comp. Sci*, Computer Society Press (1988), 68–80.
41. R. M. Karp, The transitive closure of a random digraph, *Random Struct. Alg.* **1** (1990), 73–93.
42. G. Kemkes and N. Wormald, An improved upper bound on the length of the longest cycle of a supercritical random graph (2009), arXiv:0907.3511.
43. H. Kesten, Aspects of first passage percolation, *École d' Été de Probabilités de Saint-Flour, XIV–1984*, Lecture Notes in Mathematics **1180**, Springer (1986), 125–264.

44. T. Łuczak, Component behavior near the critical point of the random graph process, *Random Struct. Alg.* **1** (1990), 287–310.
45. T. Łuczak, Cycles in a random graph near the critical point, *Random Struct. Alg.* **2** (1991), 421–439.
46. T. Łuczak, Random trees and random graphs, *Random Struct. Alg.* **13** (1998), 485–500.
47. T. Łuczak, B. G. Pittel and J. C. Wierman, The structure of a random graph near the point of the phase transition, *Trans. Amer. Math. Soc.* **341** (1994), 721–748.
48. T. Łuczak and T. G. Seierstad, The diameter behavior in the random graph process, Mittag–Leffler preprint Report No 5, 2008/2009 Spring.
49. T. Łuczak and J. C. Wierman, The chromatic number of random graphs at the double-jump threshold, *Combinatorica* **9** (1989), 39–49.
50. A. Martin-Löf, Symmetric sampling procedures, general epidemic processes and their threshold limit theorems, *J. Appl. Probab.* **23** (1986), 265–282.
51. M. Molloy and B. A. Reed, A critical point for random graphs with a given degree sequence, *Random Struct. Alg.* **6** (1995), 161–180.
52. M. Molloy and B. A. Reed, The size of the largest component of a random graph on a fixed degree sequence, *Combin. Probab. Comput.* **7** (1998), 295–306.
53. A. Nachmias and Y. Peres, Component sizes of the random graph outside the scaling window, *ALEA Latin Amer. J. Probab. Math. Stat.* **3** (2007), 133–142.
54. A. Nachmias and Y. Peres, Critical random graphs: diameter and mixing time, *Ann. Probab.* **36** (2008), 1267–1286.
55. B. Pittel, On tree census and the giant component in sparse random graphs, *Random Struct. Alg.* **1** (1990), 311–342.
56. B. Pittel, On the largest component of the random graph at a nearcritical stage, *J. Combin. Theory (B)* **82** (2001), 237–269.
57. B. Pittel and C. Wormald, Counting connected graphs inside-out, *J. Combin. Theory (B)* **93** (2005), 127–172.
58. O. Riordan and N. Wormald, The diameter of sparse random graphs, *Combin. Probab. Comput.* **19** (2010), 835–926.
59. R. W. Robinson and N. C. Wormald, Almost all cubic graphs are Hamiltonian, *Random Struct. Alg.* **3** (1992), 117–125.
60. V. E. Stepanov, Phase transitions in random graphs (Russian), *Teor. Verojatnost. i Primenen.* **15** (1970), 200–216; transl. in *Theory Probab. Appl.* **15** (1970), 55–67.
61. E. M. Wright, The number of connected sparsely edged graphs, III. Asymptotic results, *J. Graph Theory* **4** (1980), 393–407.

11

Network reliability and synthesis

F. T. BOESCH, A. SATYANARAYANA and C. L. SUFFEL

This chapter is an introduction to several graph-theoretic concepts related to the analysis and synthesis of reliable or invulnerable networks. We describe the notion of signed reliability domination of a system and review some of its applications to reliability analysis. We consider the problem of analysis and give a brief summary of the difficulty in calculating various reliability measures. We also introduce some concepts of synthesis of a most reliable network. The chapter concludes with an introduction to a non-probabilistic approach to evaluating the ability of a network to resist damage.

1. Introduction

Over the last few decades, there have been several outstanding advances in reliability theory that have utilized graph-theoretic ideas. An important problem in this area is to determine the reliability of a system from the reliability of its components. Our purpose in this chapter is to introduce graph-theorists to the basis of these applications.

There is an extensive literature on reliability theory, and some of the terms used are not standard in graph theory. Historically, network reliability has been concerned with the problem of determining whether there is a path of operating elements from one specified vertex in a network to another. A variety of measures have been proposed and a number of graph-theoretic parameters have been used to calculate network reliability. One of the most commonly used measures of performance is K-terminal reliability.

Throughout this chapter, K denotes a specified set of vertices in a graph G. Given that the elements (vertices or edges) of G may fail with known probabilities, the *K-terminal reliability* of a graph is the probability that K is contained in some connected subgraph of the graph all of whose elements are operational. A graph invariant known as the *reliability domination* is based on network reliability analysis.

The original definition of reliability domination involves the concept of a formation of a graph, which is a collection of trees with particular properties. This notion of domination possesses many interesting properties and plays a key role in network reliability theory. In this chapter we give a self-contained presentation of the reliability problem and its connection to domination. We present several results that show how this invariant is related to other graph-theoretic properties. The concept was first introduced by Satyanarayana and Prabhakar [43], and has subsequently been investigated by a number of other investigators (see, for example, Huseby [28]).

Our chapter is organized as follows. In Sections 2 and 3 we introduce the notion of the signed reliability domination of a system and survey some of its applications to reliability analysis. The analysis problem is considered in Section 4, and a brief summary is given of the difficulty of calculating various reliability measures. In Section 5 we consider the synthesis of optimal networks. We conclude, in Section 6, with an introduction to a non-probabilistic approach to evaluating the ability of a network to resist damage.

We consider both graphs and digraphs, and these may have multiple edges or multiple arcs as well as loops. Note that, as usual, V denotes the set of vertices of a graph or digraph, while E denotes the set of edges in a graph and A denotes the set of arcs in a digraph.

2. Domination in digraphs

The basic problem that we are concerned with in this section is this:

Given one specific vertex s in a digraph D and a set K of vertices that includes s, what is the probability that s has a path to each of the vertices in K when each of the vertices not in K, and each of the arcs of D may fail?

The set of elements surviving at a given point in time is referred to as a 'state' and is an 'operating' state if there is a path from s to each vertex in K. The probability that we seek is just that of the collection of operating states when the joint probabilities of the states are known. In this case (as well as in all of the reliability scenarios that we consider here), the minimal operating states are of paramount importance, and we discuss these next. In order to do this, we need some terminology and notation.

A rooted tree can naturally be thought of as having all of its edges directed away from the root, and throughout our discussion of digraphs, we assume this interpretation. Given a vertex s and a set K of vertices in a digraph D (as above), we define a

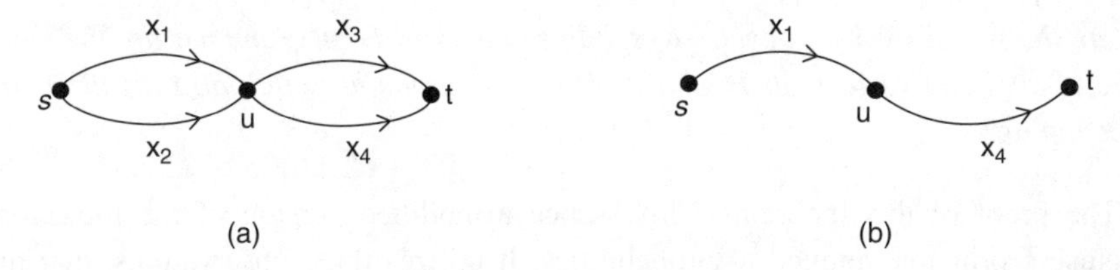

Fig. 1.

rooted tree T to be a K*-tree* if its root is s, it contains every vertex in K, and each of its sink vertices is in K. In other words, a K-tree of D is simply a minimal subdigraph of D in which there is a path from s to each vertex in K. Fig. 1 shows a digraph D and a K-tree T in D having root s, with $K = \{s, t\}$.

In what follows, for a given digraph D, we assume that K is a set of vertices that includes a specified vertex s as source. We say that D is a K*-digraph* if each of its arcs is in some K-tree. If D is such a digraph, we let $\mathcal{T}_K$ be the collection of K-trees rooted at s. A *formation* is a collection of K-trees rooted at s whose union is D, and is *odd* or *even* depending on its cardinality being odd or even. For a given set K, the *signed domination* $\sigma_K(D)$ of a digraph D is defined to be the number of odd formations in D minus the number of even formations.

We begin with the surprising fact (see [41]) that the signed domination of a K-digraph D with n vertices and m arcs is

$$\sigma_K(D) = (-1)^{m-n+1}$$

if D is an acyclic K-digraph, and is 0 otherwise. The fact that $\sigma_K(D)$ is 0 when D is not a K-digraph follows from the definition, since there are no formations in this case. The main assertion is established by induction on the number of arcs. It uses the fact that if e is the arc su, then $\sigma_K(D) = -\sigma_K(D-e)$ when the in-degree $\text{id}(u) > 1$, and $\sigma_K(D) = \sigma_{K|e}(D|e)$ when $\text{id}(u) = 1$. (Here, $K|e$ denotes the set of vertices that result in K when s and u are identified.) A detailed proof can be found in [16].

Signed domination has been used in the following directed network reliability problem. At a given time, each vertex and each arc of a digraph D are in one of two states, either *failed* or *functioning*. A vertex v can *communicate* with another vertex w if there is a directed v–w path in which all vertices and arcs are functioning. The problem of interest here, known as the *source-to-terminal reliability problem* (or the *s–K reliability problem*, for short), is to find the probability that s can communicate with every vertex in K when the joint probability distribution for the states is known. This probability is the *s–K-terminal reliability* of D and is denoted by $R_{s,K}(D)$. The following result of Satyanarayana [41] provides a simplified way of computing its value.

Theorem 2.1 *The s–K-terminal reliability of a digraph D is*

$$R_{s,K}(D) = \sum_H (-1)^{m_H - n_H + 1} \Pr(H),$$

where the sum is over all acyclic K-subdigraphs H of D, m_H and n_H are the numbers of arcs and vertices in H and $\Pr(H)$ is the probability that all arcs in H are functioning.

The proof of this theorem is in essence a modified version of the inclusion-exclusion principle applied to probabilities. It utilizes three observations: that the contribution of a subdigraph with a cycle is 0, that the contribution of a subdigraph that is not a K-digraph is also 0, and that the signed domination of an acyclic K-digraph is the expression given above.

The following example illustrates these ideas. Consider the digraph D shown in Fig. 1(a), where $K = \{s, t\}$. Each vertex in D operates with probability 1, and the arcs operate with independent probabilities $\Pr(e_i) = p_i$ for $i = 1, 2, 3, 4$. All of the acyclic K-subgraphs of D are shown in Fig. 2, the first four being K-trees. The seven formations are thus $\{H_1, H_4\}$, $\{H_2, H_3\}$, $\{H_1, H_2, H_3\}$, $\{H_1, H_2, H_4\}$, $\{H_1, H_3, H_4\}$, $\{H_2, H_3, H_4\}$ and $\{H_1, H_2, H_3, H_4\}$, so that $\sigma(D) = 4 - 3 = 1$. Also, in accordance with Theorem 2.1, we see that

$$\begin{aligned}R_{s,K} &= (-1)^{2-3+1}\Pr(H_1) + (-1)^{2-3+1}\Pr(H_2) + (-1)^{2-3+1}\Pr(H_3)\\ &\quad + (-1)^{2-3+1}\Pr(H_4) + (-1)^{3-3+1}\Pr(H_5) + (-1)^{3-3+1}\Pr(H_6)\\ &\quad + (-1)^{3-3+1}\Pr(H_7) + (-1)^{3-3+1}\Pr(H_8) + (-1)^{4-3+1}\Pr(H_9)\\ &= p_1p_2 + p_1p_4 + p_2p_3 + p_2p_4 - p_1p_2p_3 - p_1p_2p_4 - p_1p_3p_4\\ &\quad - p_2p_3p_4 + p_1p_2p_3p_4.\end{aligned}$$

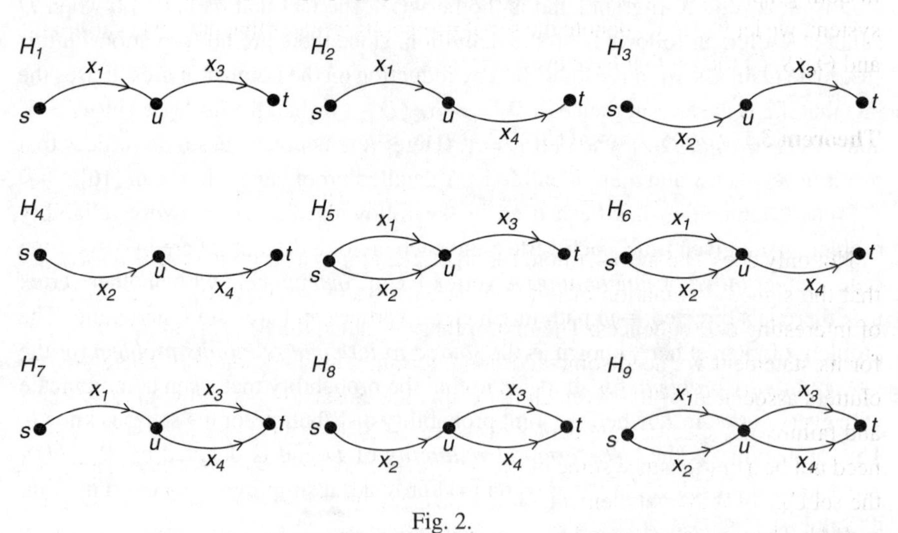

Fig. 2.

Another important application of the parameter σ_K has been (in absolute value) as a measure of the computational complexity of pivoting algorithms. This is discussed in greater detail in Section 4.

3. Coherent systems and domination in graphs

In this section, we switch our discussion from digraphs to graphs, and go into considerably greater depth. The 'K concepts' introduced for digraphs in Section 2 have straightforward counterparts in the undirected case. For example, given a set K of vertices in a graph G, a *K-tree* is a tree of G that contains all of K and has all of its leaves in K, and the graph G is a *K-graph* if each of its edges lies in some *K-tree*. The notions of *formation* and *signed domination* are likewise similar to the directed versions. There is one key difference, however, in that in the undirected case the value of the signed domination can be any integer, whereas in the directed case only the values 0, 1 and -1 are possible.

We begin with some general concepts from set theory. Given a finite non-empty set S, a non-empty collection $\mathcal{C}$ of subsets of S is called a *clutter* if none of the subsets is contained in any of the others. If $\mathcal{C}$ is a clutter, the pair $(S, \mathcal{C})$ is called a *system*, and is said to be *coherent* if each element of S is in a set in $\mathcal{C}$. Any set $X \subseteq S$ that contains a set in $\mathcal{C}$ is called an *operating state* of $(S, \mathcal{C})$. The set of all operating states of $(F, \mathcal{C})$ is denoted by $\Theta(S, \mathcal{C})$.

A collection of sets $\mathcal{A} \subseteq \mathcal{C}$ is called a *formation* of $(S, \mathcal{C})$ if $(S, \mathcal{A})$ is itself a coherent system. A formation is *odd* or *even* depending on its cardinality being odd or even. The *signed domination* $\sigma(S, \mathcal{C})$ of a formation $(S, \mathcal{C})$ is the number of odd formations minus the number of even formations.

The following theorem, due to Huseby [28], provides a characterization of signed domination for systems, and has been used to establish other results on domination in both graphs and digraphs. Given the collection $\Theta(S, \mathcal{C})$ of operating states of a system, we let $\Theta_o(S, \mathcal{C})$ denote the set of its operating states that have odd cardinality and $\Theta_e(S, \mathcal{C})$ the set that have even cardinality.

Theorem 3.1 *For any system $(S, \mathcal{C})$,*

$$\sigma(S, \mathcal{C}) = (-1)^{|S|}(|\Theta_e(S, \mathcal{C})| - (|\Theta_o(S, \mathcal{C})|).$$

Not only does Theorem 3.1 hold for all clutters (and in particular reaffirms the fact that the signed domination of a non-coherent system is always 0), it has a number of interesting consequences. The first of these is called the *domination theorem*, and for its statement we need some additional terminology and notation. There are two clutters associated with a given clutter $\mathcal{C}$ that are closely associated with the success and failure of a specified element x of the set S. Note that $\mathcal{C} - x = \{A - x : A \in \mathcal{C}\}$ need not be a clutter since some of the sets $A - x$ may be subsets of others. However, the set $\mathcal{C}_{+x}$ of maximal elements of $\mathcal{C} - x$ is a clutter, as is the set $\mathcal{C}_{-x} = \{A \in \mathcal{C} : x \notin A\}$. The two sets $\mathcal{C}_{+x}$ and $\mathcal{C}_{-x}$ are called the *minors of $\mathcal{C}$ with respect to x*. Note that if $x \in A$, then $A \in \Theta(S, \mathcal{C})$ if and only if $A - x \in \Theta(S - \{x\}, \mathcal{C}_{+x})$, while if $x \notin A$, then $A \in \Theta(S, \mathcal{C})$ if and only if $A - x \in \Theta(S - \{x\}, \mathcal{C}_{-x})$. The following three corollaries are due to Satyanarayana and Chang [42].

Corollary 3.2 (Domination theorem) If $(S, \mathcal{C})$ is a system and if $x \in S$, then

$$\sigma(S, \mathcal{C}) = \sigma(S - \{x\}, \mathcal{C}_{+x}) - \sigma(S - \{x\}, \mathcal{C}_{-x}).$$

The original version of this result (see [40]) was for all-terminal domination in graphs, which was then extended in [42] to K-terminal domination. Subsequently, Agrawal and Barlow [2] proved that it holds for all coherent systems, after which it was further extended to general clutters by Huseby [28], [29]. Because this result has a number of important consequences, we give an abbreviated version of Huseby's proof for graphs.

Proof We begin by showing that

$$|\Theta_o(S, \mathcal{C})| = |\Theta_o(S - \{e\}, \mathcal{C}_{-e})| - |\Theta_e(S - \{e\}, \mathcal{C}_{+e})|.$$

Note that, for a given subset U of S, if $e \notin U$, then $U \in \Theta_o(S, \mathcal{C})$ if and only if there is a clutter element C such that $C \subseteq U$, and so C does not contain e – that is, $U \in \Theta_o(S, \mathcal{C})$ if and only if $U \in \Theta_o(S - \{e\}, \mathcal{C}_{-e})$.

On the other hand, if $e \in U$, then $U \in \Theta_o(S, \mathcal{C})$ if and only if there is a clutter element C for which $C \subseteq U$, and so $C - \{e\} \subseteq U - \{e\}$. Thus, if $e \in U$, then $U \in \Theta_o(S, \mathcal{C})$ if and only if $U - \{e\} \in \Theta_o(S - \{e\}, \mathcal{C}_{+e})$.

The desired equality is a combination of the two cases, and the corresponding equality

$$|\Theta_e(S, \mathcal{C}|) = |\Theta_e(S - \{e\}, \mathcal{C}_{-e}|) - |\Theta_o(S - \{e\}, \mathcal{C}_{+e})|$$

can be proved in the same way. The corollary now follows from Theorem 3.1. ■

When the clutter in the system is the set of K-trees in a graph, it is straightforward to deduce the following result. We adopt the convention that if a vertex of e is in K, then $K|e$ denotes the set obtained from K by replacing that vertex by the coalesced vertex in $G|e$, and $K|e = K$ otherwise.

Corollary 3.3 *If K is a set of vertices and e is an edge in a graph G, then*

$$\sigma_K(G) = \sigma_{K|e}(G|e) - \sigma_K(G - e).$$

Using Corollary 3.3, one can show that if G is a connected graph with n vertices and m edges and K is a set of its vertices, then the sign of $\sigma_K(G)$ is $(-1)^{m-n+1}$. With $\sigma_K^*(G)$ denoting the absolute value of $\sigma_K(G)$, we have the following result.

Corollary 3.4 *If K is a set of vertices and $e = vw$ is a non-loop edge in a graph G, and if either v or w has degree* 1 *and is in K, then*

$$\sigma_K^*(G) = \sigma_{K|e}^*(G|e) + \sigma_K^*(G - e).$$

Since the conclusion to Corollary 3.4 does not hold for all edges, it is of interest to know precisely when there is equality. The next corollary answers this question.

Corollary 3.5 *If K is a set of vertices and if e is an edge in a graph G, then*

$$\sigma_K^*(G) \neq \sigma_{K|e}^*(G|e) + \sigma_K^*(G - e)$$

if and only if $G - e$ is a K-graph and either e is a loop or some end-vertex of G is on e but is not in K.

Our next corollary gives a characterization of the K-terminal domination of a graph G in terms of certain of its connected spanning subgraphs. Among the connected spanning subgraphs for which all of the vertices in K lie in the same component, we let $s_o(G, K)$ denote the number with an odd number of edges and $s_e(G, K)$ be the number with an even number of edges. The result is an immediate consequence of Theorem 3.1 and the fact that $s_o(G, K) = |\Theta_o(E, \mathcal{T}_K)|$ and $s_e(G, K) = |\Theta_e(E, \mathcal{T}_K)|$ (as before, E is the set of edges and $\mathcal{T}_K$ is the set of K-trees in G).

Corollary 3.6 *If K is any set of vertices in a graph G, then*

$$\sigma_K^*(G) = |s_o(G, K) - s_e(G, K)|.$$

All-terminal domination

When specialized to all-terminal domination (that is, when $K = V$), the previous corollary has an interesting consequence. In this case, $s_o(G, K)$ is just the number of connected spanning subgraphs with an odd number of edges, and $s_e(G, K)$ is the number with an even number of edges. We now denote by $s_o(G)$ the number of spanning connected subgraphs with odd cycle rank, and by $s_e(G)$ the number with even cycle rank. (Recall that the cycle rank of a connected graph with n vertices and m edges is $m - n + 1$.) While it is not always true that $s_o(G) = s_o(G, V)$ and $s_e(G) = s_e(G, V)$, the two pairs of numbers are always equal–that is, $\{s_o(G), s_e(G)\} = \{s_o(G, V), s_e(G, V)\}$. This equality, together with the fact that the sign of $\sigma_V(G)$ is $(-1)^{m-n+1}$, yields the result.

Corollary 3.7 *For every graph G, $\sigma_V^*(G) = s_e(G) - s_o(G)$.*

Since the all-terminal domination $\sigma_V^*(G) > 0$ for every connected graph without loops, it follows that, for such a graph, $s_e(G) > s_o(G)$.

We next look at the parity of $\sigma_V^*(G)$. It follows from Corollary 3.7 that a graph G has an odd number of connected spanning subgraphs if and only if $\sigma_V^*(G)$ is odd. The following theorem of the authors [16] says precisely when this happens.

Theorem 3.8 *The all-terminal domination of a connected graph G is odd if and only if G is bipartite.*

We note in passing that Turán's theorem implies that $K_{r,r}$ and $K_{r,r+1}$ are the only simple connected graphs with their numbers of vertices and edges that have odd all-terminal domination.

Spanning trees revisited

In the case of all-terminal domination of a graph G, the K-trees are just the spanning trees of G. However, if $K \neq V$, then not every spanning tree is a K-tree, and so there is no obvious connection between $\sigma_K^*(G)$ and the set of spanning trees. However, perhaps surprisingly, $\sigma_K^*(G)$ is equal to the number of spanning trees of a certain type.

To establish this connection, we require some preliminaries. Let G be a connected graph and let T be one of its spanning trees. If e is an edge of G in T, then the set of edges of G that join the two components of $T - e$ is called the *fundamental cutset determined by e with respect to T*. Similarly, if e is an edge of G not in T, then $T \cup \{e\}$ has exactly one cycle, called the *fundamental cycle determined by e with respect to T*.

We now assume that the set E of all edges in G has a linear order $<$. An edge e in T is called *internally active* in T if it is the first edge (under $<$) in its fundamental cutset, and an edge e not in T is called *externally active* in T if it is the first edge (under $<$) in its fundamental cycle. Note that, of the m edges in G, any number of the $n - 1$ edges in T can be internally active, and any number of the $m - n + 1$ edges not in T can be externally active.

Continuing with our notation, where $\mathcal{T}$ is the set of spanning trees of G, we let $\mathcal{T}_{i,j}$ be the set of those having i internally active edges and j that are externally active. Further, we let $\tau(G)$ and $\tau_{i,j}(G)$ denote the cardinalities of these sets.

Now consider a set K of vertices of the graph G, with $<$ an ordering of its edge-set E. Any spanning tree of G that is not already a K-tree can be reduced to a K-tree by repeatedly pruning leaves that are not in K. Thus, each spanning tree contains a unique K-tree T^K. Let $\mathcal{T}_<(G, K)$ consist of those spanning trees T of G that have no externally active edges, and for which every internally active edge is in its K-tree T^K, and let $\tau_<(G, K)$ be its cardinality. We now have the following theorem, proved in [16].

Theorem 3.9 *For any set K of vertices in a graph G with a linear order $<$ on its edge-set E,*

$$\sigma_K^*(G) = \tau_<(G, K).$$

As a special case of this theorem, we let K consist of the two vertices of the smallest edge under the ordering $<$. Then $\sigma_K^*(G)$ is the number of spanning trees of G with just one internally active edge and no externally active edges. At the other extreme, we have the following corollary in the all-terminal case. Here, $\mathcal{T}_<(G)$ denotes the set of spanning trees of G with no externally active edges, and $\tau_<(G)$ denotes its cardinality.

Corollary 3.10 *For every graph G, $\sigma_V^*(G) = \tau_<(G)$.*

It follows from these results that the number $\tau_<(G)$ of spanning trees without any externally active edges, and the number $\tau_<(G, K)$ for a given K, are both independent of the linear order $<$ chosen. Indeed, in his study of the chromatic polynomial of a graph, Tutte [49] noted this fact for all of the numbers $\tau_{i,j}(G)$. (Recall that the chromatic polynomial $P(G; \lambda)$ of a graph G gives the number of proper λ-colourings of G). As a consequence of a result of Tutte for the number of spanning trees with exactly one internally active edge and no externally active edges (as well as for the number with these properties reversed), we have the following corollary.

Theorem 3.11 *For any edge vw of a connected graph G,*

$$\sigma^*_{\{v,w\}}(G) = |P(G; \lambda)/(\lambda - 1)|_{\lambda=1} .$$

An immediate consequence of this corollary is that (with $V(e)$ as the pair of end-points of the edge e) if G is connected, then $\sigma^*_{V(e)}(G)$ has the same value for all its edges. Further, Whitney [52] showed that if $\phi_i(G)$ is the number of spanning forests of G with i components and no externally active edges, then

$$P(G; \lambda) = \sum_{i=1}^{n} (-1)^{n-i} \phi_i(G) \lambda^i .$$

(An edge of a forest is externally active if it is externally active for some tree in the forest.) Since a spanning forest is a spanning tree when $i = 1$, we have the following result for the all-terminal domination of a graph.

Theorem 3.12 *For any connected graph G,*

$$\sigma^*_V(G) = |P(G; \lambda)/\lambda|_{\lambda=0}| .$$

Satyanarayana and Tindell [46] introduced a generalization $P(G, K; \lambda)$ of the chromatic polynomial that features a given set K of vertices in a graph. As one would hope, $P(G, V; \lambda) = P(G; \lambda)$ and, like the classical polynomial, its terms always alternate in sign. It has a number of other interesting properties, but the one of most interest to us here is the following generalization of Theorem 3.12:

For any connected graph G and any subset K of its vertices,

$$\sigma^*_K(G) = |P(G, K; \lambda)/\lambda|_{\lambda=0} .$$

An interesting connection between $\sigma^*_K(G)$ and the number of orientations of G of a certain type was established by Satyanarayana and Procesi-Ciampi [44]. Given a set K of vertices in a graph G and a vertex s in K, we say that an orientation of G is a *K-orientation rooted at s* if the in-degree of s is 0 and all of the vertices of out-degree 0 are in K. Satyanarayana and Procesi-Ciampi showed that the number of such orientations always equals $\sigma^*_K(G)$, and is thus independent of the root selected from K.

4. Computational complexity of reliability

Valiant [50] proved that computing the K-terminal reliability $R_K(G)$ of a graph G is in general NP-hard, and Provan [37] proved that this is the case even for planar graphs. These results motivated the search for classes of graphs that admit polynomial algorithms for the computation of the K-terminal reliability (see, for example, [3], [20] and [36]).

A special case of the K-terminal problem arises when no edges in a graph G fail, but the vertices not in a specified subset K fail with known probabilities. The *K-terminal connectedness reliability* of G is the probability that the surviving vertices induce a subgraph in which all of the vertices in K lie in a single component. AboElFotoh and Colbourn [1] showed that this problem is also NP-hard in general, and that it remains so even for the quite restricted families of chordal graphs (those in which each cycle of length greater than 3 has a chord) and comparability graphs (those that can be oriented transitively).

Another reliability problem, called the *residual vertex connectedness reliability* of a graph, is concerned with the situation where the edges do not fail but the vertices fail independently of one another. The network is considered to be in an operating state if the surviving vertices induce a connected subgraph, and the probability of the network being in an operating state is the residual vertex reliability of the graph. We note that this problem is not a special case of any of the K-terminal reliability problems. Those K-terminal models described before constitute coherent systems. The residual vertex connectedness model is not coherent since a supergraph of a connected graph may be disconnected. Computing the residual vertex connectedness reliability is NP-hard and remains so for split graphs as well as planar and bipartite graphs [48]. Efficient algorithms for computing this reliability for restricted families of graphs are given in [21]. These restricted classes include the complements of planar graphs, trees, series-parallel graphs, graphs of tree width at most k (sometimes called partial k-trees), directed path graphs and permutation graphs.

5. Synthesis of reliable networks

In the most general sense, a network design or synthesis problem requires the construction of a network that meets certain pre-set specifications. Very often the problem takes the form of a constrained optimization problem, which may appear to be quite innocuous at first sight. For example, the problem of finding those graphs (assumed to be simple throughout this section) having given numbers of vertices and edges, but with the maximum possible connectivity, was posed by Berge [12] and solved by Harary [26]. On the other hand, the corresponding problem for the maximum number of spanning trees has thus far evaded a complete solution. Special cases of the problem have been solved, including for values of m close to the extremes of $n-1$ and $\frac{1}{2}n(n-1)$ for connected graphs of order n, as well as when m equals the number of edges in a complete multipartite graph in which the parts differ in size by at most 1.

This section concerns the variation of this question for all-terminal reliability – that is, we are interested in finding a graph G with n vertices and m edges for which the all-terminal reliability $R_V(G)$ is as large as possible. Here we have to be careful since there are some formulations of the problem that do not admit a solution. For convenience, we let $\mathcal{G}_{n,m}$ be the set of graphs with n vertices and m edges. The graphs G that we consider are those in which no vertices fail and all edges operate with the same probability p, independent of one another. Let $\eta_i(G)$ denote the number of connected spanning subgraphs of G with i edges. Then

$$R_V(G, p) = \sum_{i=0}^{m} \eta_i(G) p^i (1 - p)^{m-i}.$$

Thus, the reliability depends on p as well as on the structure of the system, and so, even though the restriction to fixed numbers of vertices and edges constrains the collection of candidates to be finite, there may be no one system that maximizes the function for all values of p.

A graph G in $\mathcal{G}_{n,m}$ for which the reliability is a maximum for all values of p is said to be *uniformly most reliable*. Clearly, if there is a graph G that maximizes each of the numbers $\eta_i(H)$ over all graphs H in $\mathcal{G}_{n,m}$, then G is uniformly most reliable. However, it is not known whether this condition is necessary. Nevertheless, most approaches taken thus far have been concerned with the optimization of the individual values of η_i. The following theorem on comparative all-terminal reliability plays an important role in applying this approach; it has a straightforward proof using calculus.

Theorem 5.1 *Let G and H be graphs with n vertices and m edges.*

- *If $\eta_j(G) > \eta_j(H)$ for some $j < m$, and if $\eta_i(G) = \eta_i(H)$ for all $i < j$, then there exists $r < 1$ such that $R_V(G, p) > R_V(H, p)$ for all p with $0 < p < r$.*
- *If $\eta_j(G) > \eta_j(H)$ for some $j < m$, and if $\eta_i(G) = \eta_i(H)$ when $j < i \leq m$, then there exists $r < 1$ such that $R_V(G, p) > R_V(H, p)$ for all p with $r < p < 1$.*

An important consequence of this theorem is the following corollary (see [9]).

Corollary 5.2 *Let G be uniformly most reliable among the graphs with n vertices and m edges, and let $l = \lfloor 2m/n \rfloor$. Then, among all graphs with n vertices and m edges, G has*

- *the maximum number of spanning trees;*
- *the maximum edge-connectivity – namely, l;*
- *the minimum number of edge-cuts of size l among graphs with this edge-connectivity.*

We continue to consider graphs with fixed numbers n of vertices and m of edges. Since any such graph with edge-connectivity $\lambda = \lfloor 2m/n \rfloor$ must have minimum degree $\delta = \lambda$, the number of edge-cuts of size λ must be at least as large as the number of vertices of minimum degree δ. Thus, there must be at least $(1 + \lfloor 2m/n \rfloor)n - 2m$ such vertices, since that is the least number with degree δ that such a graph can have. Thus, if $\lambda(G) = \lfloor 2m/n \rfloor$, and if G has this number of vertices of degree λ, then G satisfies the last two properties in Corollary 5.2.

A graph with $\lambda(G) = \lfloor 2m/n \rfloor$ is called *super-*λ if every edge-cut of size λ is the set of edges at some vertex. In [8] and [9], it is shown that when $m \geq \lfloor \frac{6}{5}n \rfloor$, such graphs always exist and, furthermore, there is one with precisely $(1+\lfloor 2m/n \rfloor)n-2m$ vertices of minimum degree. Additionally, it is shown that when $n \leq m < \lfloor \frac{6}{5}n \rfloor$, none of the graphs G with $\lambda(G) = \lfloor 2m/n \rfloor$ is super-λ. Interestingly, the structure of those graphs with $n \leq m < \lfloor \frac{6}{5}n \rfloor$ and $\lambda(G) = \lfloor 2m/n \rfloor$ that have the minimum number of edge-cuts of size λ were also given. They are the graphs obtained from a cubic graph or multigraph of order $2r$ by inserting the remaining $n - 2r$ vertices into the $3r$ edges so that the numbers on those edges are as nearly equal as possible.

Independently, Kelmans [30] and Myrvold *et al.* [34] proved that uniformly most reliable graphs do not always exist. They showed that, if p is sufficiently close to 1, then for $m = 2k^2 - 2k - 1$ when $n = 2k$ (the even case) and for $m = 2k^2 - 3$ when $n = 2k + 1$ (the odd case) with $k \geq 3$, there is a unique graph that maximizes the all-terminal reliability $R_V(G, p)$. However, among graphs with n vertices and m edges, it does not have the maximum number of spanning trees. Thus, it follows from Theorem 5.1 and Corollary 5.2 that there is no uniformly most reliable graph in these cases.

Some progress has been made on determining whether (for given n and m) there exists a uniformly most reliable graph, using results on the maximum number of spanning trees in graphs in this family. For example, Kelmans and Chelnokov [31] and Shier [47] independently showed that, for $m \geq \frac{1}{2}n(n-2)$, the graph obtained by removing a set of independent edges from K_n has the maximum number of spanning trees among all graphs in $\mathcal{G}_{n,m}$. Using a reliability-increasing operation known as *swing surgery* (first introduced by Kelmans [30]), Satyanarayana, Schoppman and Suffel [45] showed that these graphs are uniformly most reliable. For graphs with nearly as many edges – namely, for $\frac{1}{2}(n^2 - 3n + 4) \leq m < \frac{1}{2}(n^2 - 2n)$ and in some special cases when $m = \frac{1}{2}(n^2 - 3n)$ or $\frac{1}{2}(n^2 - 3n + 2)$ – graphs with the maximum number of spanning trees have been identified (see [24], [35]), although no progress has been made on the uniformly most reliable problem for these pairs.

As for sparse cases (those with between $n - 1$ and $n + 3$ edges), the graphs having the greatest number of spanning trees have been determined and shown to be uniformly most reliable (see [15], [51]). For $n - 1$ edges, they are trees; for n edges, cycles; for $n + 1$ edges, three internally disjoint paths as nearly equal in length as possible between a pair of vertices; for $n + 2$ edges, a particular subdivision of K_4; and for $n + 3$ edges, a particular subdivision of $K_{3,3}$. In all of these cases, the graphs maximize all of the η_is.

Between the sparse and nearly full extremes, there are few results. Cheng [18] and Rodriguez and Petingi [39] proved that complete multipartite graphs in which the parts differ in size by at most 1 have the most spanning trees among the graphs with those numbers of vertices and edges. However, it is not known whether these graphs are uniformly most reliable.

A fair amount of work has been done on determining graphs that maximize the reliability when p is close to 1. As was already noted, such graphs must have $\lambda = \lfloor 2m/n \rfloor$ and the minimum possible number of edge cuts of size λ. Indeed, these graphs maximize each of the coefficients η_i for $m - \lambda \leq i \leq m$.

In an attempt to determine those graphs that maximize other η_is, some researchers have concentrated on a deeper study of super-λ graphs. (Recall that, as previously noted, if $m \geq \lfloor \frac{6}{5}n \rfloor$, a graph with $\lambda = \lfloor 2m/n \rfloor$ that minimizes the number of edge-cuts of size λ must be super-λ.) To this end, Esfahanian and Hakimi [23] introduced the *restricted edge-connectivity* λ', defined to be the minimum size of an edge-cut that is not the set $E(v)$ of edges at some vertex v. It can be shown that a graph is super-λ if and only if $\lambda(G) = \delta(G) < \lambda'(G)$. Consequently, the restricted edge-connectivity quantifies the super-λ property; that is, the minimum possible value of $\lambda'(G)$ is the minimum degree of an edge (defined to be 2 less than the sum of the degrees of its endpoints). A super-λ graph is called *optimal* if $\lambda'(G)$ equals the minimum degree of an edge. Esfahanian and Hakimi showed that, for p sufficiently close to 1, the optimal super-λ graphs are the most reliable of the super-λ graphs.

Meng and Ji [32] extended these ideas to graphs that maximize the minimum size of an edge-cut that produces only components of order 3 or more; other extensions and refinements of these ideas can be found in [6], [17], [38] and [53]. We emphasize that, even in those cases in which there are no uniformly most reliable graphs, the study of the graphs that optimize the reliability for p close to 1 is of practical importance, so a continuation of this study is justified from a pragmatic point of view.

We conclude this section with a few comments on the extension of the problem of reliability to multigraphs. In the sparse cases previously discussed, it was observed by Gross and Saccoman [25] that the uniformly most reliable graphs in $\mathcal{G}_{n,m}$ remain so when all multigraphs with the same numbers of vertices and edges are considered. On the other hand, the simpler problem of determining for multigraphs whether the result of removing an independent set of edges from a complete graph has the maximum number of spanning trees remains an open question. Finally, Myrvold [33] reported that S. Trace proved the surprising result that, for certain n and m and p close to 1, the reliability over the class of all planar multigraphs is maximized by a proper multigraph and not by a simple graph.

6. Other measures of vulnerability

There are two entirely different approaches to measuring the susceptibility of a network to damage. The use of probability models is the standard approach in reliability theory. A different approach uses graph invariants as deterministic measures, which we call *vulnerability theory*. This approach is fairly well known in the graph theory literature; in fact, there is a section in Berge's classic book [12] that describes the use of connectivity as a measure of a network's susceptibility to damage. Here we give a few examples and refer to survey papers (for example, [7], [13] and [14]) for more details. In this section, all graphs are assumed to be simple.

Having decided on some 'reasonable' parameter that can be used as a measure of vulnerability, one can then define an optimization problem that relates to the design of invulnerable networks. For example, if the edge-connectivity λ is used as a measure of vulnerability, then an appropriate design objective is to create graphs G that have the maximum edge-connectivity among all graphs with given numbers of vertices and edges. There are many solutions to this particular optimization problem; for example, every union of edge-disjoint Hamiltonian cycles on a set of n vertices clearly has this property. As when designing graphs that have optimal reliability properties, we call the problem of optimizing a vulnerability parameter over a given class of graphs a *synthesis problem*. In a previous section, we discussed Harary's solution to the problem of maximizing the connectivity for given values of n and m, and the value was determined to be $\lfloor 2m/n \rfloor$. Since $\kappa(G) \leq \lambda(G) \leq \lfloor 2m/n \rfloor$ for any graph G in $\mathcal{G}_{n,m}$, it follows that maximizing the connectivity also maximizes the edge-connectivity. However, it can easily be shown that the converse is not true.

There are a number of other graph invariants that are reasonable measures of vulnerability; some of these can be found in [7], [13] and [14]. One example is the *toughness* $\tau(G)$, defined to be the minimum of all ratios $|S|/k(G-S)$, taken over all sets S of vertices in G for which $k(G-S)$, the number of components of $G-S$, is at least 2 (this invariant was introduced by Chvátal [19] in the study of Hamiltonicity). The corresponding synthesis problem of maximizing $\tau(G)$ for given n and m was partially solved by Doty [22]. Another measure of vulnerability is the *binding number* $\beta(G)$, defined as the minimum of $|N(S)|/|S|$ taken over all non-empty sets S of vertices for which the set of neighbours is not all of V. This concept was introduced by Woodall at about the same time that Chvátal introduced toughness, and these two parameters are the subject of Chapter 8 of this book.

The last vulnerability measure that we mention here is the *integrity* $\iota(G)$, defined as the minimum of $|S| + m(G-S)$, taken over all subsets S of the vertex-set V, where $m(G-S)$ is the largest order of a component of $G-S$. We note that in contrast to toughness, the sets considered need not be disconnecting sets. However, it is easily shown that, for any connected graph other than the complete graph, a minimizing set for $\iota(G)$ is in fact a disconnecting set. Integrity was first introduced as a vulnerability measure by Barefoot, Entringer and Swart [7]; see also the related work by Bagga *et al.* [4], [5]. The synthesis problem for integrity was solved by Beineke, Goddard and Lipman [11]. Other vulnerability results can be found in an overview by Beineke [10].

Acknowledgement

This chapter is dedicated to the memory of the first author, Frank Boesch, a dear friend and valued colleague, who died after the work was begun.

References

1. H. M. F. AboElFotoh and C. J. Colbourn, Computing 2-terminal reliability for radio broadcast networks, *IEEE Trans. Reliab.* **38** (1980), 538–555.
2. A. Agrawal and R. E. Barlow, A survey of network reliability and domination theory, *Oper. Res.* **32** (1984), 478–492.
3. S. Arnborg and A. Proskurowski, Linear time algorithms for NP-hard problems restricted to K-trees, *Discrete Appl. Math.* **23** (1989), 11–24.
4. K. Bagga, L. Beineke, M. Lipman and R. Pippert, On the edge integrity of graphs, *Congr. Numer.* **60** (1987), 141–144.
5. K. Bagga, L. Beineke, M. Lipman and R. Pippert, The separation sequence and network reliability, *Congr. Numer.* **66** (1988), 293–300.
6. C. Balbuena, P. Garcia-Vasquez and X. Marcote, Sufficient conditions for λ'- connectivity in graphs with girth g, *J. Graph Theory* **52** (2006), 73–86.
7. C. Barefoot, R. Entringer and H. Swart, Vulnerability graphs – a comparative survey, *J. Combin. Math. Combin. Comput.* **1** (1987), 13–22.
8. D. Bauer, F. Boesch, C. Suffel and R. Tindell, Combinatorial optimization problems in the analysis and design of probabilistic networks, *Networks* **15** (1985), 257–271.
9. D. Bauer, F. Boesch, C. Suffel and R. Van Slyke, On the validity of a reduction of a reliable network design to a graph extremal problem, *IEEE Trans. Circuits and Systems* **34** (1987), 1579–1581.
10. L. Beineke, Explorations into graph vulnerability, *Graph Theory, Combinatorics, and Applications*, Vol. 1, Kalamazoo (1988) (eds. Y. Alavi *et al.*), Wiley (1991), 159–177.
11. L. Beineke, W. Goddard and M. Lipman, Graphs with maximum edge-integrity, *Ars Combin.* **46** (1997), 119–127.
12. C. Berge, *The Theory of Graphs and its Applications*, Methuen, 1962.
13. F. Boesch, Synthesis of reliable networks – a survey, *IEEE Trans. Reliab.* **35** (1986), 240–246.
14. F. Boesch, An overview of graph theory applications to network reliability vulnerability, *Graph Theory Notes of New York* **18**, N. Y. Academy of Sciences, 1989, 29–37.
15. F. T. Boesch, X. Li and C. Suffel, On the existence of uniformly optimally reliable networks, *Networks* **21** (1991), 181–194.
16. F. T. Boesch, A. Satyanarayana and C. L. Suffel, Some alternate characterizations of reliability domination, *Probab. Engrg. Inform. Sci.* **4** (1990), 257–276.
17. P. Bonsma, N. Ueffing and L. Volkmann, Edge-cuts leaving components of order at least three, *Discrete Math.* **256** (2002), 431–439.
18. C. S. Cheng, Maximizing the total number of trees in a graph; two related problems in graph theory and design, *J. Combin. Theory (B)* **31** (1981), 240–248.
19. V. Chvátal, Tough graphs and Hamiltonian circuits, *Discrete Math.* **5** (1973), 215–228.
20. C. J. Colbourn, *The Combinatorics of Network Reliability*, Oxford Univ. Press, 1987.
21. C. J. Colbourn, A. Satyanarayana, C. Suffel and K. Sutner, Computing residual connectedness reliability for restricted networks, *Discrete Appl. Math.* **44** (1993), 221–232.
22. L. Doty, A large class of maximally tough graphs, *OR Spectrum* **13** (1991), 147–151.

23. A. Esfahanian and S. Hakimi, On computing a conditional edge connectivity of a graph, *Inform. Process. Letters* **27** (1988), 195–199.
24. B. Gilbert and W. Myrvold, Maximizing spanning trees in almost complete graphs, *Networks* **30** (1997), 23–30.
25. D. Gross and J. T. Saccoman, Uniformly optimally reliable networks, *Networks* **31** (1998), 217–225.
26. F. Harary, The maximum connectivity of a graph, *Proc. Nat. Acad. Sci. USA* **48** (1962), 1142–1146.
27. F. Harary, *Graph Theory*, Addison-Wesley, 1969.
28. A. B. Huseby, A unified theory of domination and signed domination with applications to exact reliability computations, *Statistical Research Report*, Institute of Mathematics, Univ. of Oslo, Norway, 1984.
29. A. B. Huseby, Domination theory and the Crapo β-invariant, *Networks* **19** (1989), 135–149.
30. A. K. Kelmans, On graphs with randomly deleted edges, *Acta. Math. Acad. Sci. Hungar.* **37** (1981), 77–88.
31. A. K. Kelmans and V. M. Chelnokov, A certain polynomial of a graph and graphs with an extremal number of trees, *J. Combin. Theory (B)* **16** (1974), 197–214.
32. J. Meng and Y. Ji, On a kind of restricted edge connectivity of graphs, *Discrete Appl. Math.* **117** (2002), 183–193.
33. W. Myrvold, Reliable network synthesis: Some recent developments, *Combinatorics, Graph Theory, and Algorithms*, Vol. II (1999), 651–660.
34. W. Myrvold, K. Cheung, L. Page and E. Perry, Uniformly most reliable networks do not always exist, *Networks* **21** (1991), 417–419.
35. L. Petingi, F. T. Boesch and C. L. Suffel, On the characterization of graphs with maximum number of spanning trees, *Discrete Math.* **179** (1998), 155–166.
36. T. Politof and A. Satyanarayana, A survey of different algorithms for the reliability analysis of planar networks, *IEEE Trans. Reliab.* **35** (1986), 252–259.
37. J. S. Provan, The complexity of reliability computations on planar and acyclic graphs, *SIAM J. Comput.* **15** (1986), 694–702.
38. J. Qu, Edge cuts leaving components of order at least m, *Discrete Math.* **305** (2005), 365–371.
39. J. Rodriguez and L. Petingi, Graphs with maximum number of spanning trees and optimally reliable graphs, *J. Combin. Optim.* **1** (1996), 57–68.
40. A. Satyanarayana, *Multi-terminal Network Reliability*, Technical Report ORC 80-6, OR Center, Univ. of California, Berkeley, 1980.
41. A. Satyanarayana, A unified formula for analysis of some network reliability problems, *IEEE Trans. Reliab.* **3** (1982), 23–32.
42. A. Satyanarayana and M. K. Chang, Network reliability and the factoring theorem, *Networks* **13** (1983), 107–120.
43. A. Satyanarayana and A. Prabhakar, A new topological formula algorithm for reliability analysis of complex systems, *IEEE Trans. Reliab.* **27** (1978), 82–100.
44. A. Satyanarayana and R. Procesi-Ciampi, *On Some Acyclic Orientations of a Graph*, Technical Report ORC 81-11, OR Center, Univ. of California, Berkeley, 1981.
45. A. Satyanarayana, L. Schoppman and C. L. Suffel, A reliability-improving transformation with applications to network reliability, *Networks* **22** (1992), 209–216.
46. A. Satyanarayana and R. Tindell, Chromatic polynomials and network reliability, *Discrete Math.* **67** (1987), 57–79.
47. D. R. Shier, Maximizing the number of spanning trees in a graph with n nodes and m edges, *J. Res. Nat. Bur. Standards (B)* **78** (1974), 193–196.

48. K. Sutner, A. Satyanarayana and C. Suffel, The complexity of the residual node connectedness reliability problem, *SIAM J. Comput.* **20** (1991), 149–155.
49. W. T. Tutte, A contribution to the theory of chromatic polynomials, *Canad. J. Math.* **6** (1954), 80–91.
50. L. Valiant, The complexity of enumeration and reliability problems, *SIAM J. Comput.* **8** (1979), 410–421.
51. G. Wang, A proof of Boesch's conjecture, *Networks* **24** (1994), 277–284.
52. H. Whitney, A logical expansion in mathematics, *Bull. Amer. Math. Soc.* **38** (1932), 572–579.
53. J. Xu and K. Xu, On restricted edge-connectivity of graphs, *Discrete Math.* **243** (2002), 291–298.

12

Connectivity algorithms

ABDOL-HOSSEIN ESFAHANIAN

This chapter presents an exposition of how connectivity algorithms have advanced over the years. Most of these algorithms work by making a number of calls to a max-flow subroutine. As these calls determine the bulk of the computation, attempts have been made to minimize the number of such calls.

1. Introduction

A variety of algorithms for the computation of the vertex-connectivity $\kappa(G)$ and the edge-connectivity $\lambda(G)$ of a graph G have been developed over the years. Most of these algorithms work by solving a number of maximum-flow problems (see Section 3 of the Preliminaries chapter); in other words, these algorithms compute the desired connectivity by making a number of calls to a max-flow subroutine. The major part of the computation in such an algorithm comes from these calls, and therefore attempts have been made to make the number of max-flow calls as small as possible.

Even and Tarjan [6] were among the first to present a max-flow based connectivity algorithm. Subsequent results include the work of Schnorr [25], Kleitman [21], Galil [10], [11], Esfahanian and Hakimi [3], Matula [23], Mansour and Schieber [22] and Henzinger, Rao and Gabow [17]. The problem of determining whether $\kappa(G)$ (or $\lambda(G)$) is greater than a prescribed value, without computing its actual value, has been studied by Tarjan [26], Mansour and Schieber [22] and Gabow [9].

In this chapter, we explain how the computation of connectivities can be reduced to solving a number of max-flow problems. We then give an exposition of the advancement of the connectivity algorithms over the years. A brief review of the literature is given in the later sections, along with some discussion. We begin with the edge-connectivity of graphs, then turn to the arc-connectivity of digraphs, and then the vertex-connectivity of graphs. Because the vertex-connectivity or edge-connectivity of a disconnected or trivial graph is 0, we assume throughout that our graphs are connected and non-trivial. Throughout this chapter, n denotes the order of a graph or a digraph, and m is the number of edges or arcs.

2. Computing the edge-connectivity

In this section we discuss a progression of algorithms for the edge-connectivity $\lambda(G)$ of a graph G. Initially, this is done via the local edge-connectivity $\lambda(v, w)$ of any pair of vertices v and w in G. By definition, $\lambda(v, w)$ is the minimum number of edges whose removal from G leaves v and w in different components and, by Menger's theorem, this equals the maximum number of edge-disjoint v–w paths. The edge-connectivity of a graph G, being the minimum number of edges in a set whose removal leaves a disconnected graph, is clearly the least edge-connectivity of a pair of its vertices:

$$\lambda(G) = \min\{\lambda(v, w) : v, w \in V\}.$$

The value of $\lambda(v, w)$ can be computed by solving a max-flow problem in a network (such as that described in the Preliminaries chapter). First, convert G into a network N_G by replacing each edge vw of G by the pair of arcs (v, w) and (w, v), and taking the weight of each arc to be 1. Thus, we have the following algorithm (see Even [5]).

Algorithm 1
Input: A graph $G = (V, E)$ and vertices s and t.
Output: $\lambda(s, t)$.

1. Convert G into the network N_G.
2. Assign s to be the source vertex and t the sink vertex.
3. Find a maximum-flow f in N_G.
4. Set $\lambda(s, t)$ equal to the value of the flow f.
5. Stop.

As Even [5] showed, the time complexity of the above algorithm is $O(nm)$. Provided that we have access to max-flow software, we can use this algorithm as a subroutine to compute $\lambda(v, w)$ for all pairs of vertices v and w, take the minimum of these quantities, and thus compute $\lambda(G)$. For a graph of order n, there are $\frac{1}{2}n(n-1)$ pairs of vertices for which we need to compute the edge-connectivity. It turns out, however, that we actually need to compute far fewer than this.

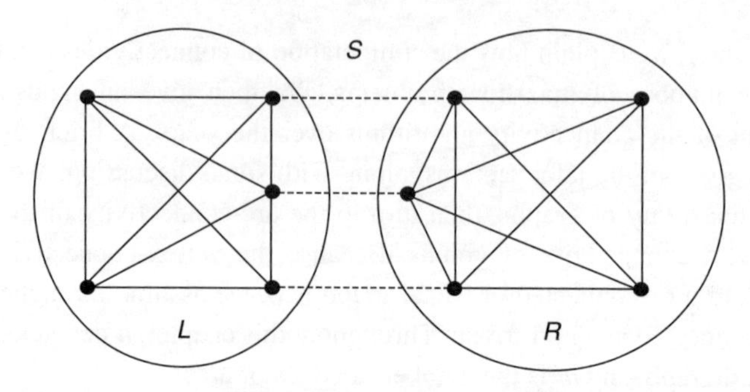

Fig. 1. A graph, a minimum edge-cut S (dotted edges), and its L and R sides

If S is a minimum edge-cut of a graph G, then $G - S$ has just two components. We arbitrarily denote the two sets of vertices by L and R (as indicated in Fig. 1) and call them the *sides* of S. A key observation is that, for any vertices v and w in different sides of S, $\lambda(v, w) = \lambda(G)$. Thus $\lambda(G)$ can be determined if we have an oracle as follows. First, select a vertex v in one side of S, and then, using the oracle, identify a vertex w in the other side. Next, compute $\lambda(v, w)$ using Algorithm 1. By the above observation, this is $\lambda(G)$. These ideas led Even and Tarjan [6] and Schnorr [25] to construct our next algorithm.

Algorithm 2
Input: A graph $G = (V, E)$.
Output: $\lambda(G)$.

1. Select a vertex $v \in V$.
2. Using Algorithm 1, compute $\lambda(v, w)$ for every $w \in V - \{v\}$.
3. Assign $\lambda(G) \leftarrow \min\{\lambda(v, w)\}$.
4. Stop.

This algorithm reduces the number of computations of $\lambda(v, w)$ from our earlier value of $\frac{1}{2}n(n-1)$ to $n - 1$, which is a significant reduction. If you keep staring at Fig. 2, you might notice that this algorithm computes $\lambda(G)$ if set V in Step 1 is replaced by any set Y that contains vertices from both L and R; such a set is called a *λ-covering* of G. Formally, a subset Y of the vertices in G, with $|Y| \geq 2$, is a *λ-covering* if Y contains a pair of vertices v and w, for which $\lambda(v, w) = \lambda(G)$. Clearly, the smaller the set, the fewer the calls to the max-flow subroutine. This observation and our next result led to new algorithms for computing the edge-connectivity.

It is well known that for any graph G, $\lambda(G) \leq \delta(G)$, the minimum vertex degree in G. What happens to the sizes of the sides of a minimum edge-cut (see Fig. 2) when $\lambda(G) < \delta(G)$? The significance of this question will become clear later. The following result (see [3]) answers that question. (Keep Fig. 2 in mind!)

Theorem 2.1 *Let G be a graph, and let L and R be the two sides of a minimum edge-cut S. If $\lambda(G) < \delta(G)$, then $|L| > \delta(G)$ and $|R| > \delta(G)$.*

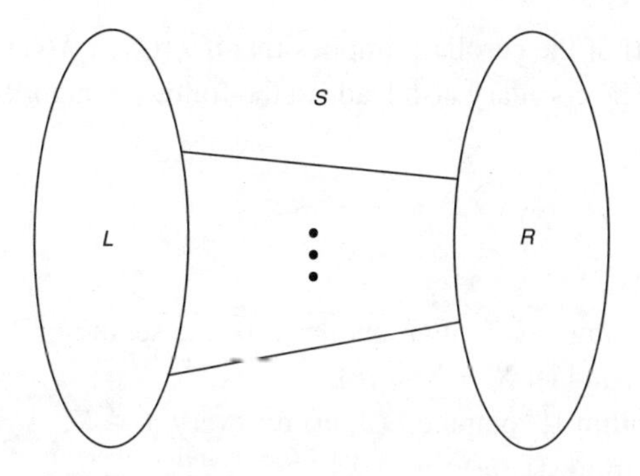

Fig. 2. A graph, a minimum edge-cut S and its L and R sides

Proof Let $L = \{v_1, v_2, \ldots, v_l\}$. On the one hand,

$$\deg v_1 + \deg v_2 + \cdots + \deg v_l \geq \delta l$$

while on the other hand,

$$\deg v_1 + \deg v_2 + \cdots + \deg v_l = 2|E(\langle L \rangle)| + |S|,$$

where $\langle L \rangle$ is the subgraph of G induced by L. Since $\langle L \rangle$ can have no more than $\frac{1}{2}l(l-1)$ edges, and by hypothesis $|S| < \delta$, it follows that $\delta l < l(l-1) + \delta$. Since $l = 1$ gives $\lambda(G) = \delta(G)$, we see that $|L| > \delta$. The same argument applies to R. ■

Before we discuss an application of this result, some observations are in order.

Corollary 2.2 *Let G be a graph with $\lambda(G) < \delta(G)$, and let L and R be the sides of some minimum edge-cut S. Then*

(a) *both L and R contain at least one vertex that is not incident to any edge in S;*
(b) *both L and R contain at least one non-leaf vertex of each spanning tree of G.* (See Fig. 3.)

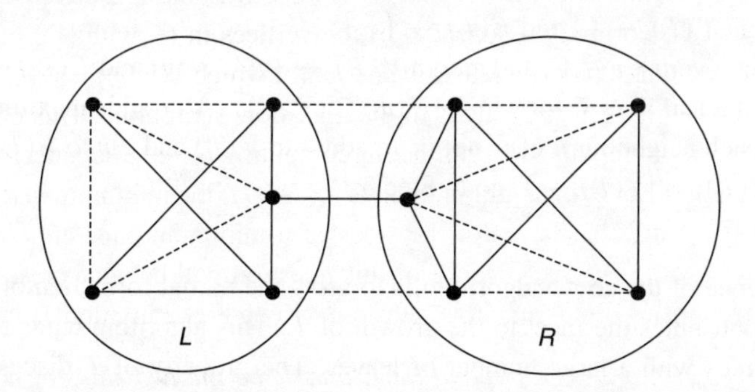

Fig. 3. A graph, a minimum edge-cut and a spanning tree (dotted edges)

Note that part (a) of the corollary implies that if $\lambda(G) < \delta(G)$, then the diameter of G is at least 3. The corollary also leads to the following algorithm.

Algorithm 3
Input: A graph $G = (V, E)$.
Output: $\lambda(G)$.

1. Select a spanning tree T of G, and let Y be the set of non-leaves of T.
2. Select $v \in Y$, and let $X = Y - \{v\}$.
3. Using Algorithm 1, compute $\lambda(v, w)$ for every $w \in X$.
4. Assign $c \leftarrow \min\{\lambda(v, w) : w \in X\}$.
5. Assign $\lambda(G) \leftarrow \min\{c, \delta(G)\}$.
6. Stop.

The correctness of this algorithm can be seen by noting that if $\lambda(G) < \delta(G)$ then c (in Step 4) equals $\lambda(G)$ and, regardless of this, Step 5 produces the correct value for $\lambda(G)$. Note also that the more leaves T has, the fewer are the calls required to Algorithm 1. However, as shown by Garey and Johnson [13], finding a spanning tree with the maximum number of leaves is NP-hard. Thus, the only savings that Algorithm 3 can guarantee is two fewer calls than Algorithm 2, since every non-trivial tree has at least two leaves.

In pursuit of even smaller λ-coverings, Esfahanian and Hakimi [3] discovered that the set of leaves of the spanning tree T produced by the next algorithm is a λ-covering of G, provided that $\lambda(G) < \delta(G)$. By part (b) of Corollary 2.2, this immediately implies that both L and R (as indicated in Fig. 2) contain both leaves and non-leaves of T. In other words, if Y is the set of all non-leaves of T, then both Y and $V - Y$ are λ-coverings of G, provided that $\lambda(G) < \delta(G)$. (For each vertex v in a graph G, $E(v)$ denotes the set of edges incident with v, and $N(v)$ denotes the set of neighbours of v.)

Algorithm 4
Input: A graph $G = (V, E)$.
Output: A spanning tree T.

1. Assign $V(T) \leftarrow \{\,\}$ and $E(T) \leftarrow \{\,\}$.
2. Select a vertex $v \in V$, and assign $V(T) \leftarrow \{v\} \cup N(v)$ and $E(T) \leftarrow E(v)$.
3. Select a leaf w in T for which $|N(w) \cap (V(G) - V(T))|$ is maximum.
4. For each neighbour u of w not in T, add u to $V(T)$ and uw to $E(T)$.
5. If $|E(T)| < |V(T)| - 1$, go to Step 3.
6. Stop.

The essence of the above algorithm is to grow the partial formation of H from a leaf that contributes the most to the growth of T. This algorithm tends to generate spanning trees with a large number of leaves. The property of T discussed above suggests the following algorithm for computing $\lambda(G)$.

Algorithm 5
Input: A graph $G = (V, E)$.
Output: $\lambda(G)$.

1. Use Algorithm 4 and generate a spanning tree T of G.
2. Let Y be the set of non-leaves of T, and let X be the smaller of Y and $V - Y$.
3. Select an arbitrary vertex v in X, and let $W = X - \{v\}$.
4. Using Algorithm 1, compute $\lambda(v, w)$ for every $w \in W$.
5. Assign $c \leftarrow \min\{\lambda(v, w) : w \in W\}$.
6. Assign $\lambda(G) \leftarrow \min\{c, \delta(G)\}$.
7. Stop.

The correctness of this algorithm should be evident from the discussion. Furthermore, since $|X| \leq \frac{1}{2}n$, it makes at most $\frac{1}{2}n$ calls to Algorithm 1.

Matula [23] improved further upon Algorithm 5 by making use of Theorem 2.1 and dominating sets; recall that a *dominating set* in a graph is a set X of vertices with the property that each vertex in the graph is either in X or adjacent to a vertex in X. The following result can be easily deduced from Theorem 1.2.

Corollary 2.3 *Let S be a minimum edge-cut in a graph G with $\lambda(G) < \delta(G)$. Then every dominating set in G is a λ-covering of G.*

This corollary suggests the following algorithm for computing the edge-connectivity.

Algorithm 6
Input: A graph $G = (V, E)$.
Output: $\lambda(G)$.

1. Select a dominating set X of G.
2. Select an arbitrary vertex $v \in X$, and let $W = X - \{v\}$.
3. Using Algorithm 1, compute $\lambda(v, w)$ for every $w \in W$.
4. Assign $c \leftarrow \min\{\lambda(v, w) : w \in W\}$.
5. Assign $\lambda(G) \leftarrow \min\{c, \delta(G)\}$.
6. Stop.

Clearly, this algorithm determines $\lambda(G)$ correctly. Furthermore, the smaller the dominating set, the fewer are the calls to Algorithm 1. While finding a smallest dominating set is NP-hard (see, for example, Garey and Johnson [13]), finding some dominating set is easy. The next algorithm provides a way of generating a 'small' dominating set; recall that the *neighbourhood* $N(X)$ of a set X of vertices is the union of the neighbourhoods of the vertices in X.

Algorithm 7
Input: A connected non-trivial graph $G = (V, E)$.
Output: A dominating set X.

1. Select a vertex $v \in V$, and let $X = \{v\}$.
2. If $V - (X \cup N(X)) = \{\,\}$ then Stop.
3. Select $w \in V - (X \cup N(X))$.
4. Assign $X \leftarrow X \cup \{w\}$.
5. Go to Step 2.

By using a dominating set X, as produced by this algorithm, and amortizing the cost of computing $\lambda(v, w)$ for the vertices in X, Matula [23] was able to bring down the overall complexity of computing $\lambda(G)$ to $O(nm)$. His algorithm is the fastest one known for determining the edge-connectivity of a graph.

3. Computing the arc-connectivity

We now turn our attention to computing the arc-connectivity of digraphs, and for simplicity we assume that D is a strongly connected and non-trivial digraph. Consider the representation of such a digraph D in Fig. 4, with an arbitrary minimum arc-cut S of minimum size.

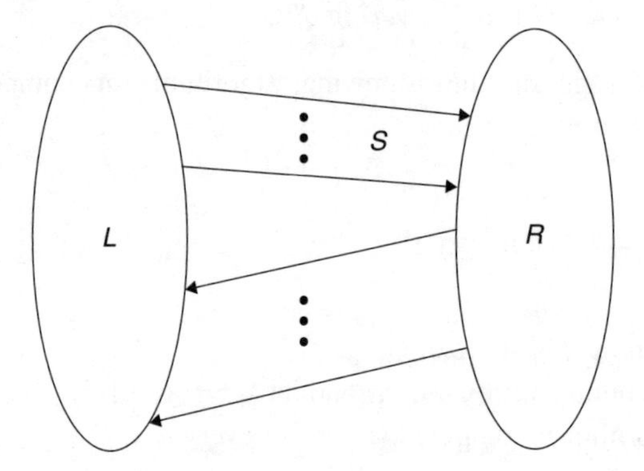

Fig. 4. A digraph, minimum arc-cut S, and its L and R sides

Note that, for any vertices $v \in L$ and $w \in R$, $\lambda(D) = \lambda(v, w)$. However, for general vertices v and w, it may be that $\lambda(D) \neq \lambda(w, v)$. Consequently, one cannot use Algorithm 2 directly to compute $\lambda(D)$, since the vertex selected in Step 1 of the algorithm may belong to R. This situation was remedied by the following result of Schnorr [25].

Theorem 3.1 *Let D be a digraph, S be a minimum arc-cut with left and right sides L and R and $W = \{w_1, w_2, \ldots, w_l\}$ be a λ-covering of D for S. Then*

$$\lambda(D) = \min\{\lambda(w_1, w_2), \lambda(w_2, w_3), \ldots, \lambda(w_{l-1}, w_l), \lambda(w_l, w_1)\}.$$

Proof Clearly, we may assume that the vertices in W are ordered so that $w_1 \in L$. Let r be the smallest index for which vertex w_r is in R; such a vertex must exist, since W is a λ-covering. It follows that $\lambda(D) = \lambda(w_{r-1}, w_r)$. ■

Based on this result, Schnorr [25] devised the following algorithm for computing the arc-connectivity of a digraph.

Algorithm 8
Input: A strongly connected digraph $D = (V, E)$.
Output: $\lambda(D)$.

1. Let $V = \{v_1, v_2, \ldots, v_n\}$.
2. Using Algorithm 1, compute $\lambda(v_i, v_{i+1})$ for $i = 1, 2, \ldots, n-1$, and $\lambda(v_n, v_1)$.
3. Assign $\lambda(D) \leftarrow \min\{\lambda(v_1, v_2), \lambda(v_2, v_3), \ldots, \lambda(v_{n-1}, v_n), \lambda(v_n, v_1)\}$.
4. Stop.

This algorithm reduces the number of calls from $n(n-1)$, as discussed earlier, to just n. Further improvements have been made, based on techniques similar to others used in computing the edge-connectivity of a graph; for example, there is a version of Theorem 2.1 for digraphs [3]. The existence of a λ-covering with at most half the vertices was also shown for every graph G for which $\lambda(G) < \delta(G)$ [3]. Mansour and Schieber [22] used the notion of dominating sets (as developed by Matula) to create two algorithms for computing the arc-connectivity. A combination of their algorithms yields one of order $O(\min\{mn, n\lambda^2\})$.

4. Computing the vertex-connectivity

In this section we cover some of the basic ideas involved in computing the connectivity (also called the vertex-connectivity) of a graph. Because the ideas are so similar for digraphs, we focus on the undirected case here. We begin by explaining how the computation of the connectivity can be reduced to solving a number of maximum-flow problems, which are discussed in the Preliminaries chapter.

Recall that $\kappa(G)$ is the minimum number of vertices whose removal from graph G leaves a disconnected or a trivial graph, while for non-adjacent vertices v and w, $\kappa(v, w)$ is the minimum number of vertices whose removal leaves a graph in which v and w lie in different components. (See Chapter 1 for a survey of graph connectivity.) Just as there is a connection between global and local edge-connectivities,

$$\lambda(G) = \min\{\lambda(v, w) : v, w \in V\},$$

so there is a similar one for vertex-connectivities,

$$\kappa(G) = \min\{\kappa(v, w) : v, w \in V\},$$

provided that when v and w are adjacent, we define $\kappa(v, w)$ in G to be one more than its value in $G - vw$. To avoid trivial cases, we consider only non-complete but connected graphs.

Even [5] showed that, for non-adjacent vertices v and w, the local connectivity $\kappa(v, w)$ can be found by solving a maximum-flow problem in a related network. Given a graph G and a pair s and t of non-adjacent vertices, let the network N_G be defined as follows. First, replace each edge vw of G by an arc from v to w and an arc from w to v to form the digraph D_G; arcs coming into s or going away from t should be removed, as they play no role in finding a max-flow. Then form N_G from D_G by successively replacing each vertex v_i other than s and t by a pair of vertices u_i and w_i, adding an arc from u_i to w_i, and replacing each arc into v_i with an arc into u_i and each arc out of v_i with an arc out of w_i (with the corresponding other ends). Finally, assign each arc weight 1. An example is shown in Fig. 5.

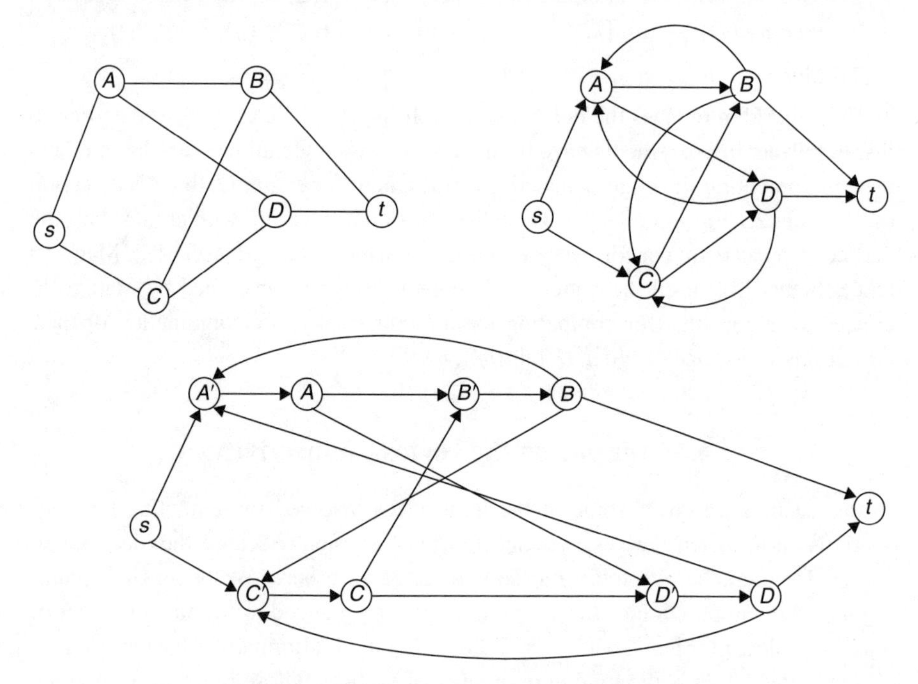

Fig. 5. A graph G, its digraph D_G and its associated network N_G

Here is Even's algorithm.

Algorithm 9

Input: A graph $G = (V, E)$ with non-adjacent vertices s and t.

Output: $\kappa(s, t)$.

1. Form the associated network N_G.
2. Find a maximum-flow function f for N_G.
3. Set $\kappa(s, t)$ equal to the value of f.
4. Stop.

Even [5] showed that the time complexity of his algorithm is $O(mn^{2/3})$. The algorithm can be used as a subroutine to compute $\kappa(v, w)$ for all pairs of non-adjacent vertices v and w. Just as Algorithm 1, a local result, leads to global versions (such as Algorithm 2), so Algorithm 9 leads to an analogous global algorithm.

Algorithm 10
Input: A graph $G = (V, E)$.
Output: $\kappa(G)$.

1. Using Algorithm 9, compute $\kappa(v, w)$ for every pair of non-adjacent vertices v and w.
2. Assign $\kappa(G) \leftarrow \min\{\kappa(v, w)\}$.
3. Stop.

This algorithm requires $\frac{1}{2}n(n-1) - m$ calls to Algorithm 9. However, there are algorithms for computing κ that require fewer calls to a maximum-flow algorithm, one of which we next describe.

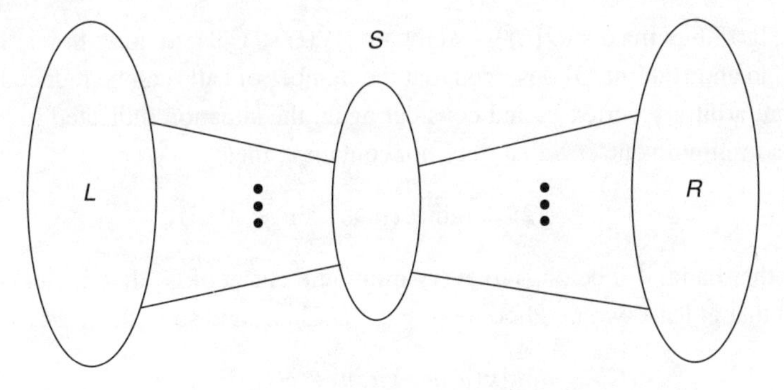

Fig. 6. A graph and a minimum cutset

Consider the representation of a graph G shown in Fig. 6, where S is a minimum cutset. (As before, we assume that G is connected and not complete.) Further, L is the set of vertices in one component of $G - S$ and R consists of the remaining vertices. Clearly, for vertices $v \in L$ and $w \in R$, $\kappa(v, w) = \kappa(G)$, and so one might be tempted to use the same idea as was used in Algorithm 2, by choosing an arbitrary vertex v and finding the minimum connectivity between v and a vertex not adjacent to it.

However, for this to work, there must be a minimum cutset that does not contain v. This can be achieved in the following way. Recall that in any graph G, $\kappa(G) \leq \delta(G)$. Hence, if S is a minimum cutset of G and X is any set of more than $\delta(G)$ vertices, then X has at least one vertex that is not in S. Therefore, $\kappa(G)$ can be computed as

$$\kappa(G) = \min_{v \in X}\{\min\{\kappa(v, w) : w \neq v\}\}.$$

Even and Tarjan [6] observed that, if we keep track of the minimum of the connectivities $\kappa(v, w)$ as these values are computed, then a set X of order $\kappa(G)+1$ suffices. Here is their algorithm.

Algorithm 11
Input: A graph $G = (V, E)$.
Output: $\kappa(G)$.

1. Assign $i \leftarrow 1, k \leftarrow n - 1$, and let $V = \{v_1, v_2, \ldots, v_n\}$.
2. For $j = i + 1, i + 2, \ldots, n$,
   2.1 If $i > k$, go to Step 4;
   2.2 If v_i and v_j are not adjacent, compute $\kappa(v_i, v_j)$ using Algorithm 9, and assign $k \leftarrow \min\{k, \kappa(v_i, v_j)\}$.
3. Assign $i \leftarrow i + 1$, and go to Step 2.
4. Assign $\kappa(G) \leftarrow k$.
5. Stop.

This algorithm makes $O((n - \delta(G) - 1)\kappa(G))$ calls to max-flow. However, Esfahanian and Hakimi [3] observed that the number of calls can be reduced further.

Take an arbitrary vertex v, and consider again the situation indicated in Fig. 6. If there is a minimum cutset S that does not contain v, then

$$\kappa(G) = \min\{\kappa(v, w) : w \neq v\}.$$

On the other hand, if v belongs to every minimum cutset of G, then it can be shown (see [3]) that at least two neighbours of v are not in S, and so in this case,

$$\kappa(G) = \min\{\kappa(u, w) : u, w \in N(v), u \neq w\}.$$

Since we do not know which of these situations applies to our arbitrary vertex v, both must be considered. This yields our next algorithm.

Algorithm 12
Input: A graph $G = (V, E)$.
Output: $\kappa(G)$.

1. Select a vertex v of minimum degree.
2. Compute $k_1 = \min\{\kappa(v, w) : w \neq v\}$.
3. Compute $k_2 = \min\{\kappa(u, w) : u, w \in N(v), u \neq w\}$.
4. Assign $\kappa(G) \leftarrow \min\{k_1, k_2\}$.
5. Stop.

This algorithm makes $O(n + \delta^2)$ calls to the maximum-flow algorithm; for a further refinement, see [3].

5. Concluding remarks

In this chapter, we have presented some of the key developments that have arisen in the pursuit of fast algorithms for computing graph connectivities. All of these algorithms are max-flow-based, but researchers have tried other methods as well. For example, Henzinger and Rao [16] developed a randomized algorithm for computing the connectivity. Algorithms have also been developed for deciding whether a graph is k-connected or l-edge-connected, some of which are not max-flow based. We conclude with a table that summarizes facts (for a graph G with n vertices and m edges) concerning algorithms related to connectivity.

Decision	Authors	Year	Complexity	Comments
		Edge-connectivity or Arc-connectivity		
$\lambda = 2, 3$	Tarjan [26]	1972	$O(m+n)$	uses depth-first search
λ (general)	Even and Tarjan [6]	1975	$O(\min\{m^{3/2}n, mn^{5/3}\})$	n calls to max-flow
λ (digraph)	Schnorr [25]	1979	$O(\lambda mn)$	n calls to max-flow
λ (general)	Esfahanian and Hakimi [3]	1984	$O(\lambda mn)$	at most $\frac{1}{2}n$ calls to max-flow
λ (digraph)	Esfahanian and Hakimi [3]	1984	$O(\lambda mn)$	at most $\frac{1}{2}n$ calls to max-flow
λ (general)	Matula [23]	1987	$O(mn)$	uses dominating sets
$\lambda = l$	Matula [23]	1987	$O(ln^2)$	
λ (digraph)	Mansour and Schieber [22]	1989	$O(mn)$	
$\lambda = l$	Gabow [9]	1991	$O(m + l^2 n \log(n/l))$	uses matroids
		Vertex-connectivity		
$\kappa = 2$	Tarjan [26]	1972	$O(m+n)$	uses depth-first search
$\kappa = 3$	Hopcroft and Tarjan [18]	1973	$O(m+n)$	uses 3-connected components
κ (general)	Even and Tarjan [6]	1975	$O(\kappa(n-\delta-1)mn^{2/3})$	max-flow based
$\kappa = k$	Even [4]	1975	$O(kn^3)$	max-flow based
κ (general)	Galil [10]	1980	$O(mn \min\{\kappa, n^{2/3}\})$	max-flow based
$\kappa = k$	Galil [10]	1980	$O(kmn \min\{k, n^{1/2}\})$	max-flow based
κ (general)	Esfahanian and Hakimi [3]	1984	$O((n-1 + \frac{1}{2}\delta(\delta-3))mn^{2/3})$	max-flow based
$\kappa = 4$	Kanevsky and Ramachandran [20]	1991	$O(n^2)$	—
κ (general)	Henzinger and Rao [16]	1996	$O(\kappa mn \log n)$	randomized algorithm

References

1. M. Becker, W. Degenhardt, J. Doenhardt, S. Hertel, G. Kaninke and W. Keber, A probabilistic algorithm for vertex connectivity of graphs, *Inf. Proc. Letters* **15** (1982), 135–136.
2. E. A. Dinic, Algorithm for solution of a problem of maximum flow in a network with power estimation, *Soviet Math. Dokl.* **11** (1970), 1277–1280.
3. A.-H. Esfahanian and S. L. Hakimi, On computing the connectivities of graphs and digraphs, *Networks* **14** (1984), 355–366.
4. S. Even, An algorithm for determining whether the connectivity of a graph is at least k, *SIAM J. Comput.* **4** (1975), 393–396.
5. S. Even, *Graph Algorithms*, Computer Science Press, 1979.
6. S. Even and R. E. Tarjan, Network flow and testing graph connectivity, *SIAM J. Comput.* **4** (1975), 507–518.
7. H. Frank and W. Chou, Connectivity considerations in the design of survivable networks, *IEEE Trans. on Circuit Theory* **CT-17** (1970), 486–490.
8. G. N. Frederickson, Ambivalent data structures for dynamic 2-edge-connectivity and k smallest spanning trees, *SIAM J. Comput.* **26** (1997), 484–538.
9. H. Gabow, A matroid approach to finding edge connectivity and packing arborescences, *J. Computer and System Science* **50** (1995), 259–273.
10. Z. Galil, Finding the vertex connectivity of graphs, *SIAM J. Comput.* **9** (1980), 197–199.
11. Z. Galil and G. F. Italiano, Reducing edge connectivity to vertex connectivity, *ACM SIGACT News* **22** (1991), 57–61.
12. Z. Galil and G. F. Italiano, Fully dynamic algorithms for 2-edge connectivity, *SIAM J. Comput.* **21** (1992), 1047–1069.
13. M. R. Garey and D. S. Johnson, *Computers and Intractability, A Guide to the Theory of NP-Completeness*, W. H. Freeman, 1979.
14. R. E. Gomory and T. C. Hu, Multi-terminal network flows, *J. Social and Industrial and Applied Math.* **9** (1961), 551–570.
15. D. Gusfield, Optimal mixed graph augmentation, *SIAM J. Comput.* **16** (1987), 599–612.
16. M. R. Henzinger and S. Rao, Faster vertex connectivity algorithms, *Proc. 37th IEEE Symp. on Foundations of Computer Science* (1996), 1–15.
17. M. R. Henzinger, S. Rao and H. N. Gabow, Computing vertex connectivity: new bounds from old techniques, *Proc. 37th IEEE Symp. on Foundations of Computer Science* (1996), 462–471.
18. J. Hopcroft and R. E. Tarjan, Dividing a graph into triconnected components, *SIAM J. Comput.* **2** (1973), 135–158.
19. T. Hsu, Undirected vertex-connectivity structure and smallest four-vertex-connectivity augmentation, *Proc. 6th ISAAC LNCS* **1004** (1995), 274–283.
20. A. Kanevsky and V. Ramachandran, Improved algorithms for graph four-connectivity, *J. Computer and System Science* **42** (1991), 288–306.
21. D. J. Kleitman, Methods for investigating connectivity of large graphs, *IEEE Trans. Circuit Theory* **16** (1969), 232–233.
22. Y. Mansour and B. Schieber, Finding the edge connectivity of directed graphs, *J. Algorithms* **10** (1989), 76–85.
23. D. W. Matula, Determining edge connectivity in $O(nm)$, *Proc. 28th Symp. on Foundations of Computer Science* (1987), 249–251.

24. H. Nagamochi and T. Ibaraki, Computing edge connectivity in multigraphs and capacitated graphs, *SIAM J. Discrete Math.* **5** (1992), 54–66.
25. C. P. Schnorr, Bottlenecks and edge connectivity in unsymmetrical networks, *SIAM J. Comput.* **8** (1979), 265–274.
26. R. E. Tarjan, Depth first search and linear graph algorithms, *SIAM J. Comput.* **1** (1972), 146–160.

13

Using graphs to find the best block designs

R. A. BAILEY and PETER J. CAMERON

A statistician designing an experiment wants to get as much information as possible from the data gathered. Often this means the most precise estimate possible (that is, an estimate with minimum possible variance) of the unknown parameters. If there are several parameters, this can be interpreted in many ways: do we want to minimize the average variance, or the maximum variance, or the volume of a confidence region for the parameters?

In the case of block designs, these optimality criteria can be calculated from the concurrence graph of the design, and in many cases from its Laplacian eigenvalues. The Levi graph can also be used. The various criteria turn out to be closely connected with other properties of the graph as a network, such as the number of spanning trees, isoperimetric number, or the sum of the resistances between pairs of vertices when the graph is regarded as an electrical network.

1. What makes a block design good?

Experiments are designed in many ways: for example, Latin squares, block designs or split-plot designs. Combinatorialists, on the other hand, have a much more specialized usage of the term 'design', as we remark later. We are concerned here with incomplete-block designs, more special than the statistician's designs and more general than the mathematician's.

To a statistician, a *block design* has two components. There is an underlying set Ω of experimental units, partitioned into b blocks of size k. There is a further set $\mathcal{T}$ of v treatments, and also a function f from units to treatments, specifying which treatment is allocated to which experimental unit – that is, $f(\omega)$ is the treatment allocated to experimental unit ω. Thus each block defines a subset, or maybe a multi-subset, of the treatments.

In a *complete-block design*, we have $k = v$ and each treatment occurs once in every block. Here we assume that blocks are *incomplete*, in the sense that $k < v$.

We assume that the purpose of the experiment is to find out about the treatments, and the differences between them. The blocks are an unavoidable nuisance, an inherent feature of the experimental units. In an agricultural experiment the experimental units may be field plots and the blocks may be fields or plough-lines; in a clinical trial the experimental units may be patients and the blocks hospitals; in process engineering the experimental units may be runs of a machine that is recalibrated each day and the blocks days. See [5] for further examples.

In all of these situations, the values of b, k and v are given. Given these values, not all incomplete-block designs are equally good. This chapter describes some criteria that can be used to choose between them.

For example, Fig. 1 shows two block designs with $v = 15$, $b = 7$ and $k = 3$. We use the convention that the treatments are labelled $1, 2, \ldots, v$, that columns represent blocks, and that the order of the entries in each column is not significant. Where necessary, we use the notation Γ_j to refer to the block which is shown as the jth column, for $j = 1, 2, \ldots, b$.

1	1	2	3	4	5	6
2	4	5	6	10	11	12
3	7	8	9	13	14	15

(a)

1	1	1	1	1	1	1
2	4	6	8	10	12	14
3	5	7	9	11	13	15

(b)

Fig. 1. Two block designs with $v = 15$, $b = 7$ and $k = 3$

The *replication* r_i of treatment i is $\left|f^{-1}(i)\right|$, the number of experimental units to which it is allocated. For the design in Fig. 1(a), r_i is 1 or 2 for all i. As we see later, statisticians tend to prefer designs in which all the replications are as equal as possible. If all treatments have the same replication, then the design is *equireplicate*: the common value of r_i is usually written as r, and $vr = bk$.

The design in Fig. 1(b) is a *queen-bee design* because there is (at least) one treatment that occurs in every block. Scientists tend to prefer such designs because they have been taught to compare every treatment to one distinguished treatment, which may be called a *control treatment*.

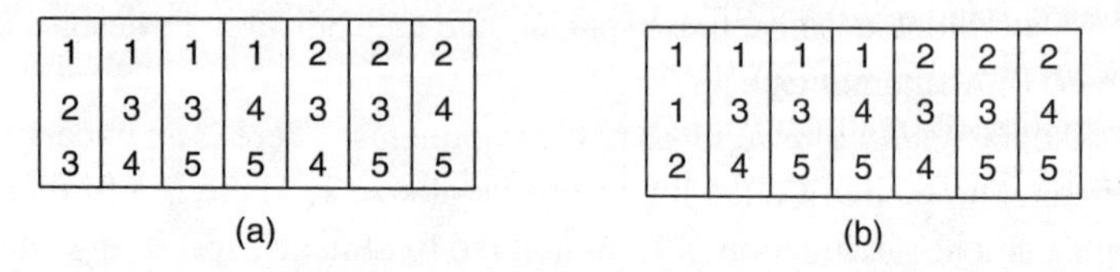

Fig. 2. Two block designs with $v=5$, $b=7$ and $k=3$

Fig. 2 shows two block designs with $v=5, b=7$ and $k=3$. The design in Fig. 2(b) shows a new feature: treatment 1 occurs on two experimental units in block Γ_1. A block design is *binary* if $f(\alpha) \neq f(\omega)$ whenever α and ω are experimental units in the same block. The design in Fig. 2(a) is binary. It seems to be obvious that binary designs must be better than non-binary ones, but we shall see later that this is not necessarily so. However, if there is any block on which f is constant, then that block provides no information about treatments, so we assume from now on that there are no such blocks.

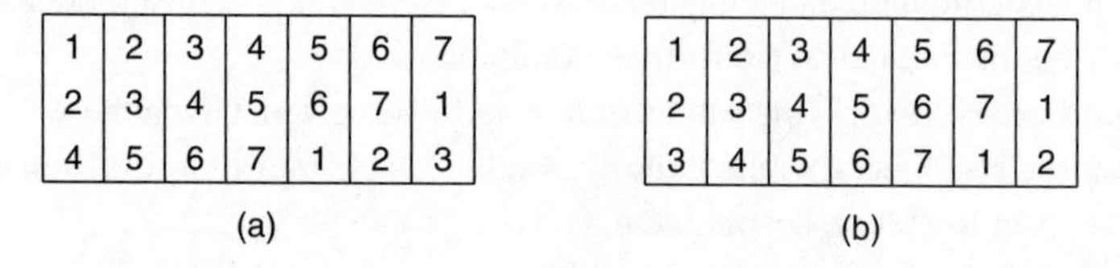

Fig. 3. Two block designs with $v=7$, $b=7$ and $k=3$

Fig. 3 shows two equireplicate binary block designs with $v=7$, $b=7$ and $k=3$. A binary design is *balanced* if every pair of distinct treatments occurs together in the same number of blocks. If that number is λ, then $r(k-1)=(v-1)\lambda$. Such designs are also called 2-*designs* or *BIBDs*. The design in Fig. 3(a) is balanced with $\lambda=1$; the design in Fig. 3(b) is not balanced.

Pure mathematicians usually assume that, if they exist, balanced designs are better than non-balanced ones; indeed, many do not call a structure a 'design' unless it is balanced. As we shall show in Section 4, this assumption is correct for all the criteria considered here. However, for given values of v and k, a non-balanced design with a larger value of b may produce more information than a balanced design with a smaller value of b.

2. Graphs from block designs

In this section, we introduce two graphs that we will use to represent a block design, and discuss the Laplacian matrix of a graph in general, specializing to the two graphs of interest.

The Levi graph

A simple way of representing a block design is its *Levi graph*, or *incidence graph*, introduced in [40]. This graph has $v+b$ vertices, one for each block and one for each treatment. There are bk edges, one for each experimental unit. If experimental unit ω is in block j and $f(\omega)=i$, then the corresponding edge $\tilde{e}_\omega$ joins vertices i and j. Thus the graph is bipartite, with one partite set consisting of block vertices and the other consisting of treatment vertices. Moreover, the graph has multiple edges if the design is not binary. Fig. 4 gives the Levi graph of the design in Fig. 2(b).

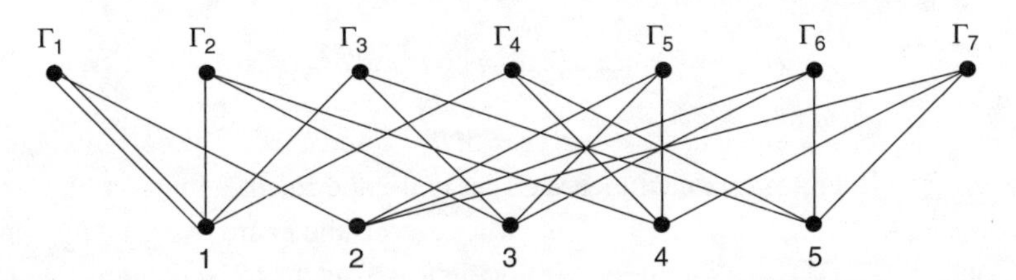

Fig. 4. The Levi graph of the design in Fig. 2(b)

We regard two block designs as the same if one can be obtained from the other by permuting the experimental units within each block. Since the vertices of the Levi graph are labelled, there is a bijection between block designs and their Levi graphs.

Let n_{ij} be the number of edges from treatment-vertex i to block-vertex j – that is, treatment i occurs on n_{ij} experimental units in block j. The $v \times b$ matrix $\mathbf{N} = (n_{ij})$ is the *incidence matrix* of the block design. If the rows and columns of $\mathbf{N}$ are labelled, we can recover the block design from its incidence matrix.

The concurrence graph

In a binary design, the *concurrence* λ_{ij} of treatments i and j is r_i if $i=j$, and otherwise is the number of blocks in which i and j both occur. For non-binary designs we have to count the number of occurrences of the pair $\{i, j\}$ in blocks according to multiplicity, so that λ_{ij} is the (i, j)-entry of $\mathbf{\Lambda}$, where $\mathbf{\Lambda} = \mathbf{N}\mathbf{N}^t$. The matrix $\mathbf{\Lambda}$ is called the *concurrence matrix* of the design.

The *concurrence graph* of the design has the treatments as vertices. There are no loops. If $i \neq j$, then there are λ_{ij} edges between vertices i and j. Each such edge corresponds to a pair $\{\alpha, \omega\}$ of experimental units in the same block, with $f(\alpha)=i$ and $f(\omega)=j$: we denote this edge by $e_{\alpha\omega}$. (This edge does not join the experimental units α and ω; it joins the treatments applied to these units.) It follows that the degree d_i of vertex i is given by

$$d_i = \sum_{j \neq i} \lambda_{ij}.$$

Figs. 5 and 6 show the concurrence graphs of the designs in Figs. 1 and 2, respectively.

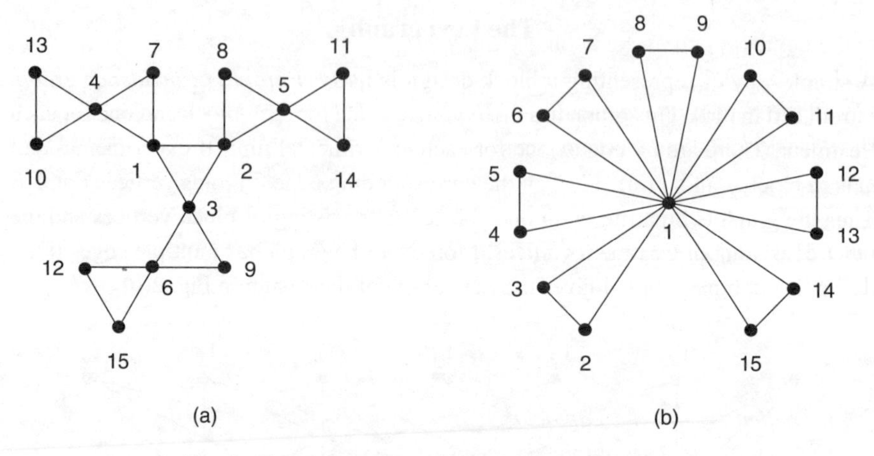

Fig. 5. The concurrence graphs of the designs in Fig. 1

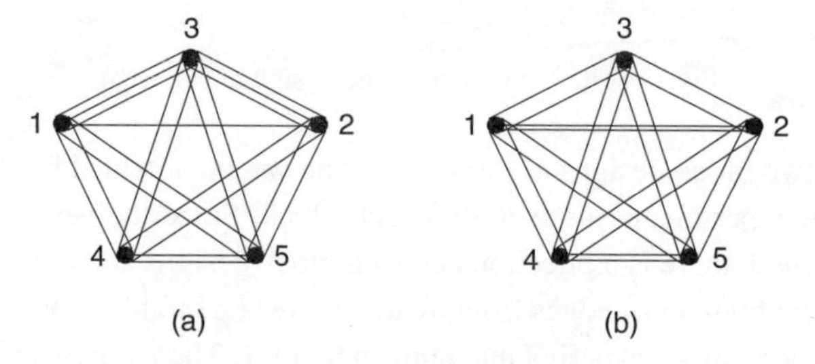

Fig. 6. The concurrence graphs of the designs in Fig. 2

If $k = 2$, then the concurrence graph is effectively the same as the block design. Although the block design cannot be recovered from the concurrence graph for larger values of k, we shall see in Section 3 that the concurrence graphs contain enough information to decide between two block designs on any of the usual statistical criteria. They were introduced as *variety concurrence graphs* in [44], but are so useful that they may have been considered earlier.

Throughout the rest of this chapter, we denote the concurrence graph and the Levi graph of any design Δ under consideration by G and $\tilde{G}$, respectively.

The Laplacian matrix of a graph

Let H be an arbitrary graph with n vertices: it may have multiple edges, but no loops. The *Laplacian matrix* $\mathbf{L}$ of H is defined to be the square matrix with rows and columns indexed by the vertices of H whose (i, i)-entry L_{ii} is the degree of vertex i and whose (i, j)-entry L_{ij} is the negative of the number of edges between vertices i and j if $i \neq j$. Then $L_{ii} = \sum_{j \neq i} L_{ij}$ for $1 \leq i \leq n$, and so the row sums of $\mathbf{L}$ are all 0. It follows that $\mathbf{L}$ has eigenvalue 0 on the all-1 vector; this is called the *trivial eigenvalue* of $\mathbf{L}$. We show below that the multiplicity of the zero eigenvalue is equal to the number of connected components of H. Thus the multiplicity is 1 if and only if H is connected.

Call the remaining eigenvalues of $\mathbf{L}$ *non-trivial*. They are all non-negative, as we show in the following theorem (see [7]).

Theorem 2.1

(a) *If* $\mathbf{L}$ *is a Laplacian matrix, then* $\mathbf{L}$ *is positive semi-definite.*
(b) *If* $\mathbf{L}$ *is a Laplacian matrix of order* n *and* $\mathbf{x}$ *is any vector in* $\mathbb{R}^n$*, then*

$$\mathbf{x}^t\mathbf{L}\mathbf{x} = \sum_{\text{edges } ij} (x_i - x_j)^2.$$

(c) *If* $\mathbf{L}_1$ *and* $\mathbf{L}_2$ *are the Laplacian matrices of graphs* H_1 *and* H_2 *with the same vertices, and if* H_2 *is obtained from* H_1 *by inserting one extra edge, then* $\mathbf{L}_2 - \mathbf{L}_1$ *is positive semi-definite.*
(d) *If* $\mathbf{L}$ *is the Laplacian matrix of the graph* H*, then the multiplicity of the zero eigenvalue of* $\mathbf{L}$ *is equal to the number of connected components of* H*.*

Proof Each edge between vertices i and j defines an $n \times n$ matrix whose entries are all 0, apart from the following submatrix:

$$\begin{array}{c} \\ i \\ j \end{array} \begin{array}{c} \begin{array}{cc} i & j \end{array} \\ \begin{bmatrix} 1 & -1 \\ -1 & 1 \end{bmatrix} \end{array}.$$

The Laplacian is the sum of these matrices, which are all positive semi-definite. This proves (a), (b) and (c).

From (b), the vector $\mathbf{x}$ is in the null-space of the Laplacian if and only if $\mathbf{x}$ takes the same value on both vertices of each edge, which happens if and only if it takes a constant value on each connected component. This proves (d). ■

Theorem 2.1 shows that the smallest non-trivial eigenvalue of a connected graph is positive; this eigenvalue is sometimes called the *algebraic connectivity* of the graph. The statistical importance of this is shown in Section 3.

In Section 3 we shall also need the Moore–Penrose generalized inverse $\mathbf{L}^-$ of $\mathbf{L}$ (see [45]). Put $\mathbf{P}_0 = n^{-1}\mathbf{J}_n$, where $\mathbf{J}_n$ is the $n \times n$ matrix whose entries are all 1, so that $\mathbf{P}_0$ is the matrix of orthogonal projection onto the space spanned by the all-1 vector. If H is connected then $\mathbf{L} + \mathbf{P}_0$ is invertible, and

$$\mathbf{L}^- = (\mathbf{L} + \mathbf{P}_0)^{-1} - \mathbf{P}_0,$$

so that $\mathbf{L}\mathbf{L}^- = \mathbf{L}^-\mathbf{L} = \mathbf{I}_n - \mathbf{P}_0$, where $\mathbf{I}_n$ is the identity matrix of order n.

Laplacians of the concurrence and Levi graphs

There is a relationship between the Laplacian matrices of the concurrence and Levi graphs of a block design Δ. Let $\mathbf{N}$ be the incidence matrix of the design, and $\mathbf{R}$ be the diagonal matrix (with rows and columns indexed by treatments) whose (i, i) entry is the replication r_i of treatment i. If the design is equireplicate, then $\mathbf{R} = r\mathbf{I}_v$, where r is the replication number.

For the remainder of this chapter, we use $\mathbf{L}$ for the Laplacian matrix of the concurrence graph G of Δ, and $\tilde{\mathbf{L}}$ for the Laplacian matrix of the Levi graph $\tilde{G}$ of Δ. Then it is straightforward to show that

$$\mathbf{L} = k\mathbf{R} - \mathbf{N}\mathbf{N}^t \quad \text{and} \quad \tilde{\mathbf{L}} = \begin{bmatrix} \mathbf{R} & -\mathbf{N} \\ -\mathbf{N}^t & k\mathbf{I} \end{bmatrix}.$$

The Levi graph is connected if and only if the concurrence graph is connected; thus 0 is a simple eigenvalue of $\tilde{\mathbf{L}}$ if and only if it is a simple eigenvalue of $\mathbf{L}$, which in turn occurs if and only if all contrasts between treatment parameters are estimable (see Section 3). A block design with this property is itself called *connected*: we consider only connected block designs.

In the equireplicate case, the above expressions for $\mathbf{L}$ and $\tilde{\mathbf{L}}$ give a relationship between their Laplacian eigenvalues, as follows. Let $\mathbf{x}$ be an eigenvector of $\mathbf{L}$ with eigenvalue $\phi \neq rk$. Then, for each of the two solutions θ of the quadratic equation

$$rk - \phi = (r - \theta)(k - \theta),$$

there is a unique vector $\mathbf{z}$ in $\mathbb{R}^b$ for which $[\mathbf{x}^t \quad \mathbf{z}^t]^t$ is an eigenvector of $\tilde{\mathbf{L}}$ with eigenvalue θ. Conversely, any eigenvalue $\theta \neq k$ of $\tilde{\mathbf{L}}$ arises in this way.

The Laplacian matrices of the concurrence graphs in Fig. 6 are shown in Table 1.

$$\begin{bmatrix} 8 & -1 & -3 & -2 & -2 \\ -1 & 8 & -3 & -2 & -2 \\ -3 & -3 & 10 & -2 & -2 \\ -2 & -2 & -2 & 8 & -2 \\ -2 & -2 & -2 & -2 & 8 \end{bmatrix} \qquad \begin{bmatrix} 8 & -2 & -2 & -2 & -2 \\ -2 & 8 & -2 & -2 & -2 \\ -2 & -2 & 8 & -2 & -2 \\ -2 & -2 & -2 & 8 & -2 \\ -2 & -2 & -2 & -2 & 8 \end{bmatrix}$$

(a) (b)

Table 1. The Laplacian matrices of the concurrence graphs in Fig. 6

3. Statistical issues

In this section, we explain various optimality criteria that have been introduced to identify designs in which estimators have small variance. The criteria we consider, known as A, D and E, can (in the case of a block design) be expressed in terms of the Laplacian eigenvalues of its concurrence graph.

Estimation and variance

As part of the experiment, we measure the response Y_ω on each experimental unit ω. If ω is in block Γ, then we assume that

$$Y_\omega = \tau_{f(\omega)} + \beta_\Gamma + \varepsilon_\omega; \tag{1}$$

here, τ_i is a constant depending on treatment i, β_Γ is a constant depending on block Γ, and ε_ω is a random variable with expectation 0 and variance σ^2. Furthermore, if $\alpha \neq \omega$, then ε_α and ε_ω are uncorrelated.

It is clear that we can add a constant to every block parameter, and subtract that constant from every treatment parameter, without changing (1). It is therefore impossible to estimate the individual treatment parameters. However, if the design is connected, then we can estimate all *contrasts* in the treatment parameters – that is, all linear combinations of the form $\sum_i x_i \tau_i$ for which $\sum_i x_i = 0$. In particular, we can estimate all the simple treatment differences $\tau_i - \tau_j$.

An *estimator* is a function of the responses Y_ω, so it is itself a random variable. An estimator of a value is *unbiased* if its expectation is equal to the true value, and is *linear* if it is a linear function of the responses. Amongst linear unbiased estimators, the *best* one (the so-called BLUE) is the one with the least variance. Let V_{ij} be the variance of the BLUE for $\tau_i - \tau_j$.

If all the experimental units form a single block, then the BLUE of $\tau_1 - \tau_2$ is just the difference between the average responses for treatments 1 and 2. It follows that

$$V_{12} = \left(\frac{1}{r_1} + \frac{1}{r_2}\right)\sigma^2.$$

When $v = 2$, this variance is minimized (for a given number of experimental units) when $r_1 = r_2$. Moreover, if the responses are normally distributed, then the length of the 95% confidence interval for $\tau_1 - \tau_2$ is proportional to $\mathrm{t}(r_1 + r_2 - 2, 0.975)\sqrt{V_{12}}$, where $\mathrm{t}(d, p)$ is the $100p$-th percentile of the t-distribution on d degrees of freedom. The smaller the confidence interval, the more likely is our estimate to be close to the true value. This length can be made smaller by increasing $r_1 + r_2$, decreasing $|r_1 - r_2|$ or decreasing σ^2.

However, matters are not so simple when $k < v$ and $v > 2$. The following result can be found in any statistical textbook about block designs (see the section on further reading).

Theorem 3.1 *Let* $\mathbf{L}$ *be the Laplacian matrix of the concurrence graph of a connected block design. If* $\sum_i x_i = 0$*, then the variance of the BLUE of* $\sum_i x_i \tau_i$ *is equal to* $(\mathbf{x}^t\mathbf{L}^-\mathbf{x})k\sigma^2$*. In particular, the variance* V_{ij} *of the BLUE of the simple difference* $\tau_i - \tau_j$ *is given by* $V_{ij} = (L^-_{ii} + L^-_{jj} - 2L^-_{ij})k\sigma^2$*.*

Optimality criteria

We want all of the V_{ij} to be as small as possible, but this is a multi-dimensional problem if $v > 2$. Let $\bar{V}$ be the average of the variances V_{ij} over all treatments i, j with $i \neq j$. Theorem 3.1 shows that, for each fixed i,

$$\begin{aligned}\sum_{j\neq i} V_{ij} &= \sum_{j\neq i}(L^-_{ii} + L^-_{jj} - 2L^-_{ij})k\sigma^2 \\ &= [(v-1)L^-_{ii} + (\mathrm{Tr}(\mathbf{L}^-) - L^-_{ii}) + 2L^-_{ii}]k\sigma^2 \\ &= [vL^-_{ii} + \mathrm{Tr}(\mathbf{L}^-)]k\sigma^2,\end{aligned}$$

because the row sums and column sums of $\mathbf{L}^-$ are all 0. It follows that $\bar{V} = 2k\sigma^2\,\mathrm{Tr}(\mathbf{L}^-)/(v-1)$.

Let $\theta_1, \theta_2, \ldots, \theta_{v-1}$ be the non-trivial eigenvalues of $\mathbf{L}$, now listed according to multiplicity and in non-decreasing order. Then

$$\mathrm{Tr}(\mathbf{L}^-) = \frac{1}{\theta_1} + \frac{1}{\theta_2} + \cdots + \frac{1}{\theta_{v-1}},$$

and so

$$\bar{V} = 2k\sigma^2 \times \frac{1}{\text{harmonic mean of } \theta_1, \theta_2, \ldots, \theta_{v-1}}.$$

A block design is defined to be *A-optimal* (in some given class of designs with the same values of b, k and v) if it minimizes the value of $\bar{V}$; here 'A' stands for 'average'. Thus a design is A-optimal if and only if it maximizes the harmonic mean of $\theta_1, \theta_2, \ldots, \theta_{v-1}$.

For $v > 2$, the generalization of a confidence interval is a confidence ellipsoid centred at the point $(\hat{\tau}_1, \hat{\tau}_2, \ldots, \hat{\tau}_v)$, which gives the estimated value of $(\tau_1, \tau_2, \ldots, \tau_v)$ in the $(v-1)$-dimensional subspace of $\mathbb{R}^v$ for which $\sum \tau_i = 0$. A block design is called *D-optimal* if it minimizes the volume of this confidence ellipsoid. Since this volume is proportional to $\sqrt{\det(\mathbf{L}^- + \mathbf{P}_0)}$, a design is D-optimal if and only if it maximizes the geometric mean of $\theta_1, \theta_2, \ldots, \theta_{v-1}$; here, 'D' stands for 'determinant'.

Rather than looking at averages, we might consider the worst case. If all the entries in the vector $\mathbf{x}$ are multiplied by a constant c, then the variance of the estimator of $\sum x_i\tau_i$ is multiplied by c^2. Thus, those contrast vectors $\mathbf{x}$ that give the largest variance relative to their own length are those that maximize $\mathbf{x}^t\mathbf{L}^-\mathbf{x}/\mathbf{x}^t\mathbf{x}$; these are precisely the eigenvectors of $\mathbf{L}$ with eigenvalue θ_1. A design is defined to be *E-optimal* if it maximizes the value of θ_1; here, 'E' stands for 'extreme'.

More generally, for p in $(0, \infty)$, a design is called *Φ_p-optimal* if it minimizes

$$\left(\frac{\sum_{i=1}^{v-1}\theta_i^{-p}}{v-1}\right)^{1/p}.$$

Thus A-optimality corresponds to $p=1$, D-optimality corresponds to the limit as $p \to 0$, and E-optimality corresponds to the limit as $p \to \infty$.

Let $\mathbf{L}_1$ and $\mathbf{L}_2$ be the Laplacian matrices of the concurrence graphs of block designs Δ_1 and Δ_2 for v treatments in blocks of size k. If $\mathbf{L}_2 - \mathbf{L}_1$ is positive semi-definite, then Δ_2 is at least as good as Δ_1 on all the Φ_p-criteria. Theorem 2.1(c) shows that adding an extra block to a design cannot decrease its performance on any Φ_p-criterion.

There are even more general classes of optimality criteria (see [28] and [48] for details). Here we concentrate on A-, D- and E-optimality.

Questions and an example

A first obvious question to ask is: do these criteria agree with each other?

Our optimality properties are all functions of the concurrence graph. What features of this graph should we look for if we are searching for optimal, or near-optimal, designs? Symmetry? (Nearly) equal degree? (Nearly) equal numbers of edges between pairs of vertices? Distance-regularity? Large girth (ignoring cycles within a block)? Small numbers of short cycles (ditto)? High connectivity? Non-trivial automorphism group? Also, is it more useful to look at the Levi graph rather than the concurrence graph?

Example 4 Fig. 7 shows the values of the A- and D-criteria for all equireplicate block designs with $v=8$, $b=12$ and $k=2$: these are just regular graphs with 8 vertices and degree 3. The harmonic mean is shown on the A-axis, and the geometric mean on the D-axis. (Note that this figure includes some designs that were omitted from Fig. 3 of [4].) The rankings on these two criteria are not exactly the same, but

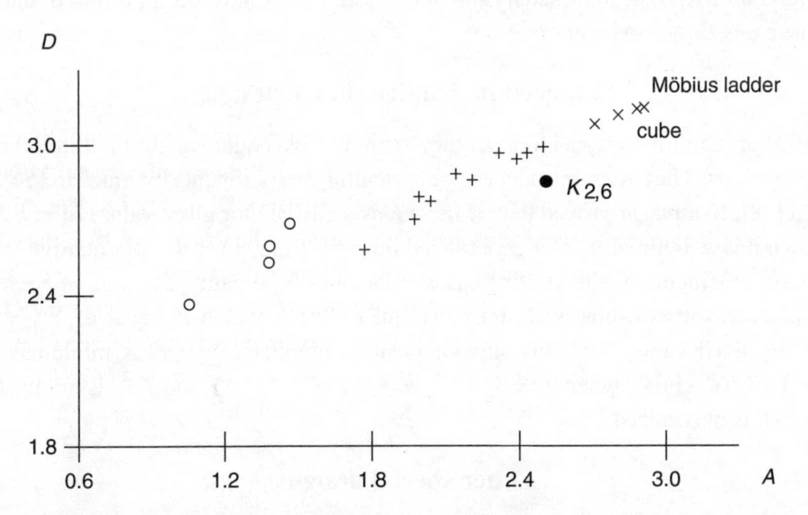

Fig. 7. Values of two optimality criteria for all equireplicate block designs with $v=8$, $b=12$ and $k=2$, and for $K_{2,6}$

they do agree at the top end, where it matters. The second-best graph on both criteria is the cube; the best is the Möbius ladder, whose vertices are the elements of $\mathbb{Z}_8$ and whose edges are $\{i, i+1\}$ and $\{i, i+4\}$ for i in $\mathbb{Z}_8$. These two graphs are so close on both criteria that, for practical purposes, they can be regarded as equally good.

The plotting symbols show the edge-connectivities of the graphs: edge-connectivities of 3, 2, 1 are shown as $\times$, $+$, $\circ$, respectively. This suggests that the higher the edge-connectivity the better is the design on the A- and D- criteria. This is intuitively reasonable: if $k = 2$, then the edge-connectivity is the minimum number of blocks whose removal disconnects the design. In this context, it has been called the *breakdown number* (see [39]).

The four graphs with edge-connectivity 3 have no double edges, so concurrences differ by at most 1. The only other regular graph with no double edges is ranked eighth (amongst regular graphs) by the A-criterion. This suggests that (near-)equality of concurrences is not sufficient to give a good design.

The symbol $\bullet$ shows the non-regular graph $K_{2,6}$, which also has eight vertices and twelve edges. It is not as good as the regular graphs with edge-connectivity 3, but it beats many of the other regular graphs.

This pattern is typical of the block designs investigated by statisticians for most of the 20th century. The A- and D-criteria agree closely at the top end. High edge-connectivity appears to show good designs. Many of the best designs have a high degree of symmetry.

4. Highly patterned block designs

Balanced incomplete-block designs, when they exist, are optimal; in cases where there is no BIBD, designs satisfying similar conditions have been proposed, and in some cases shown to be optimal.

Balanced incomplete-block designs

BIBDs are intuitively appealing, as they seem to give equal weight to all treatment comparisons. They were introduced for agricultural experiments by Yates in [54].

In [38], Kshirsagar proved that, if there exists a BIBD for given values of v, b and k, then it is A-optimal. Kiefer generalized this in [35] to cover Φ_p-optimality for all p in $(0, \infty)$, including the limiting cases of D- and E-optimality. The core of Kiefer's proof is as follows: binary designs maximize $\mathrm{Tr}(\mathbf{L})$, which is equal to $\sum_{i=1}^{v-1} \theta_i$; for any fixed value T of this sum of positive numbers, $\sum \theta_i^{-p}$ is minimized at $(\mathrm{v}-1)[T/(v-1)]^{-p}$ when $\theta_1 = \theta_2 = \cdots = \theta_{v-1} = T/(v-1)$; and T^{-p} is minimized when T is maximized.

Other special designs

Of course, it frequently occurs that the values of b, v and k available for an experiment are such that no BIBD exists. (Necessary conditions for the existence of a BIBD include the well-known divisibility conditions $v \mid bk$ and $v(v-1) \mid bk(k-1)$, which follow from the elementary results in Section 1, and *Fisher's inequality* asserting that $b \geq v$.)

In the absence of a BIBD, various other special types of design have been considered, and some of these have been proved optimal. Here is a short sample.

A design is *group-divisible* if the treatments can be partitioned into 'groups' all of the same size, so that the number of blocks containing two treatments is λ_1 if they belong to the same group, and λ_2 otherwise. Chêng [16], [17] showed that if there is a group-divisible design with two groups and with $\lambda_2 = \lambda_1 + 1$ in the class of designs with given v, b and k, then it is Φ_p-optimal for all p, and in particular it is A-, D- and E-optimal.

A *regular-graph design* is a binary equireplicate design with two possible concurrences, λ and $\lambda+1$. It is easily proved that, in such a design, the number of treatments lying in $\lambda + 1$ blocks with a given treatment is constant; so the graph H whose vertices are the treatments, with two vertices joined whenever they lie in $\lambda + 1$ blocks, is regular.

Chêng [17] showed that a group-divisible design with $\lambda_1 = 0$ and $\lambda_2 = 1$, if one exists, is Φ_p-optimal in the class of regular-graph designs for all p. Cheng and Bailey [20] showed that a regular-graph design for which the graph is *strongly regular* (see [15]) and which has singular concurrence matrix is Φ_p-optimal, for all p, among binary equireplicate designs with given v, b and k.

Designs with the property described here are particular examples of *partially balanced designs* with respect to an association scheme (see Bailey [3]).

Another class which turns out to be optimal in many cases, but whose definition is less combinatorial, consists of the *variance-balanced designs*, which we consider later in the chapter.

5. D-optimality

D-optimal block designs are those whose concurrence graphs contain the largest number of spanning trees.

Spanning trees of the concurrence graph

Let G be the concurrence graph of a connected block design, and let $\mathbf{L}$ be its Laplacian matrix. A *spanning tree* for G is a spanning subgraph that is a tree. Kirchhoff's famous *matrix-tree theorem* in [36] states the following.

Theorem 5.1 *If G is a connected graph with v vertices and Laplacian matrix* $\mathbf{L}$, *then the product of the non-trivial eigenvalues of* $\mathbf{L}$ *is equal to v multiplied by the number of spanning trees in G.*

Thus we have a test for D-optimality:

A design is D-optimal if and only if its concurrence graph has the maximum number of spanning trees.

Note that Theorem 5.1 gives an easy proof of Cayley's theorem on the number of spanning trees of the complete graph K_v. The non-trivial eigenvalues of its Laplacian matrix are all equal to v, so Theorem 5.1 shows that it has v^{v-2} spanning trees.

If G is sparse, then it may be much easier to count the number of spanning trees than to compute the eigenvalues of $\mathbf{L}$. For example, if G has a single cycle, of length s, then the number of spanning trees is s, irrespective of the remaining edges in G.

In [26], Gaffke discovered the importance of Kirchhoff's theorem for optimal block designs. Chêng followed this up in papers such as [16], [17], [18] and [19]. Particularly intriguing is the following theorem from [21]. A design is *nearly balanced* if

- no two replications differ by more than 1;
- for each fixed i, no two concurrences λ_{ij} and λ_{ik} differ by more than 1.

Theorem 5.2 *Consider block designs with $k=2$ (connected graphs). For each given v there is a threshold b_0 such that, if $b \geq b_0$, then any D-optimal design for v treatments in b blocks of size 2 is nearly balanced.*

In fact, there is no known example where the threshold b_0 is greater than v.

Spanning trees of the Levi graph

In [27] Gaffke stated the following relationship between the numbers of spanning trees in the concurrence graph and in the Levi graph.

Theorem 5.3 *Let G and $\tilde{G}$ be the concurrence graph and Levi graph of a connected incomplete-block design for v treatments in b blocks of size k. Then the number of spanning trees in $\tilde{G}$ is equal to k^{b-v+1} times the number of spanning trees in G.*

Thus, an alternative test for D-optimality is to count the number of spanning trees in the Levi graph. For binary designs, the Levi graph has fewer edges than the concurrence graph if and only if $k \geq 4$.

6. A-optimality

A-optimal block designs minimize the average pairwise resistance of the concurrence graph, when regarded as an electrical network where each edge has unit resistance.

The concurrence graph as an electrical network

We can consider any connected graph as an electrical network with a 1-ohm resistance in each edge. Connect a 1-volt battery between two specified vertices i and j. Then current flows in the network, according to these rules:

Ohm's law: In every edge, the voltage drop is the product of the current and the resistance.

Kirchhoff's voltage law: The total voltage drop from one vertex to any other vertex is the same, no matter which path we take from one to the other.

Kirchhoff's current law: At each vertex not connected to the battery, the total incoming current is equal to the total outgoing current.

We find the total current from i to j, and then use Ohm's law to define the effective resistance R_{ij} between i and j as the reciprocal of this current. It is a standard result of electrical network theory that the linear equations implicitly defined above for the currents and voltage differences have a unique solution.

Now let G be the concurrence graph of a block design with treatment set $\mathcal{T}$ and set of experimental units Ω. Current flows in each edge $e_{\alpha\omega}$, where α and ω are experimental units in the same block which receive different treatments; let $I(\alpha, \omega)$ be the current from $f(\alpha)$ to $f(\omega)$ in this edge. Thus I is a function $I\colon \Omega \times \Omega \mapsto \mathbb{R}$ for which

(a) $I(\alpha, \omega) = 0$ if $f(\alpha) = f(\omega)$, or if α and ω are in different blocks.
(b) $I(\alpha, \omega) = -I(\omega, \alpha)$ for (α, ω) in $\Omega \times \Omega$.

This defines a further function $I_{\text{out}}\colon \mathcal{T} \mapsto \mathbb{R}$ by

$$I_{\text{out}}(l) = \sum_{\alpha: f(\alpha)=l} \sum_{\omega\in\Omega} I(\alpha, \omega), \qquad \text{for } l \text{ in } \mathcal{T}.$$

Voltage is another function $V\colon \mathcal{T} \mapsto \mathbb{R}$. The following two conditions ensure that Ohm's and Kirchhoff's laws are satisfied.

(c) If there is any edge in G between $f(\alpha)$ and $f(\omega)$, then
$$I(\alpha, \omega) = V(f(\alpha)) - V(f(\omega)).$$
(d) If $l \neq i$ and $l \neq j$, then $I_{\text{out}}(l) = 0$.

If G is connected, and if different voltages $V(i)$ and $V(j)$ are given for a pair of distinct treatments i and j, then there are unique functions I and V satisfying conditions (a)–(d). Moreover, $I_{\text{out}}(j) = -I_{\text{out}}(i) \neq 0$. Then R_{ij} is defined by

$$R_{ij} = \frac{V(i) - V(j)}{I_{\text{out}}(i)}.$$

It can be shown that the value of R_{ij} does not depend on the choice of values for $V(i)$ and $V(j)$, so long as these are different. In practical examples, it is usually convenient to take $V(i) = 0$ and to let I take integer values.

What has all of this got to do with block designs? The following theorem, which is a standard result from electrical engineering, gives the answer.

Theorem 6.1 *If* $\mathbf{L}$ *is the Laplacian matrix of a connected graph G, then the effective resistance* R_{ij} *between vertices i and j is given by*

$$R_{ij} = \left(L^-_{ii} + L^-_{jj} - 2L^-_{ij}\right).$$

Comparing this with Theorem 3.1, we see that $V_{ij} = R_{ij} \times k\sigma^2$. We thus have the following test for A-optimality:

A design is A-optimal if and only if its concurrence graph, regarded as an electrical network, minimizes the sum of the pairwise effective resistances between all pairs of vertices.

Effective resistances are easy to calculate without matrix inversion if the graph is sparse.

Fig. 8 shows the concurrence graph of a block design with $v = 12$, $b = 6$ and $k = 3$. Only vertices i and j are labelled. Otherwise, numbers beside arrows denote currents and numbers in square brackets denote voltages. It is straightforward to check that the above conditions (a)–(d) are satisfied. Now $V(i) - V(j) = 47$ and $I_{\text{out}}(i) = 36$, and so $R_{ij} = \frac{47}{36}$. Therefore $V_{ij} = \frac{47}{12}\sigma^2$. Moreover, for graphs consisting of b triangles arranged in a cycle like this, it is clear that the average effective resistance, and hence the average pairwise variance, can be calculated as a function of b.

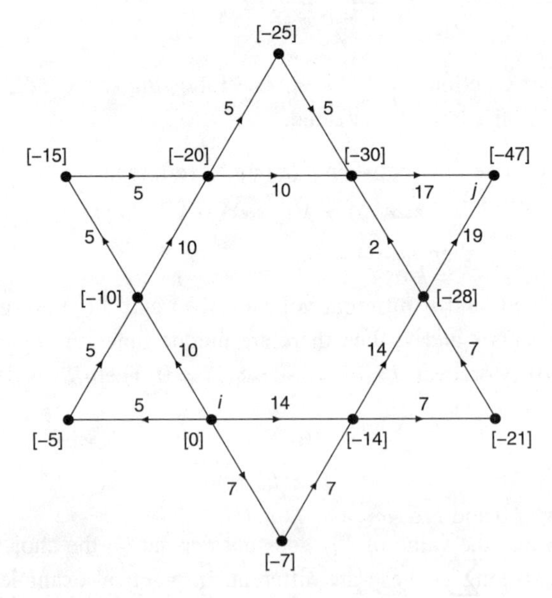

Fig. 8. The current between vertices i and j in a concurrence graph

The Levi graph as an electrical network

The Levi graph $\tilde{G}$ of a block design can also be considered as an electrical network. Denote by $\mathcal{B}$ the set of blocks. Now the currents are defined on the ordered edges of $\tilde{G}$. Recall that, if ω is an experimental unit in block Γ, then the edge $\tilde{e}_\omega$ joins Γ to $f(\omega)$. Thus the currents are defined on $(\Omega \times \mathcal{B}) \cup (\mathcal{B} \times \Omega)$ and the voltages are defined on $\mathcal{T} \cup \mathcal{B}$. The above conditions (a)–(d) need to be modified appropriately.

The next theorem shows that a current–voltage pair (I, V) in the concurrence graph G can be transformed into a current–voltage pair $(\tilde{I}, \tilde{V})$ in the Levi graph $\tilde{G}$. In $\tilde{G}$, the current $\tilde{I}(\alpha, \Gamma)$ flows in the edge $\tilde{e}_\alpha$ from vertex $f(\alpha)$ to vertex Γ, where $\alpha \in \Gamma$. The pairwise variance V_{ij} can thus be calculated from the effective resistance $\tilde{R}_{ij}$ in the Levi graph.

Theorem 6.2 *Let G be the concurrence graph and let $\tilde{G}$ be the Levi graph of a connected block design with block size k. If i and j are two distinct treatments, let R_{ij} and $\tilde{R}_{ij}$ be the effective resistances between vertices i and j in the electrical networks defined by G and $\tilde{G}$, respectively. Then $\tilde{R}_{ij} = kR_{ij}$, and so $V_{ij} = \tilde{R}_{ij}\sigma^2$.*

Proof Let (I, V) be a current–voltage pair on G. For $(\alpha, \Gamma) \in \Omega \times \mathcal{B}$, put

$$\tilde{I}(\alpha, \Gamma) = -\tilde{I}(\Gamma, \alpha) = \sum_{\omega \in \Gamma} I(\alpha, \omega)$$

if $\alpha \in \Gamma$; otherwise, put $\tilde{I}(\alpha, \Gamma) = \tilde{I}(\Gamma, \alpha) = 0$. Put $\tilde{V}(i) = kV(i)$ for all i in $\mathcal{T}$, and let

$$\tilde{V}(\Gamma) = \sum_{\omega \in \Gamma} V(f(\omega))$$

for all Γ in $\mathcal{B}$. It is clear that $\tilde{I}$ satisfies the analogues of conditions (a) and (b).

If $\alpha \in \Gamma$, then

$$\begin{aligned}\tilde{I}(\alpha, \Gamma) &= \sum_{\omega \in \Gamma} I(\alpha, \omega) = \sum_{\omega \in \Gamma} [V(f(\alpha)) - V(f(\omega))] \\ &= kV(f(\alpha)) - \tilde{V}(\Gamma) = \tilde{V}(f(\alpha)) - \tilde{V}(\Gamma),\end{aligned}$$

so the analogue of condition (c) is satisfied.

If $\Gamma \in \mathcal{B}$, then

$$\tilde{I}_{\text{out}}(\Gamma) = \sum_{\alpha \in \Gamma} \tilde{I}(\Gamma, \alpha) = -\sum_{\alpha \in \Gamma} \sum_{\omega \in \Gamma} I(\alpha, \omega) = 0,$$

because $I(\alpha, \alpha) = 0$ and $I(\alpha, \omega) = -I(\omega, \alpha)$. If $l \in \mathcal{T}$, then

$$\tilde{I}_{\text{out}}(l) = \sum_{\alpha : f(\alpha) = l} \sum_{\Gamma \in \mathcal{B}} \tilde{I}(\alpha, \Gamma) = \sum_{\alpha : f(\alpha) = l} \sum_{\omega \in \Omega} I(\alpha, \omega) = I_{\text{out}}(l).$$

In particular, $\tilde{I}_{\text{out}}(l) = 0$ if $l \neq i$ and $l \neq j$, which shows that the analogue of condition (d) is satisfied. It follows that $(\tilde{I}, \tilde{V})$ is the current–voltage pair on $\tilde{G}$ defined by $\tilde{V}(i)$ and $\tilde{V}(j)$.

Now

$$\tilde{R}_{ij} = \frac{\tilde{V}(i) - \tilde{V}(j)}{\tilde{I}_{\text{out}}(i)} = \frac{k(V(i) - V(j))}{I_{\text{out}}(i)} = kR_{ij}.$$

Then Theorems 3.1 and 6.1 show that $V_{ij} = R_{ij}\sigma^2$. ■

When $k = 2$, it seems to be easier to use the concurrence graph than the Levi graph because it has fewer vertices, but for larger values of k the Levi graph may be better since it does not have all the within-block cycles that the concurrence graph has. Fig. 9 gives the Levi graph of the block design whose concurrence graph is in Fig. 8, with the same two vertices i and j attached to the battery. This gives $\tilde{R}_{ij} = \frac{47}{12}$, which is in accordance with Theorem 6.2.

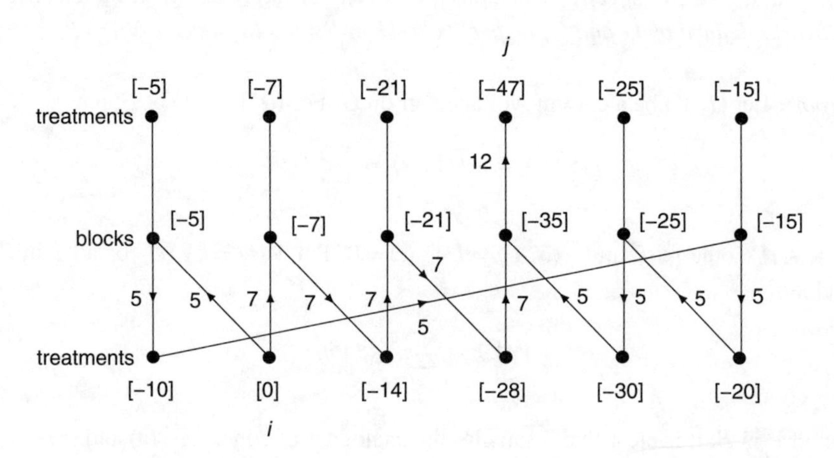

Fig. 9. The current between vertices i and j for the Levi graph corresponding to the concurrence graph in Fig. 8

Here is another way of visualizing Theorem 6.2. From the block design we construct a graph G_0 with vertex-set $\mathcal{T} \cup \Omega \cup \mathcal{B}$: the edges are $\{\alpha, \Gamma\}$ for $\alpha \in \Gamma \in \mathcal{B}$, and $\{\alpha, f(\alpha)\}$ for $\alpha \in \Omega$. Let (I_0, V_0) be a current–voltage pair on G_0 for which both battery vertices are in $\mathcal{T}$. We obtain the Levi graph $\tilde{G}$ from G_0 by ignoring the vertices in Ω. Thus the resistance in each edge of $\tilde{G}$ is twice that in each edge in G_0, so this step multiplies each effective resistance by 2.

Because none of the battery vertices is in $\mathcal{B}$, we can now obtain G from $\tilde{G}$ by replacing each path of the form (i, Γ, j) by an edge (i, j). There is no harm in scaling all the voltages by the same amount, so we can obtain (I, V) on G from $(\tilde{I}, \tilde{V})$ on $\tilde{G}$ by putting $V(i) = \tilde{V}(i)/k$ for i in $\mathcal{T}$, and $I(\alpha, \omega) = V(f(\alpha)) - V(f(\omega))$ for α, ω in the same block. If Γ is a block, then

$$0=\sum_{\alpha\in\Gamma}\tilde{I}(\alpha,\Gamma)=\sum_{\alpha\in\Gamma}[\tilde{V}(f(\alpha))-\tilde{V}(\Gamma)]=k\sum_{\alpha\in\Gamma}V(f(\alpha))-k\tilde{V}(\Gamma),$$

and so $\tilde{V}(\Gamma)=\sum_{\alpha\in\Gamma}V(f(\alpha))$. Also, if $\alpha\in\Gamma$, then

$$\begin{aligned}\sum_{\omega\in\Gamma}I(\alpha,\omega)&=\sum_{\omega\in\Gamma}[V(f(\alpha))-V(f(\omega))]\\&=kV(f(\alpha))-\sum_{\omega\in\Gamma}V(f(\omega))\\&=\tilde{V}(f(\alpha))-\tilde{V}(\Gamma)=\tilde{I}(\alpha,\Gamma).\end{aligned}$$

Thus, this transformation reverses the one used in the proof of Theorem 6.2.

There is yet another way of obtaining Theorem 6.2. If we use the responses Y_ω to estimate the block parameters β_Γ in (1) as well as the treatment parameters τ_i (see Section 3), then the standard theory of linear models shows that, if the design is connected, then we can estimate linear combinations of the form $\sum_{i=1}^{v}x_i\tau_i+\sum_{j=1}^{b}z_j\beta_j$ so long as $\sum x_i=\sum z_j$. Moreover, the variance of the BLUE of this linear combination is

$$[\ \mathbf{x}^t\quad \mathbf{z}^t\ ]\mathbf{C}^-\begin{bmatrix}\mathbf{x}\\ \mathbf{z}\end{bmatrix}\sigma^2,\quad \text{where } \mathbf{C}=\begin{bmatrix}\mathbf{R} & \mathbf{N}\\ \mathbf{N}^t & k\mathbf{I}_b\end{bmatrix}$$

and $\mathbf{R}$ is the diagonal matrix of replications.

If we reparametrize equation (1) by replacing β_j by $-\gamma_j$ for $j=1,2,\ldots,b$, then the estimable quantities are the contrasts in $\tau_1,\tau_2,\ldots,\tau_v,\gamma_1,\gamma_2,\ldots,\gamma_b$. The so-called *information matrix* $\mathbf{C}$ must be modified by multiplying the last b rows and the last b columns by -1: this gives precisely the Laplacian $\tilde{\mathbf{L}}$ of the Levi graph $\tilde{G}$. Just as for $\mathbf{L}$, but unlike $\mathbf{C}$, the null-space is spanned by the all-1 vector.

Spanning thickets

We have seen that the value of the D-criterion is a function of the number of spanning trees of the concurrence graph G. It turns out that the closely related notion of a spanning thicket enables us to calculate the A-criterion (more precisely, the value of each pairwise effective resistance in G).

A *spanning thicket* for the graph is a spanning subgraph that consists of two trees (one of which may be an isolated vertex).

Theorem 6.3 *If i and j are distinct vertices of G, then*

$$R_{ij}=\frac{\textit{number of spanning thickets with } i,\ j \textit{ in different parts}}{\textit{number of spanning trees}}.$$

This is also rather easy to calculate directly when the graph is sparse.

Summing all the R_{ij} and using Theorem 6.3 gives the following result from [49]. If F is a spanning thicket of G, we denote the sets of vertices in its two trees by F_1 and F_2.

Theorem 6.4

$$\sum_{i<j} R_{ij} = \frac{\sum_{\text{spanning thickets } F} |F_1|\,|F_2|}{\text{number of spanning trees}},$$

where the sum on the right is over all spanning thickets F of G.

Random walks and electrical networks

It was first pointed out by Kakutani in 1945 that there is a very close connection between random walks and electrical networks. In a simple random walk, a single step works as follows: starting at a vertex, we choose an edge containing the vertex at random, and move along it to the other end. This definition accommodates multiple edges, and is easily adapted to graphs with edge weights (where the probability of moving along an edge is proportional to the weight of the edge).

If we are thinking of an edge-weighted graph as an electrical network, we take the weights to be the conductances of the edges (the reciprocals of the resistances).

The connection is simple to state.

Theorem 6.5 *Let i and j be distinct vertices of the connected edge-weighted graph G. Apply voltages of 1 at i and 0 at j. Then the voltage at a vertex l is equal to the probability that the random walk, starting at l, reaches i before it reaches j.*

From this theorem, one can derive a formula for the effective resistance between two vertices. Here are two such formulas. Given two vertices i and j, let $P_{\mathrm{esc}}(i \to j)$ be the probability that a random walk starting at i reaches j before returning to i, and let $S_i(i,j)$ be the expected number of times that a random walk starting at i visits i before reaching j. Then the effective resistance between i and j is given by either of the two expressions

$$\frac{1}{d_i P_{\mathrm{esc}}(i \to j)} \quad \text{and} \quad \frac{S_i(i,j)}{d_i},$$

where d_i is the degree of i. (If the edge resistances are not all 1, then the term d_i should be replaced by the sum of the reciprocals of the resistances of all edges incident with vertex i.)

The random-walk approach gives alternative proofs of some of the main results about electrical networks. We discuss this in Section 11.

Foster's formula and generalizations

In 1948, Foster [25] discovered that the sum of the effective resistances between all *adjacent* pairs of vertices of a connected graph on v vertices is equal to $v - 1$. Thirteen years later, he found a similar formula for pairs of vertices at distance 2:

$$\sum_{i \sim h \sim j} \frac{R_{ij}}{d_h} = v - 2.$$

Further extensions have been found, but these require a stronger condition on the graph. The sum of resistances between all pairs of vertices at distance at most m can be written down explicitly if the graph is *walk-regular up to distance* m: this means that, for $k \leq m$, the number of closed walks of length k starting and finishing at a vertex i is independent of i. The formula was discovered by Emil Vaughan, to whom this part of the chapter owes a debt.

In particular, if the graph is distance-regular (see [14]), then the value of the A-criterion can be written down in terms of the so-called *intersection array* of the graph.

Distance

At first sight it seems obvious that pairwise variance should decrease as concurrence increases, but there are many counter-examples to this. However, the following theorem is proved in [3].

Theorem 6.6 *If the Laplacian matrix* **L** *has precisely two distinct non-trivial eigenvalues, then pairwise variance is a decreasing linear function of concurrence.*

It does appear that effective resistance, and hence pairwise variance, generally increases with distance in the concurrence graph. In [7, Question 5.1], we pointed out that this is not always exactly so, and asked whether it is nevertheless true that the maximum value of R_{ij} is achieved for some pair of vertices $\{i, j\}$ whose distance apart in the graph is maximum. Here is a counter-example.

Example Let $k = 2$, so that the block design is the same as its concurrence graph. Take $v = 10$ and $b = 14$. The graph consists of a cube, with two extra vertices 1 and 2 attached as leaves to vertex 3. The vertex antipodal to 3 in the cube is labelled 4. It is straightforward to check (either using an electrical network, or by using the fact that the cube is distance-regular) that the effective resistances between a pair of cube vertices are $\frac{7}{12}$, $\frac{3}{4}$ and $\frac{5}{6}$ for vertices at distances 1, 2 and 3. Hence $R_{1j} \leq \frac{11}{6}$ for all cube vertices j, while $R_{12} = 2$. On the other hand, the distance between vertices 1 and 2 is only 2, while that between either of them and vertex 4 is 4.

There are some 'nice' graphs where the pairwise variance does indeed increase with distance. The following result is proved in [6]. Biggs gave the equivalent result for effective resistances in [12].

Theorem 6.7 *Suppose that a block design has just two distinct concurrences, and that the pairs of vertices corresponding to the larger concurrence form the edges of a distance-regular graph H. Then the pairwise variance increases with distance in H.*

7. E-optimality

There is no precise graph-theoretic interpretation of E-optimality similar to those we discussed for A- and D-optimality, but it is closely connected with isoperimetric properties of the concurrence graph. This connection has been known to graph theorists since the work of Alon and Milman [2].

Measures of bottlenecks

A 'good' graph (for use as a network) is one without bottlenecks: any set of vertices should have many edges joining it to its complement. For any subset S of vertices, we let $\partial(S)$ (the *boundary* of S) be the set of edges which have one vertex in S and the other in its complement, and then define the *isoperimetric number* $\iota(G)$ by

$$\iota(G) = \min\left\{|\partial S|/|S| : S \subseteq V(G),\ 0 < |S| \leq \tfrac{1}{2}v\right\}.$$

The next result shows that the isoperimetric number is related to the E-criterion. It is useful not so much for identifying the E-optimal designs as for easily showing that large classes of designs cannot be E-optimal: any design whose concurrence graph has low isoperimetric number performs poorly on the E-criterion.

Theorem 7.1 (Cutset lemma 1) *Let G have an edge-cutset of size c whose removal separates the graph into parts S and $G \setminus S$, with m and n vertices respectively, where $0 < m \leq n$. Then*

$$\theta_1 \leq c\left(\frac{1}{m} + \frac{1}{n}\right) \leq \frac{2\,|\partial S|}{|S|}.$$

Proof We know that θ_1 is the minimum of $\mathbf{x}^t\mathbf{L}\mathbf{x}/\mathbf{x}^t\mathbf{x}$, taken over real vectors $\mathbf{x}$ with $\sum_i x_i = 0$. Put

$$x_i = \begin{cases} n & \text{if } i \in S, \\ -m & \text{otherwise.} \end{cases}$$

Then $\mathbf{x}^t\mathbf{x} = nm(m+n)$ and

$$\mathbf{x}^t\mathbf{L}\mathbf{x} = \sum_{\text{edges } ij} (x_i - x_j)^2 = c(m+n)^2.$$

Hence,

$$\theta_1 \le \frac{\mathbf{x}^t \mathbf{L} \mathbf{x}}{\mathbf{x}^t \mathbf{x}} = \frac{c(m+n)^2}{nm(m+n)} = c\left(\frac{1}{m} + \frac{1}{n}\right) \le \frac{2c}{m} = \frac{2\,|\partial S|}{|S|},$$

as required. ■

Corollary 7.2 *Let θ_1 be the smallest non-trivial eigenvalue of the Laplacian matrix* $\mathbf{L}$ *of the connected graph G. Then $\theta_1 \le 2\iota(G)$.*

There is also an upper bound for the isoperimetric number in terms of θ_1, which is loosely referred to as a 'Cheeger-type inequality'; for details, see Section 11.

We also require a result about vertex-cutsets.

Theorem 7.3 (Cutset lemma 2) *Let G have a vertex-cutset C of size c whose removal separates the graph into parts S and T with m, n vertices, respectively (so $nm > 0$). Let m' and n' be the number of edges from vertices in C to vertices in S, T respectively. Then*

$$\theta_1 \le \frac{m'n^2 + n'm^2}{nm(m+n)}.$$

In particular, if there are no multiple edges at any vertex of C then $\theta_1 \le c$, with equality if and only if every vertex in C is joined to every vertex in $S \cup T$.

Proof Let

$$x_i = \begin{cases} n & \text{if } i \in S, \\ -m & \text{if } i \in T, \\ 0 & \text{otherwise.} \end{cases}$$

Then $\mathbf{x}^t \mathbf{x} = nm(m+n)$ and $\mathbf{x}^t \mathbf{L} \mathbf{x} = m'n^2 + n'm^2$, and so

$$\theta_1 \le \frac{m'n^2 + n'm^2}{nm(m+n)}.$$

If there are no multiple edges at any vertex in C, then $m' \le cm$ and $n' \le cn$ and the result follows. ■

Variance balance

A block design is *variance-balanced* if all the concurrences λ_{ij} are equal for $i \ne j$. In such a design, all of the pairwise variances V_{ij} are equal. Morgan and Srivastav proved the following result in [43].

Theorem 7.4 *If the constant concurrence λ of a variance-balanced design satisfies $(v-1)\lambda = \lfloor (bk/v) \rfloor (k-1)$, then the design is E-optimal.*

A block with k different treatments contributes $\frac{1}{2}k(k-1)$ edges to the concurrence graph. Define the *defect* of a block to be

$$\tfrac{1}{2}k(k-1) - (\text{the number of edges that it contributes to the graph}).$$

The following result is proved in [7].

Theorem 7.5 *If $k < v$, then a variance-balanced design with v treatments is E-optimal if the sum of the block defects is less than $\frac{1}{2}v$.*

Table 1(b) (on page 288) shows that the design in Fig. 2(b) is variance-balanced. Block Γ_1 has defect 1, and each other block has defect 0, so the sum of the block defects is certainly less than $\frac{5}{2}$, and Theorem 7.5 shows that the design is E-optimal. It is rather counter-intuitive that the non-binary design in Fig. 2(b) can be better than the design in Fig. 2(a); in fact, in his contribution to the discussion of Tocher's paper [51], which introduced this design, David Cox said, 'I suspect that . . . balanced ternary designs are of no practical value'.

Computation shows that the design in Fig. 2(a) is Φ_p-better than the one in Fig. 2(b) if $p < 5.327$. In particular, it is A-better and D-better.

8. Some history

As we have seen, if the experimental units form a single block and there are only two treatments, then it is best for their replications to be as equal as possible. Statisticians know this so well that it is hard for us to imagine that more information may be obtained about *all* treatment comparisons if replications differ by more than 1.

In agriculture, or in any area with qualitative treatments, A-optimality is the natural criterion. If treatments are quantities of different substances, then D-optimality is preferable, as the ranking on this criterion is invariant to change of measurement units. Thus industrial statisticians have tended to prefer D-optimality, although E-optimality has become popular among chemical process engineers. Perhaps the different camps have not talked to each other as much as they should have.

For most of the 20th century, it was normal practice in field experiments to have all treatments replicated three or four times. Where incomplete blocks were used, they typically had size from 3 to 20. Yates introduced his *square lattice designs* with $v = k^2$ in [55]. He used uniformity data and two worked examples to show that these designs can give lower average pairwise variance than a design using a highly replicated control treatment, but both of his examples were equireplicate, with $r = 3$ and $r = 4$.

In the 1930s, 1940s and 1950s, analysis of the data from an experiment involved inverting the Laplacian matrix without a computer: this is easy for BIBDs, and only slightly harder if the Laplacian matrix has only two distinct non-trivial eigenvalues. The results in [35] and [38] encouraged the beliefs that the optimal designs, on all

Φ_p-criteria, are as equireplicate as possible, with concurrences as nearly equal as possible, and that the same designs are optimal, or nearly so, on all of these criteria.

Three short papers in the same journal in 1977–1982 demonstrate the beliefs at that time. In [29], John and Mitchell did not even consider designs with unequal replication. They conjectured that, if there exist any regular-graph designs for given values of v, b and k, then the A- and D-optimal designs are regular-graph designs. For the parameter sets which they had examined by computer search, the same designs were optimal on the A- and D-criteria. In [33], Jones and Eccleston reported the results of various computer searches for A-optimal designs without the constraint of equal replication. For $k = 2$ and $b = v = 10$, 11 or 12 (but not $v = 9$), their A-optimal design is almost a queen-bee design, and their designs are D-worse than those in [29]. The belief in equal replication was so ingrained that some readers assumed that there was an error in their program.

John and Williams followed this with a paper [30] on conjectures for optimal block designs with given values of v, b and k. Their conjectures included:

- the set of regular-graph designs always contains one that is optimal without this restriction;
- among regular-graph designs, the same designs are optimal on the A- and D-criteria.

They endorsed Cox's dismissal of non-binary designs, strengthening it to the statement that they 'are inefficient', and declared that the three unequally replicated A-optimal designs in [33] were 'of academic rather than of practical interest'. These conjectures and opinions seemed quite reasonable to people who had been finding good designs for the sizes needed in agricultural experiments.

At the end of the 20th century, there was an explosion in the number of experiments in genomics, using microarrays. Simplifying the story greatly, these are effectively block designs with $k = 2$, and biologists wanted A-optimal designs, but they did not know the vocabulary of 'block' or 'A-optimal', 'graph' or 'cycle'. Computers were now much more powerful than in 1980, and researchers in genomics could simply undertake computer searches without the benefit of any statistical theory. In 2001, Kerr and Churchill [34] published the results of a computer search for A-optimal designs with $k = 2$ and $v = b \leq 11$. For $v = 10$ or 11, their results were completely consistent with those in [33], which they did not cite. They called cycles *loop designs*.

Mainstream statisticians began to get involved. In 2005, Wit, Nobile and Khanin published a paper [53] giving the results of a computer search for A- and D-optimal designs with $k = 2$ and $v = b$; the results are shown in Fig. 10. The A-optimal designs differ from the D-optimal designs when $v \geq 9$, but are consistent with those found in [34].

What is going on here? Why are the designs so different when $v \geq 9$? Why is there such a sudden large change in the A-optimal designs? We explain this in the next section.

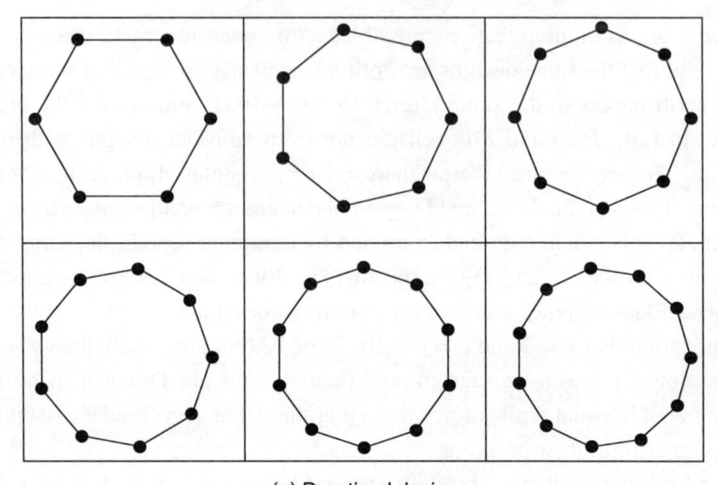

(a) D-optimal designs

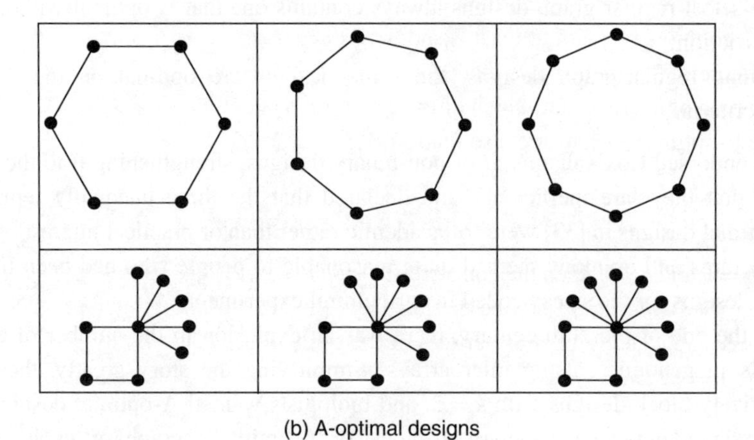

(b) A-optimal designs

Fig. 10. D- and A-optimal designs with $k=2$ and $6 \leq v=b \leq 11$

9. Block size 2

An incomplete-block design with $k=2$ is essentially the same as its concurrence graph. This leads to some simplifications.

Least replication

If $k=2$, then the design 'is' its concurrence graph, and connectivity requires that $b \geq v-1$. If $b=v-1$, then all connected designs are trees, such as those in Fig. 11. Theorem 5.1 shows that the D-criterion does not differentiate between them.

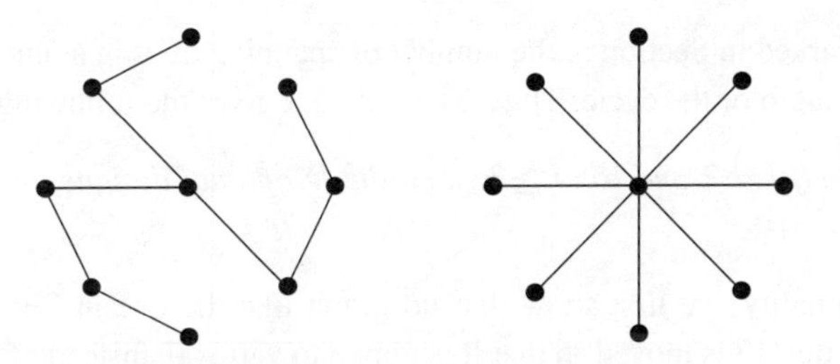

Fig. 11. Two trees with $v=9$, $b=8$ and $k=2$

In a tree, the effective resistance R_{ij} is just the length of the unique path between vertices i and j. Theorems 3.1 and 6.1 show that the only A-optimal designs are the star graphs, such as the graph on the right of Fig. 11.

In a star graph with v vertices, the contrast between any two leaves is an eigenvector of the Laplacian matrix $\mathbf{L}$ with eigenvalue 1, while the contrast between the central vertex and all the other vertices is an eigenvector with eigenvalue v. If $v \geq 5$ and the graph is not a star, then there is an edge whose removal splits the graph into two components with at least 2 and 3 vertices, respectively. The first cutset lemma (Theorem 7.1) then shows that $\theta_1 \leq \frac{5}{6} < 1$. The only other tree which is not a star is the path of length 3, for which direct calculation shows that $\theta_1 = 2 - \sqrt{2} < 1$. Hence, the E-optimal designs are also the stars.

One fewer treatment

If $b = v$ and $k = 2$, then the concurrence graph G contains a single cycle: such graphs are called *unicyclic*. Let s be the length of the cycle. All the remaining vertices are in trees attached to various vertices of the cycle. Fig. 12 shows two unicyclic graphs with $v = 12$ and $s = 6$.

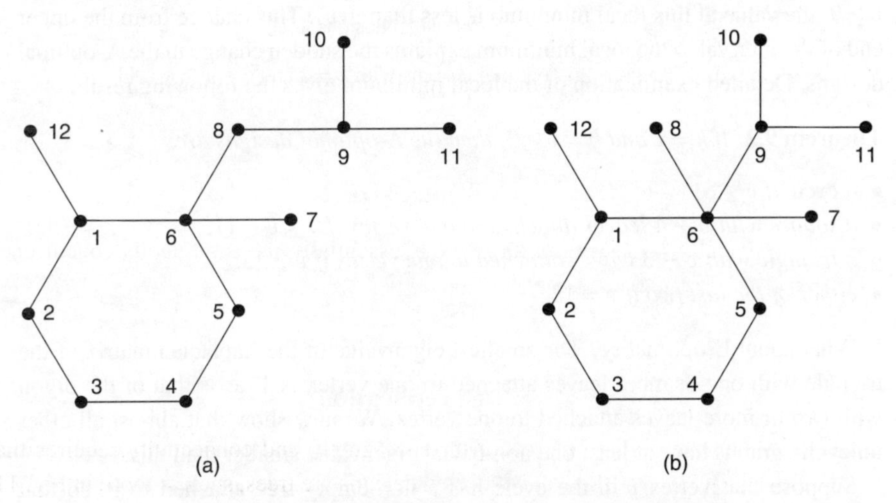

Fig. 12. Two unicyclic graphs with $b = v = 12$ and $s = 6$

As we remarked in Section 5, the number of spanning trees in a unicyclic graph is equal to the length of the cycle. Thus, Theorem 5.1 gives the following result.

Theorem 9.1 *If $k=2$ and $b=v\geq 3$, then the D-optimal designs are precisely the cycles.*

For A-optimality, we first show that no graph like the one in Fig. 12(a) can be optimal. If vertex 12 is moved so that it is joined to vertex 6, instead of vertex 1, then the sum of the variances $V_{i,12}$ for i in the cycle is unchanged and the variances $V_{i,12}$ for the remaining vertices i are all decreased. This argument shows that all the trees must be attached to the same vertex of the cycle.

Now consider the tree with vertices 6, 8, 9, 10 and 11 in Fig. 12(a). If the two edges incident with vertex 8 are modified to those in Fig. 12(b), then the set of variances between these five vertices are unchanged, as are all others involving vertex 8, but those between vertices 9, 10, 11 and vertices outside this tree are all decreased. This argument shows that, for any given length s of the cycle, the only candidate for an A-optimal design has $v-s$ leaves attached to a single vertex of the cycle.

The effective resistance between a pair of vertices at distance d in a cycle of length s is $d(s-d)/s$, while that between a leaf and the cycle vertex to which it is attached is 1. A short calculation shows that the sum of the pairwise effective resistances is equal to $\frac{1}{12}g(s)$, where

$$g(s)=-s^3+2vs^2+13s-12sv+12v^2-14v.$$

Now $\bar{V}/\sigma^2=g(s)/3v(v-1)$, and we seek the minimum of $g(s)$ for integers s in the interval $[2,v]$.

Fig. 13 plots $g(s)/3v(v-1)$ for s in $[2,v]$ and $6\leq v\leq 13$. When $v\leq 7$, the function g is monotonic decreasing, and so it attains its minimum on $[2,v]$ at $s=v$. For all larger values of v, the function g has a local minimum in the interval $[3,5]$: when $v\geq 9$, the value at this local minimum is less than $g(v)$. This change from the upper end of the interval to the local minimum explains the sudden change in the A-optimal designs. Detailed examination of the local minimum gives the following result.

Theorem 9.2 *If $k=2$ and $b=v\geq 3$, then the A-optimal designs are:*

- *a cycle if $v\leq 8$;*
- *a square with $v-4$ leaves attached to one vertex if $9\leq v\leq 11$;*
- *a triangle with $v-3$ leaves attached to one vertex if $v\geq 13$;*
- *either of the last two if $v=12$.*

What about E-optimality? The smallest eigenvalue of the Laplacian matrix of the triangle with one or more leaves attached to one vertex is 1, as is that of the digon with two or more leaves attached to one vertex. We now show that almost all other unicyclic graphs have at least one non-trivial eigenvalue smaller than this.

Suppose that vertex i in the cycle has a non-empty tree attached to it, so that $\{i\}$ is a vertex-cutset. If $s\geq 3$, then there are no double edges, so the second

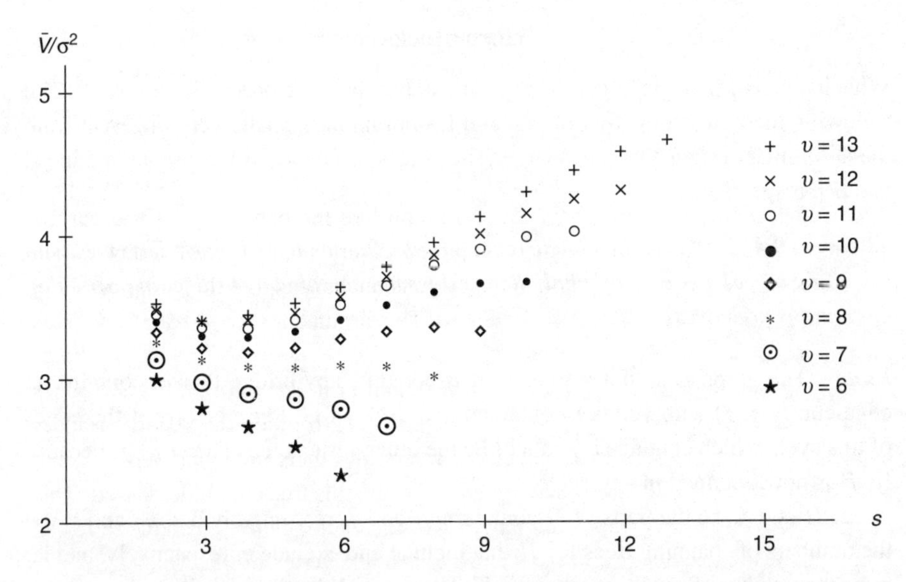

Fig. 13. Average pairwise variance, in a unicylic graph with v vertices, as a function of the length s of the cycle

cutset lemma (Theorem 7.3) shows that $\theta_1 < 1$, unless all vertices are joined to i, in which case $s = 3$. If $s = 2$ and there are trees attached to both vertices of the digon, then applying the theorem at each of these vertices shows that $\theta_1 < 1$, unless $v = 4$ and there is one leaf at each vertex of the digon: for this graph, we have $\theta_1 = 2 - \sqrt{5} < 1$. A digon with leaves attached to one vertex is just a star with one edge doubled.

The cycle of size v is a cyclic design. The smallest eigenvalue of its Laplacian matrix is $2(1 - \cos(2\pi/v))$, which is greater than 1 when $v \leq 5$, is equal to 1 when $v = 6$ and is less than 1 when $v \leq 7$. When $v = 3$ it is equal to 3, which is greater than $3 - \sqrt{3}$, which is the smallest Laplacian eigenvalue of the digon with one leaf.

Putting all of this together proves the following result.

Theorem 9.3 *If $k = 2$ and $b = v \geq 3$, then the E-optimal designs are:*

- *a cycle, if $v \leq 5$;*
- *a triangle with $v - 3$ leaves attached to one vertex, or a star with one edge doubled, if $v \geq 7$;*
- *any of the above, if $v = 6$.*

Thus, for $v \geq 9$, the ranking on the D-criterion is essentially the opposite of the ranking on the A- and E-criteria. The A- and E-optimal designs are far from equireplicate. The change is sudden, not gradual. These findings were initially quite shocking to statisticians.

More blocks

What happens when b is larger than v, but still has the same order of magnitude? The following theorems show that the A- and E-optimal designs are very different from the D-optimal designs when v is large. The proofs of Theorems 9.5 and 9.6 are in [4] and [7], respectively.

Theorem 9.4 *Let G be the concurrence graph of a connected block design Δ with $k=2$ and $b \geq v$. If Δ is D-optimal, then G does not contain any bridge: in particular, G contains no leaves.*

Proof The graph G is not a tree, so if it contains any bridge then it contains an edge-cutset $\{i, j\}$ with vertex i contained in a cycle of G. Let e be one of the edges of this cycle which contains i, and let l be the other vertex on e. Then $l \neq j$, because $\{i, j\}$ is not contained in any cycle.

Let H and K be the parts of G containing i and j, respectively. Let n_1 and n_2 be the numbers of spanning trees for H that include and exclude e, respectively, and let m be the number of spanning trees for K. Then $n_2 \geq 1$, because e is in a cycle. Every spanning tree for G consists of spanning trees for H and K together with the edge $\{i, j\}$. Hence G has $(n_1 + n_2)m$ spanning trees.

Form G' from G by removing edge e and inserting the edge e', where $e' = \{l, j\}$. Let T and T' be spanning trees for H and K, respectively. If T does not contain e, then $T \cup \{\{i, j\}\} \cup T'$ and $T \cup \{e'\} \cup T'$ are both spanning trees for G'. If T contains e, then $(T \setminus \{e\}) \cup \{\{i, j\}\} \cup \{e'\} \cup T'$ is a spanning tree for G'. Hence the number of spanning trees for G' is at least $(2n_2 + n_1)m$, which is greater than $(n_1 + n_2)m$ because $n_2 \geq 1$. Hence G does not have the maximal number of spanning trees and so Δ is not D-optimal. ∎

Theorem 9.5 *Let c be a positive integer. Then there is a positive integer v_c such that, if $b - v = c$ and $v \geq v_c$, then all A-optimal designs with $k=2$ contain leaves.*

Theorem 9.6 *If $20 \leq v \leq b \leq \frac{5}{4}v$, then the concurrence graph for any E-optimal design with $k=2$ contains leaves.*

Note that, to obtain a BIBD when $k=2$, b needs to be a quadratic function of v. What happens if b is merely a linear function of v? In [7] we conjectured that, if $b=cv$ for some constant c, then there is a threshold result like the one in Theorem 9.5. However, current work by Johnson and Walters [32] suggests something much more interesting – that there is a constant C with $3 < C < 4$ such that, if $b \geq Cv$ and $k=2$, then all A-optimal designs are (nearly) equireplicate, and that random graphs (in a suitable model) are close to A-optimal with high probability. On the other hand, if $b \leq Cv$ and b is not too small, then a graph consisting of a large almost equireplicate part (all degrees 3 and 4, with average degree close to Cv), together with a suitable number of leaves joined to a single vertex, is strictly better than any queen-bee design.

A little more history

The earlier results on D- and A-optimality in this section were proved in [4], partly to put to rest mutterings that the results of [33], [34] and [53] found by computer search were incorrect. The results on E-optimality are in [7].

In spite of the horror with which these results were greeted, it transpired that they were not new. The D- and E-optimal designs for $b = (v-1)/(k-1)$ were identified in [11] in 1991. The A-optimal designs for $k = 2$ and $b = v - 1$ had been given in [41] in 1991. Also in 1991, Tjur gave the A-optimal designs for $k = 2$ and $b = v$ in [52]: his proof used the Levi graph as an electrical network.

A fairly common response to these unexpected results was 'It seems to be just block size 2 that is a problem'. Perhaps those of us who usually deal with larger blocks had simply not thought that it was worthwhile to investigate block size 2 before the introduction of microarrays.

However, as we sketch in the next section, the problem is not block size 2, but very low average replication. The proofs there are similar to those in this section; they are given in more detail in [8] and [46]. Once again, it turns out that these results are not all new. The D-optimal designs for $v/(k-1)$ blocks of size k were given by Balasubramanian and Dey [10] in 1996, but their proof uses a version of Theorem 5.3 with the wrong value of the constant. The A-optimal designs for $v/(k-1)$ blocks of size k were published by Krafft and Schaefer in [37] in 1997, but those authors are not blameless either, because they apparently had not read [52]!

Our best explanation is that agricultural statisticians are so familiar with average replication being at least 3, that when they saw these papers they decided that these had no applicability and so forgot them.

10. Low average replication

In this section we once again consider general block size k. A block design is connected if and only if its Levi graph is connected. The Levi graph has $v + b$ vertices and bk edges, so connectedness implies that $bk \geq b + v - 1$ (that is, $b(k-1) \geq v - 1$).

Least replication

If $b(k-1) = v - 1$ and the design is connected, then the Levi graph $\tilde{G}$ is a tree and the concurrence graph G looks like those in Fig. 5. Hypergraph-theorists do not seem to have an agreed name for such designs but some graph theorists call these graphs (0, 2)-trees.

For both D- and A-optimality, it turns out to be convenient to use the Levi graph. Since all the Levi graphs are trees, Theorem 5.3 shows that the D-criterion does not distinguish among connected designs.

By Theorem 6.2, $V_{ij} = \tilde{R}_{ij}\sigma^2$. When $\tilde{G}$ is a tree, $\tilde{R}_{ij} = 2$ when i and j are in the same block; otherwise, $\tilde{R}_{ij} = 4$ if any block containing i has a treatment in common with any block containing j; and otherwise $\tilde{R}_{ij} \geq 6$. The queen-bee designs are the only ones for which $\tilde{R}_{ij} \leq 4$ for all i and j, and so they are the A-optimal designs.

The non-trivial eigenvalues of a queen-bee design are 1, k and v, with multiplicities $b-1$, $b(k-2)$ and 1, respectively. If the design is not a queen-bee design, then there is a treatment i that is in more than one block but not in all blocks. Thus, vertex i forms a cutset for the concurrence graph G which is not joined to every other vertex of G. Theorem 7.3 shows that $\theta_1 < 1$. Hence the E-optimal designs are also the queen-bee designs.

One fewer treatment

If $b(k-1) = v$, then the Levi graph $\tilde{G}$ has bk edges and bk vertices, and so it contains a single cycle which must be of some even length $2s$. If $2 \leq s \leq b$, then the design is binary. If $s = 1$, then there is a single non-binary block, whose defect is 1 (see before Theorem 7.5 for the definition); in this case, $k \geq 3$, because each block must have more than one treatment.

For $2 \leq s \leq b$, let $\mathcal{C}(b, k, s)$ be the class of designs constructed as follows. Start with a loop design for s treatments. Insert $k-2$ extra treatments into each block. The remaining $b-s$ blocks all contain the same treatment from the loop design, together with $k-1$ extra treatments. Figs. 8 and 9 show the concurrence graph and Levi graph, respectively, of a design in $\mathcal{C}(6, 3, 6)$.

For $k \geq 4$, the designs in $\mathcal{C}(b, k, 1)$ have one treatment which occurs twice in one block and once in all other blocks, with the remaining treatments all replicated once. The class $\mathcal{C}(b, 3, 1)$ contains all such designs, and also those in which the treatment in every block is the one that occurs only once in the non-binary block.

Theorem 10.1 *If $b(k-1) = v$, then the D-optimal designs are those in $\mathcal{C}(b, k, b)$.*

Proof The Levi graph $\tilde{G}$ is unicyclic, so its number of spanning trees is maximized when the cycle has maximal length. Theorem 5.3 shows that the D-optimal designs are precisely those with $s = b$. ∎

Theorem 10.2 *If $b(k-1) = v$, then the A-optimal designs are those in $\mathcal{C}(b, k, s)$, where the value of s is given in Table 2.*

k \ b	2	3	4	5	6	7	8	9	10	11	12	13
2	2	3	4	5	6	7	8	4	4	4	3 or 4	3
3	2	3	4	5	6	3	3	3	3	3	2	2
4	2	3	4	5	3	2	2	2	2	2	2	2
5	2	3	4	5	2	2	2	2	2	2	2	2
6	2	3	4	2	2	2	2	2	2	2	2	2

Table 2. Value of s for A-optimal designs for $b(k-1)$ treatments in b blocks of size k: see Theorem 10.2

Proof The Levi graph $\tilde{G}$ has one cycle, whose length is $2s$, where $1 \leq s \leq b$. A similar argument to the one used at the start of the proof of Theorem 9.2 shows that this cannot be A-optimal unless the design is in $\mathcal{C}(b, k, s)$. If $s \geq 2$ or $k \geq 4$, then each block-vertex in the cycle has $k - 2$ treatment-vertices attached as leaves; all other block-vertices are joined to the same single treatment-vertex in the cycle, and each has $k - 1$ treatment vertices attached as leaves. In $\mathcal{C}(b, 3, 1)$ the first type of design has a Levi graph like this, and the other type has the same multiset of effective resistances between treatment vertices, because their concurrence graphs are identical. The following calculations use the first type.

Let $\mathcal{V}_1$ be the set of treatment-vertices in the cycle, $\mathcal{V}_2$ the set of other treatment-vertices joined to blocks in the cycle and $\mathcal{V}_3$ the set of remaining treatment-vertices. For $1 \leq i \leq j \leq 3$, denote by $\mathcal{R}_{ij}$ the sum of the pairwise resistances between vertices in $\mathcal{V}_i$ and $\mathcal{V}_j$. Put

$$R_1 = \sum_{d=1}^{s-1} \frac{2d(2s - 2d)}{2s} = \tfrac{1}{3}(s^2 - 1)$$

and

$$R_2 = \sum_{d=0}^{s-1} \frac{(2d + 1)(2s - 2d - 1)}{2s} = \tfrac{1}{6}(2s^2 + 1).$$

Then

$$\begin{aligned}
\mathcal{R}_{11} &= \tfrac{1}{2} s R_1, \\
\mathcal{R}_{12} &= s(k - 2)(R_2 + s), \\
\mathcal{R}_{13} &= (b - s)(k - 1)(R_1 + 2s), \\
\mathcal{R}_{22} &= s(k - 2)(k - 3) + \tfrac{1}{2} s(k - 2)^2[R_1 + 2(s - 1)], \\
\mathcal{R}_{23} &= (b - s)(k - 1)(k - 2)(R_2 + 3s), \\
\mathcal{R}_{33} &= (b - s)(k - 1)(k - 2) + 2(b - s)(b - s - 1)(k - 1)^2.
\end{aligned}$$

Hence, the sum of the pairwise effective resistances between treatment-vertices in the Levi graph is $\frac{1}{6} g(s)$, where

$$g(s) = -(k - 1)^2 s^3 + 2b(k - 1)^2 s^2 - (6bk(k - 1) - 4k^2 + 2k - 1)s + c$$

and $c = b(k - 1)(12b(k - 1) - 5k - 4)$.

If $s = 1$, then the design is non-binary. However,

$$g(1) - g(2) = (3k - 9 + 6b)(k - 1) - 3,$$

which is positive, because $k \geq 2$ and $b \geq 2$. Therefore the non-binary designs are never A-optimal.

Direct calculation shows that $g(2) > g(3)$ when $b = 3$, and that $g(2) > g(3) > g(4)$ when $b = 4$. These inequalities hold for all values of k, even though g is not decreasing on the interval $[2, 4]$ for large k when $b = 4$.

If $b = 5$ and $k \geq 6$, then $g(3) > g(2)$ and $g(5) > g(2)$. Thus the local minimum of g occurs in the interval $(1, 3)$ and is the overall minimum of g on the interval $[1, 5]$.

Differentiation gives

$$g'(b) = b(k-1)\big((b-6)(k-1) - 6\big) + 4k^2 - 2k + 1.$$

If $g'(b) > 0$, then g has a local minimum in the interval $(1, b)$. If, in addition, $g(3) > g(2)$, then the minimal value for integer s occurs at $s = 2$. These conditions are both satisfied if $k = 3$ and $b \geq 12$, $k = 4$ and $b \geq 8$, $k \geq 5$ and $b \geq 7$, or $k \geq 9$ and $b \geq 6$.

Given Theorem 9.2, there remain only a finite number of pairs (b, k) to be checked individually to find the smallest value of $g(s)$. The results are in Table 2. ■

Theorem 10.3 *If $b(k-1) = v$, $b \geq 3$ and $k \geq 3$, then the E-optimal designs are those in $\mathcal{C}(b, k, b)$ if $b \leq 4$, and those in $\mathcal{C}(b, k, 2)$ and $\mathcal{C}(b, k, 1)$ if $b \geq 5$.*

Proof If $2 < s < b$, then the concurrence graph G has a cut vertex that is not joined to all other vertices; moreover, G has no multiple edges. Theorem 7.3 shows that $\theta_1 < 1$.

Direct calculation shows that $\theta_1 = 1$ if $s = 1$ or $s = 2$. For $k \geq 4$, all contrasts between singly replicated treatments in the same block are eigenvectors of the Laplacian matrix L with eigenvalue k. When $k \geq 3$ and $s = b$, the contrast between singly and doubly replicated treatments has eigenvalue $2(k-1)$. For $s = b$, a straightforward calculation shows that the remaining eigenvalues of L are

$$k - \cos\left(\frac{2\pi n}{b}\right) \pm \sqrt{(k-1)^2 - \sin^2\left(\frac{2\pi n}{b}\right)}$$

for $1 \leq n \leq b - 1$. The smallest of these is

$$k - \cos(2\pi/b) - \sqrt{(k - 1^2) - \sin^2(2\pi/b)},$$

which is greater than 1 if $b = 3$ or $b = 4$, but less than 1 if $k \geq 3$ and $b \geq 5$. ■

11. Further reading

The Laplacian matrix of a graph, and its eigenvalues, are widely used, especially in connection with network properties such as connectivity, expansion and random walks. A good introduction to this material can be found in the textbook by Bollobás [13], especially Chapters II (electrical networks) and IX (random walks). Connections between the smallest non-zero eigenvalue and connectivity are described in surveys by de Abreu [1] and Mohar [42]. In this terminology, a version of Theorem 9.3 is in [24].

The basic properties of electrical networks can be found in textbooks of electrical engineering – for example, Balabanian and Bickart [9]. A treatment related to the multivariate Tutte polynomial appears in Sokal's survey [50]. Bollobás [13] describes several approaches to the theory, including the fact (which we have not used) that the current flow minimizes the power consumed in the network, and explains the interactions between electrical networks and random walks in the network. See also Deo [22].

The connections with optimal design theory were discussed in detail by the authors in their survey [7]. Further reading on optimal design can be found in John and Williams [31], Schwabe [47] and Shah and Sinha [48]. For general principles of experimental design, see Bailey [5].

Acknowledgement

This chapter was written at the Isaac Newton Institute for Mathematical Sciences, Cambridge, UK, during the 2011 programme on Design and Analysis of Experiments.

References

1. N. M. M. de Abreu, Old and new results on algebraic connectivity of graphs, *Linear Algebra and its Applications*, **423** (2007), 53–73.
2. N. Alon and V. Milman, λ_1, isoperimetric inequalities for graphs, and superconcentrators, *J. Combin. Theory* (*B*) **38** (1985), 73–88.
3. R. A. Bailey, *Association Schemes: Designed Experiments, Algebra and Combinatorics*, Cambridge Studies in Advanced Mathematics **84**, Cambridge University Press, 2004.
4. R. A. Bailey, Designs for two-colour microarray experiments, *Applied Statistics* **56** (2007), 365–394.
5. R. A. Bailey, *Design of Comparative Experiments*, Cambridge Series in Statistical and Probabilistic Mathematics **25**, Cambridge University Press, 2008.
6. R. A. Bailey, Variance and concurrence in block designs, and distance in the corresponding graphs, *Michigan Math. J.* **58** (2009), 105–124.
7. R. A. Bailey and P. J. Cameron, Combinatorics of optimal designs, *Surveys in Combinatorics 2009* (eds. S. Huczynska, J. D. Mitchell and C. M. Roney-Dougal), London Math. Soc. Lecture Notes **365**, Cambridge University Press (2009), 19–73.
8. R. A. Bailey and A. Sajjad, Optimality in nearly minimal block designs, in preparation.
9. N. Balabanian and T. A. Bickart, *Electrical Network Theory*, Wiley, 1969.
10. K. Balasubramanian and A. Dey, D-optimal designs with minimal and nearly minimal number of units, *J. Statist. Planning and Inference* **52** (1996), 255–262.
11. R. B. Bapat and A. Dey, Optimal block designs with minimal number of observations, *Statist. Probab. Letters* **11** (1991), 399–402.
12. N. L. Biggs, Potential theory on distance-regular graphs, *Combin. Probab. Comput.* **2** (1993), 107–119.
13. B. Bollobás, *Modern Graph Theory*, Springer, 1998.

14. A. E. Brouwer, A. M. Cohen and A. Neumaier, *Distance-Regular Graphs*, Ergebnisse der Mathematik und ihrer Grenzgebiete (3) **18**, Springer, 1989.
15. P. J. Cameron, Strongly regular graphs, *Topics in Algebraic Graph Theory* (eds. L. W. Beineke and R. J. Wilson), Cambridge University Press (2004), 203–221.
16. C.-S. Chêng, Optimality of certain asymmetrical experimental designs, *Ann. Statist.* **6** (1978), 1239–1261.
17. C.-S. Chêng, Maximizing the total number of spanning trees in a graph: two related problems in graph theory and optimum design theory, *J. Combin. Theory* (*B*) **31** (1981), 240–248.
18. C.-S. Cheng, Graph and optimum design theories – some connections and examples, *Bull. Internat. Statistical Institute* **49** (Proc. 43rd Session, Buenos Aires) (1981), 580–590.
19. C.-S. Cheng, On the optimality of (M.S)-optimal designs in large systems, *Sankhyā* **54** (1992), 117–125.
20. C.-S. Cheng and R. A. Bailey, Optimality of some two-associate-class partially balanced incomplete-block designs, *Ann. Statist.* **19** (1991), 1667–1671.
21. C.-S. Cheng, J. C. Masaro and C. S. Wong, Do nearly balanced multigraphs have more spanning trees?, *J. Graph Theory* **8** (1985), 342–345.
22. N. Deo, *Graph Theory with Applications to Engineering and Computer Science*, Prentice Hall, 1980.
23. J. Dodziuk, Difference equations, isoperimetric inequality and transience of certain random walks, *Trans. Amer. Math. Soc.* **284** (1984), 787–794.
24. S. M. Fallat, S. Kirkland and S. Pati, Maximizing algebraic connectivity over unicyclic graphs, *Linear and Multilinear Algebra* **3** (2003), 221–241.
25. R. M. Foster, The average impedance of an electrical network, *Reissner Anniversary Volume, Contributions to Applied Mathematics* (ed. J. W. Edwards), Ann Arbor (1948), 333–340.
26. N. Gaffke, *Optimale Versuchsplanung für linear Zwei-Faktor Modelle*, Ph.D. thesis, Rheinisch–Westfälische Technische Hochschule, Aachen, 1978.
27. N. Gaffke, D-optimal block designs with at most six varieties, *J. Statist. Planning and Inference* **6** (1982), 183–200.
28. R. Harman, Minimal efficiency of designs under the class of orthogonally invariant information criteria, *Metrika* **60** (2004), 137–153.
29. J. A. John and T. J. Mitchell, Optimal incomplete block designs, *J. Royal Statist. Soc.* (*B*) **39** (1977), 39–43.
30. J. A. John and E. R. Williams, Conjectures for optimal block designs, *J. Royal Statist. Soc.* (*B*) **44** (1982), 221–225.
31. J. A. John and E. R. Williams, *Cyclic and Computer Generated Designs* (2nd edn.), Monographs on Statistics and Applied Probability **38**, Chapman and Hall, 1995.
32. J. R. Johnson and M. Walters, Optimal resistor networks, in preparation.
33. B. Jones and J. A. Eccleston, Exchange and interchange procedures to search for optimal designs, *J. Royal Statist. Soc.* (*B*) **42** (1980), 238–243.
34. M. K. Kerr and G. A. Churchill, Experimental design for gene expression microarrays, *Biostatistics* **2** (2001), 183–201.
35. J. Kiefer, Construction and optimality of generalized Youden designs, *A Survey of Statistical Design and Linear Models* (ed. J. N. Srivastava), North-Holland (1975), 333–353.
36. G. Kirchhoff, Über die Auflösung der Gleichenung, auf welche man bei der Untersuchung der linearen Verteilung galvanischer Ströme gefürht wird, *Ann. Phys. Chem.* **72** (1847), 497–508.

37. O. Krafft and M. Schaefer, A-optimal connected block designs with nearly minimal number of observations, *J. Statist. Planning and Inference* **65** (1997), 357–386.
38. A. M. Kshirsagar, A note on incomplete block designs, *Ann. Math. Statist.* **29** (1958), 907–910.
39. A. H. M. M. Latif, F. Bretz and E. Brunner, Robustness considerations in selecting efficient two-color microarray designs, *Bioinformatics* **25** (2009), 2355–2361.
40. F. W. Levi, *Finite Geometrical Systems*, University of Calcutta, 1942.
41. N. K. Mandal, K. R. Shah and B. K. Sinha, Uncertain resources and optimal designs: problems and perspectives, *Calcutta Statist. Assoc. Bull.* **40** (1991), 267–282.
42. B. Mohar, The Laplacian spectrum of graphs, *Graph Theory, Combinatorics and Applications*, Vol. 2 (Kalamazoo, MI, 1988), Wiley–Interscience (1991), 871–898.
43. J. P. Morgan and S. K. Srivastav, The completely symmetric designs with blocksize three, *J. Statist. Planning and Inference* **106** (2002), 21–30.
44. H. D. Patterson and E. R. Williams, Some theoretical results on general block designs, *Congr. Numer.* **15** (1976), 489–496.
45. R. M. Pringle and A. A. Rayner, *Generalized Inverse Matrices with Applications to Statistics*, Griffin's Statistical Monographs and Courses **28**, Griffin, 1971.
46. A. Sajjad, *Optimality in sparse block designs*, Ph.D. thesis, Quaid-i-Azam University, Islamabad, 2011.
47. R. Schwabe, *Optimum Designs for Multi-Factor Models*, Lecture Notes in Statistics **113**, Springer, 1996.
48. K. R. Shah and B. K. Sinha, *Theory of Optimal Designs*, Lecture Notes in Statistics **54**, Springer-Verlag, 1989.
49. L. W. Shapiro, An electrical lemma, *Math. Mag.* **60** (1987) 36–38.
50. A. D. Sokal, The multivariate Tutte polynomial (alias Potts model) for graphs and matroids, *Surveys in Combinatorics 2005* (ed. B. S. Webb), London Math. Soc. Lecture Notes **327**, Cambridge University Press (2005), 173–226.
51. K. D. Tocher, The design and analysis of block experiments, *J. Royal Statist. Soc.* (*B*) **14** (1952), 45–100.
52. T. Tjur, Block designs and electrical networks, *Ann. Statist.* **19** (1991), 1010–1027.
53. E. Wit, A. Nobile and R. Khanin, Near-optimal designs for dual channel microarray studies, *Applied Statist.* **54** (2005), 817–830.
54. F. Yates, Incomplete randomized designs, *Ann. Eugenics* **7** (1936), 121–140.
55. F. Yates, A new method of arranging variety trials involving a large number of varieties, *J. Agricultural Sci.* **26** (1936), 424–455.

Notes on contributors

Ian Anderson [*iananderson2357@gmail.com*] is an Honorary Research Fellow at the University of Glasgow, where he taught mathematics for 40 years and published research papers on graph theory and designs. He is the author of a number of books on combinatorial mathematics, including *Combinatorics of Finite Sets* and *Combinatorial Designs and Tournaments.* He has a particular interest in the history of combinatorics, and co-authored the historical survey 'Design theory: antiquity to 1950' in the second edition of the CRC *Handbook of Combinatorial Designs.*

Kiyoshi Ando [*ando@ice.uec.ac.jp*] received his Ph.D. degree from the University of Tokyo. He is currently a Professor of Informatics and Engineering at The University of Electro-Communications. His research interests include graph theory and algebraic coding theory.

R. A. Bailey [*r.a.bailey@qmul.ac.uk*] obtained a D.Phil. degree in group theory from the University of Oxford. She worked at the Open University, and then held a post-doctoral research fellowship in Statistics at the University of Edinburgh. This was followed by ten years in the Statistics Department at Rothamsted Experimental Station, which at that time came under the auspices of the Agriculture and Food Research Council. She returned to university life as Professor of Mathematical Sciences at Goldsmiths' College, University of London, and has been Professor of Statistics at Queen Mary, University of London, since 1994.

Lowell W. Beineke [*beineke@ipfw.edu*] is Schrey Professor of Mathematics at Indiana University–Purdue University Fort Wayne, where he has been since receiving his Ph.D. degree from the University of Michigan under the guidance of Frank Harary. His graph theory interests are broad and include topological graph theory, line graphs, tournaments, decompositions and vulnerability. He has published more than a hundred papers in graph theory and served as Editor of the *College Mathematics Journal*. In addition to two previous volumes in this series, *Topics in Algebraic Graph Theory* and *Topics in Topological Graph Theory*, he and Robin Wilson have edited *Selected Topics in Graph Theory* (three volumes), *Applications of Graph Theory* and *Graph Connections*.

F. T. Boesch [*deceased*] served as Department Head and the Charles Bachelor Professor of Electrical Engineering and Computer Science, and also as Dean of Faculty, at Stevens Institute of Technology, USA. Prior to coming to Stevens he worked at Bell Telephone Laboratories, where among his professional achievements he was co-recipient of the patent for the touch-tone receiver. At Stevens, he was a co-recipient of the Jess H. Davis Research Award and a recipient of the Henry Morton Distinguished Teaching Professor Award. He was the Editor-in-Chief of the international journal *Networks*, authored over 100 scholarly works, and was listed in various *Who's Who*s.

Béla Bollobás [*b.bollobas@dpmms.cam.ac.uk*] was born in Hungary and now works in Cambridge, England, where he has been a fellow of Trinity College since 1970; he has also held a joint appointment at the University of Memphis. His first doctorate was in discrete geometry, with László Fejes Tóth and Paul Erdõs as supervisors. Since then he has worked in a wide range of areas, including extremal graph theory, random graphs, graph polynomials and percolation, and has written over 350 research papers and several books. He won the senior Whitehead prize of the London Mathematical Society in 2007 and was elected to a Fellowship of the Royal Society in 2011.

Peter J. Cameron [*p.j.cameron@qmul.ac.uk*] is Professor of Mathematics at Queen Mary, University of London, where he has been since 1986, following a position as tutorial fellow at Merton College, Oxford. Since his D.Phil. degree in Oxford, he has been interested in a variety of topics in algebra and combinatorics, especially their interactions. He has held visiting positions at the University of Michigan, California Institute of Technology and the University of Sydney. He is currently chair of the British Combinatorial Committee.

Keith Edwards [*kjedwards@dundee.ac.uk*] has been at the University of Dundee since completing a D.Phil. degree in Mathematics at the University of Oxford in 1986, and is currently in the School of Computing. His research interests are in graph theory and combinatorial algorithms and complexity – in particular, detachments, colouring, planarization and fragmentability. He has been treasurer of the British Combinatorial Committee since 1999.

Abdol-Hossein Esfahanian [*esfahanian@cse.msu.edu*] is Associate Professor of Computer Science at Michigan State University. His research interests include applied graph theory, computer networks, fault-tolerant computing and web mining. He has published articles in journals such as the *IEEE Transactions on Computers*, *Networks*, *Discrete Applied Mathematics*, *Graph Theory* and *Parallel and Distributed Computing*. From 1996 to 1999, he was an Associate Editor of *Networks*.

Graham Farr [*graham.farr@monash.edu*] teaches in the Clayton School of Information Technology at Monash University, Australia. He did his first degree at the same university, before completing a D.Phil. degree in Mathematics from the University of Oxford. His interests include graph theory, enumeration (especially Tutte–Whitney polynomials), algorithms, complexity, information theory and computer history. Since 2008 he has led Computer History Tours of Melbourne. In 2011 he received a Vice-Chancellor's Award for Excellence in Postgraduate Supervision at Monash.

Ralph J. Faudree [*rfaudree@memphis.edu*] is Provost at the University of Memphis and Professor in the Department of Mathematical Sciences. His research interests and publications are in extremal graph theory, and more specifically in graphical Ramsey theory, Turán extremal theory and Hamiltonian theory of graphs. He published extensively with the prolific mathematician Paul Erdős, and has co-authored papers with many combinatorial mathematicians. He has served as a managing editor of the *Journal of Graph Theory* and was awarded the Euler Medal in 2005 by the Institute of Combinatorics.

Michael Ferrara [*michael.ferrara@ucdenver.edu*] is a graduate of Emory University in Atlanta and is an assistant professor in the Department of Mathematical and Statistical Sciences at the University of Colorado at Denver. His research interests include structural graph theory, specifically the study of paths, cycles and subdivisions, extremal problems related to degree sequences and the study of graph saturation. In his 'spare' mathematical time, he also greatly enjoys sharing exciting and fun mathematics with schoolchildren and teachers.

Ronald J. Gould [*rg@mathcs.emory.edu*] is a graduate of Western Michigan University. He has taught for more than 30 years at Emory University in Atlanta, where he is Goodrich C. White Professor of Mathematics and Computer Science. He has published extensively in graph theory and has written two books, the first being *Graph Theory* and the more recent being *Mathematics in Games, Sports and Gambling*.

Matthias Kriesell [*kriesell@imada.sdu.dk*] studied mathematics and computer science at the Leibniz University, Hannover. For many years, he was managing and technical director of a computer game company specializing in medieval trading simulations. After a two-year scholarship at the DFG graduate school on Algorithmic Discrete Mathematics at the University of Berlin, he received his Ph.D. degree from the Technical University Berlin, and held postdoctoral positions in Enschede, Hannover (habilitation), Hamburg and Odense. In 2010, he was appointed by the University of Southern Denmark, where he is now Professor of Combinatorics and Discrete Mathematics at the Department of Mathematics and Computer Science IMADA.

Dirk Meierling [*meierling@math2.rwth-aachen.de*] received his doctoral degree in mathematics from RWTH Aachen University, Germany, in 2007, where he has been a lecturer ever since. He is author or co-author of more than two dozen scientific papers. His research interests include tournaments and their generalizations, domination and connectivity in graphs, algorithmic aspects of games and mechanism design.

Ortrud Oellermann [*o.oellermann@uwinnipeg.ca*] is a professor in the Department of Mathematics and Statistics at the University of Winnipeg. After obtaining her Ph.D. degree in 1986 she held a two-year assistant professor position at Western Michigan University, following which she returned to her native South Africa where she taught at the University of Natal until she moved to Canada in November 1992. Her primary research interests are in graph connectivity, Steiner distances and abstract convexity. She co-authored *Applied and Algorithmic Graph Theory* and served as managing editor for *Ars Combinatoria*. Recently she has been active on the executive committees for the SIAM conferences on Discrete Mathematics and the newly formed CanaDAM meetings.

Dieter Rautenbach [*dieter.rautenbach@uni-ulm.de*] is the head of the Institute for Optimization and Operations Research at the University of Ulm. His main research interests lie in Combinatorial Optimization and Graph Theory. He received his education at the RWTH Aachen and moved to Ulm in 2010. Previously he held positions as associate professor at the Research Institute for Discrete Mathematics of the University of Bonn and full professor at the Institute for Mathematics of the TU Ilmenau.

Bruce Reed [*breed@cs.mcgill.ca*] holds a Canada Research Chair in the Combinatorics of Complex Networks in the School of Computer Science of McGill University in Montreal, Canada. His main research interests lie in algorithmic graph theory and the intersection of probability and combinatorics. He gave an invited talk at the 2000 ICM in Beijing and was made a Fellow of the Royal Society of Canada in 2009.

Oliver Riordan [*riordan@maths.ox.ac.uk*] was an undergraduate, graduate student and research fellow at Trinity College, Cambridge. He then held a Royal Society University Research Fellowship, partly at Trinity, and later at King's College, Cambridge. In 2007 he moved to Oxford as Professor of Discrete Mathematics and a tutorial fellow of St Edmund Hall. His main research interests are random graphs and percolation.

Appajosyula Satyanarayana [*asatya@stevens.edu*] is professor emeritus at Stevens Institute of Technology, USA, having served in the Electrical Engineering and Computer Science Department from 1984 to 2007. He authored and co-authored over 60 research papers, and mentored a significant number of Ph.D. students. Among the honours that he received at Stevens Institute were a Jess H. Davis Research Award, Alexander Humphrey and Henry Morton Distinguished Teaching Professor Awards and a Master of Engineering degree *honoris causa*. He also served as Associate Editor of *Networks* for over 15 years.

Charles L. Suffel [*csuffel@stevens.edu*] is Dean of Graduate Academics at Stevens Institute of Technology, USA. His research interests include graph theory, with an emphasis on network reliability and combinatorics. The author of more than 50 publications, he was the managing editor of the international journal *Networks* for 22 years. At Stevens, he was twice named Outstanding Teacher, was the co-recipient of a Jess H. Davis Research Award and received a Henry Morton Distinguished Teaching Professor Award. He was a participating scholar in the Scientist-in-Residence Program of the New York Academy of Sciences and was also Scholar in Residence at Purdue University, where he is now based.

Lutz Volkmann [*volkm@math2.rwth-aachen.de*] is an emeritus professor of mathematics at RWTH Aachen University, Germany. He has published extensively on functions of a complex variable and combinatorics – in particular, on graph theory. His publications include various books, the most recent being *Graphen an allen Ecken und Kanten* (German). He is co-author, with Paul Butzer, of a biography of Otto Blumenthal, published in *The Journal of Approximation Theory*.

Robin J. Wilson [*r.j.wilson@open.ac.uk*] is an emeritus professor of pure mathematics at the Open University, UK, and emeritus professor of geometry at Gresham College, London. After graduating from Oxford, he received his Ph.D. degree in number theory from the University of Pennsylvania. He has written and edited many books on graph theory and the history of mathematics, including *Introduction to Graph Theory* and *Four Colours Suffice*. His research interests formerly included graph colourings and now focus on the history of combinatorics. He has won a Lester Ford Award and a George Pólya Award from the MAA for his expository writing.

Index